Scientific and Technological Aspects of Titanium Alloys

Scientific and Technological Aspects of Titanium Alloys

Edited by Keith Liverman

www.clanryeinternational.com

Clanrye International,
750 Third Avenue, 9th Floor,
New York, NY 10017, USA

ISBN: 978-1-64726-651-6

Cataloging-in-Publication Data

Scientific and technological aspects of titanium alloys / edited by Keith Liverman.
p. cm.
Includes bibliographical references and index.
ISBN 978-1-64726-651-6
1. Titanium alloys. 2. Titanium. 3. Chemical engineering. I. Liverman, Keith.
TN693.T5 S35 2023
620.189 322--dc23

For information on all Clanrye International publications visit our website at www.clanryeinternational.com

Contents

Preface

It is often said that books are a boon to mankind. They document every progress and pass on the knowledge from one generation to the other. They play a crucial role in our lives. Thus I was both excited and nervous while editing this book. I was pleased by the thought of being able to make a mark but I was also nervous to do it right because the future of students depends upon it. Hence, I took a few months to research further into the discipline, revise my knowledge and also explore some more aspects. Post this process, I begun with the editing of this book.

Titanium alloys are made up of titanium and other chemical elements. These alloys have high tensile strength and toughness. They are light-weight, have excellent corrosion resistance, and can sustain high temperatures. They are used in medical devices, jewelry, aircraft, and spacecraft. They are also used in highly stressed components of sports cars as well as some premium consumer electronics and sports equipment. Titanium alloys are commonly divided into four categories which include alpha alloys, near-alpha alloys, alpha and beta alloys, and beta and near beta alloys. These alloys are heat treated for a variety of purposes, the most important of which is to improve strength by solution treatment and aging for optimizing specific properties, like fatigue strength, fracture toughness, and high temperature creep strength. This book provides significant information to help develop a good understanding of titanium alloys. It explores all the important scientific and technological aspects of these alloys in the modern day. Researchers and students in this field will be assisted by this book.

I thank my publisher with all my heart for considering me worthy of this unparalleled opportunity and for showing unwavering faith in my skills. I would also like to thank the editorial team who worked closely with me at every step and contributed immensely towards the successful completion of this book. Last but not the least, I wish to thank my friends and colleagues for their support.

Editor

1

Grain Refinement of Ti-15Mo-3Al-2.7Nb-0.2Si Alloy with the Rotation of TiB Whiskers by Powder Metallurgy and Canned Hot Extrusion

Jiabin Hou [1,2], Lin Gao [1], Guorong Cui [1,*], Wenzhen Chen [1], Wencong Zhang [1] and Wenguang Tian [3]

1. School of Materials Science and Engineering, Harbin Institute of Technology, Weihai 264209, China; houjiabinwh@163.com (J.H.); gaolinhit@hotmail.com (L.G.); nclwens@hit.edu.cn (W.C.); zwinc@hitwh.edu.cn (W.Z.)
2. Naval Architecture and Marine Engineering College, Shandong Jiaotong University, Weihai 264209, China
3. Oriental Bluesky Titanium Technology Co., LTD, Yantai 264003, China; tianwenguang@obtc.cn

* Correspondence: cuiguorong2010@126.com

Abstract: In situ synthesized TiB whiskers (TiBw) reinforced Ti-15Mo-3Al-2.7Nb-0.2Si alloys were successfully manufactured by pre-sintering and canned hot extrusion via adding TiB_2 powders. During pre-sintering, most TiB_2 were reacted with Ti atoms to produce TiB. During extrusion, the continuous dynamic recrystallization (CDRX) of β grains was promoted with the rotation of TiBw, and CDRXed grains were strongly inhibited by TiBw with hindering dislocation motion. Eventually, the grain sizes of composites decreased obviously. Furthermore, the stress transmitted from the matrix to TiBw for strengthening in a tensile test, besides grain refinement. Meanwhile, the fractured TiBw and microcracks around them contributed to fracturing.

Keywords: TiB whiskers; dynamic recrystallization; grain refinement; strengthening

---

1. Introduction

Metastable β titanium alloys such as Ti-15Mo-3Al-2.7Nb-0.2Si are a promising candidate applied in aerospace and automotive industries, which have the advantages of high specific strength, excellent hot and cold workability, deep hardenability and oxidation resistance [1–3]. However, during hot working, metastable β titanium alloys coarsen rapidly at elevated temperatures, which could weaken their thermal stability and mechanical properties [4]. Therefore, it is necessary to reduce the grain sizes of metastable β titanium alloys and restrict their growth. In order to overcome the drawback, lots of studies about beta-titanium alloys reinforced by intermetallic particles (or eutectic structures) were carried out [5–12]. The Ni and Co elements were partially segregated to the interdendritic region to form TiNi and TiCo intermetallic phases, resulting in fine interdendritic precipitates or eutectic structures to obtain high strength [5–7,13]. Meanwhile, TiBw was considered to be one of the best reinforcements for Ti matrix with high elastic modulus, clean bonding interface and similar thermal expansion coefficient with matrix [14]. Huang et al. [15] reported that the accumulation of strain around TiBw provided a nucleation site for dynamic recrystallization (DRX) in hot deformation process, which results in some small and fine α of Ti60. Moreover, Feng et al. [16] reported that the growth of the recrystallization primary β was strongly restricted by the TiBw during the extrusion process, which reduced the grain sizes of Ti64. Okulov et al. [17] reported that multicomponent Ti alloys was refined by adding boron during casting. The TiB needle-shape particles distributed along primary β-Ti dendrites, and reduced the secondary dendrite arm spacing of the β-Ti phase, resulting in grain refinement. At present, there are fewer scholars researching metastable β titanium alloys reinforced by

TiBw (TiBw/metastable β titanium) in the hot deformation process, particularly the grain refinement mechanism of TiBw/metastable β titanium.

The grain-refinement mechanism of TiBw/Ti-15Mo-3Al-2.7Nb-0.2Si is worthy of exploring. Above all, composites were fabricated by low energy milling, pre-sintering and canned hot extrusion via adding 2.6 vol % TiB_2.

2. Materials and Methods

The spherical Ti-15Mo-3Al-2.7Nb-0.2Si (alloy) powders (β transus ~ 827 °C) and prismatic TiB_2 powders were chosen as raw materials to produce the as-extruded bars of TiBw reinforced alloys (composites) and alloys. Alloy powders were approximately 120 μm, and TiB_2 powders were approximately 3 μm. Alloy powders (97.4 vol %) and TiB_2 powders (2.6 vol %) were mixed at a speed of 100 rpm for a period of 6 h by low energy milling (LEM) in a planetary ball mill under the protective atmosphere of argon. The weight ratio between balls and powder mixture was 5:1. The mixed powders and alloy powders were weld-sealed into a 45# steel cup (with dimensions of 52 mm in outside diameter, 40 mm in inside diameter, 50 mm in height) [16]. Then billets were pre-sintered in high-temperature box furnace at 1100 °C for 1 h and then cooled by air to room temperature. The pre-sintered billets were heated at 800, 900 and 1000 °C for 30 min and then subsequently extruded by hydraulic machine, with the extrusion ratio of 10.6:1. The as-extruded bars with a diameter of 16 mm and a length of 500 mm were obtained.

The Phases of composites were identified by X-ray diffraction (XRD) (Rigaku Corporation, Tokyo, Japan). The microstructure was observed by scanning electron microscopy (SEM) (Zeiss-MERLIN, Zürich, Switzerland) and equipped with electron back scattered diffraction (EBSD) system. The tensile specimens with gauge lengths of 15 mm, widths of 4 mm and thickness of 2 mm were cut along the extrusion direction (ED) and then polished by metallographic sandpaper. All specimens were tested with a speed of 0.5 mm/min, and the ductility was measured by extensometer.

3. Results and Discussion

In order to identify the phases clearly, the SEM and XRD of the pre-sintered and as-extruded composites are shown in Figure 1. According to XRD in Figure 1a,b, the composites contained TiB, α-Ti and β-Ti. Noting that no TiB_2 was detected, indicating that most of TiB_2 were reacted with Ti to produce TiB. Ma et al. [18] reported that when the ratio of B atoms below 49% to 50 at %, Ti atom would react with TiB_2 to form TiB. The phases composition, morphology and distribution were studied by the SEM in Figure 1c–f. According to the SEM images, TiBw, β-Ti and α-Ti were identified. During pre-sintering, TiBw emerged in the β grains boundaries. After extrusion, the sizes of TiBw increased and distributed along the ED. Meanwhile, the average sizes of β grains increased with the increasing extrusion temperature.

The inverse pole figure (IPF) mapping of as-extruded alloys and composites are shown in Figure 2. The black lines represent high-angle grain boundaries (HAGBs with misorientations > 15°), and the white lines represent low-angle grain boundaries (LAGBs with misorientations between 2° and 15°) [16,19]. The red lines distributed around phases, including α phase and TiBw. The coarse elongated β grains distributed along ED. Meanwhile, a large volume fraction of fine equiaxed β grains distributed inhomogeneously between elongated β grains, indicating dynamic recrystallization (DRX) occurred during extrusion [20,21]. Some β gains were distributed around TiBw and elongated α grains, typically as indicated by the black ellipse. Moreover, the necklace structure was distributed along the grain boundaries of elongated β grains, consisting of fine β grains and bulged HAGBs, as shown in the white ellipse. During deformation, the bulging grain boundaries severed as the nucleus of recrystallized grains, and recrystallized grains grew with the migration of HAGBs, resulting in necklace structure (DRX process on the bulged grain boundaries), that is discontinuous dynamic recrystallization (DDRX) [22].

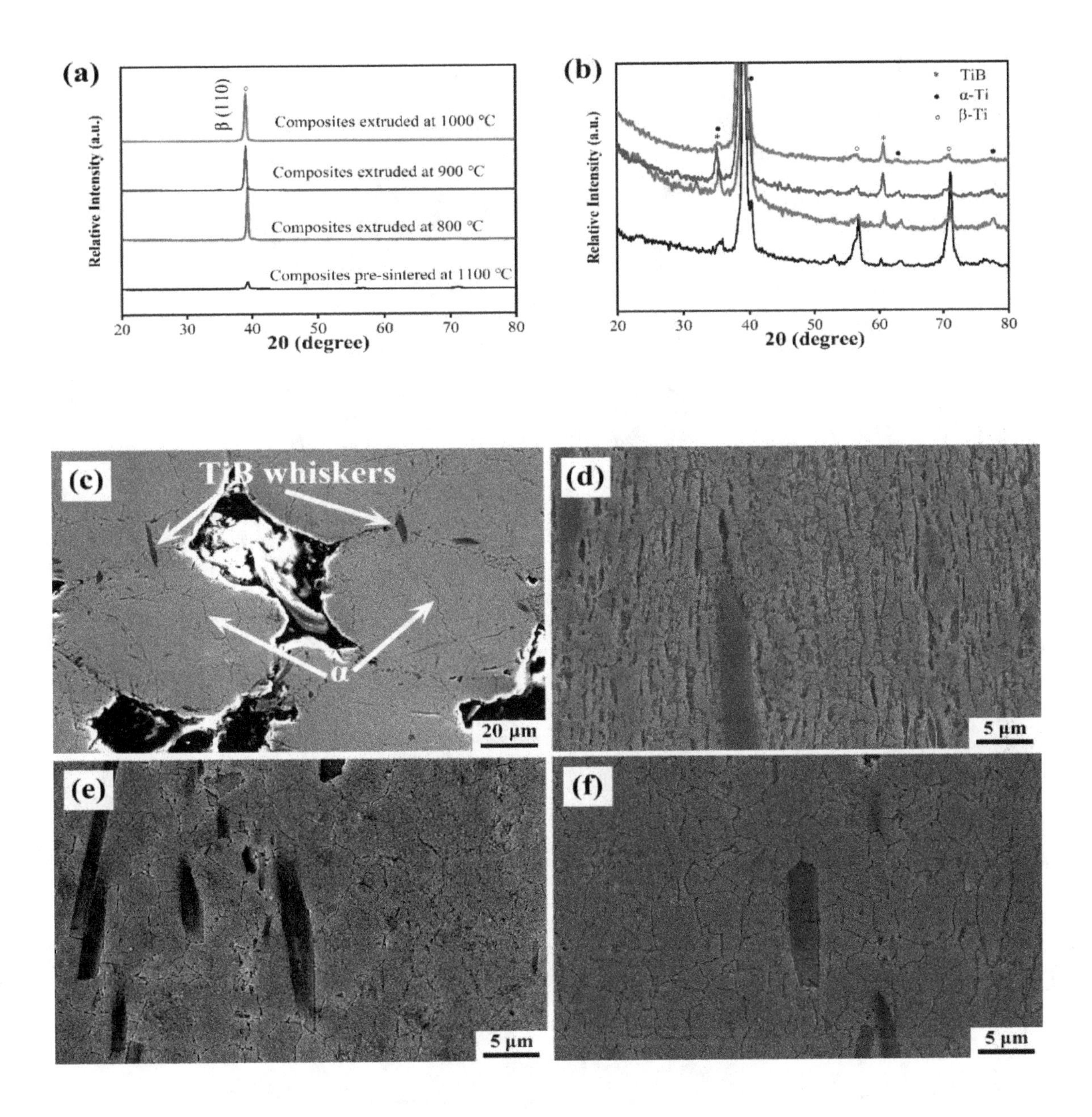

Figure 1. The XRD of pre-sintered and as-extruded composites (**a**), high magnification of XRD (**b**); SEM images of pre-sintered composites (**c**), composites extruded at 800 °C (**d**), composites extruded at 900 °C (**e**), composites extruded at 1000 °C (**f**).

In order to detect the orientation and texture of as-extruded alloys and composites, the discrete plot and texture along ED of alloys and composites extruded at 800 °C were shown in Figure 3a,b. The basal planes {0001} of α grains (⟨$2\bar{1}\bar{1}0$⟩ (point A1) and ⟨$10\bar{1}0$⟩ (point A3)) almost paralleled to ED, which contributed to ⟨$2\bar{1}\bar{1}0$⟩ α texture in Figure 3a,b and ⟨$10\bar{1}0$⟩ α texture in Figure 3b. During deformation, the prismatic glide of primary α grains (α_p) is {$10\bar{1}0$} ⟨$11\bar{2}0$⟩ [16]. Therefore, the basal planes {0001} of α_p would turn to parallel to ED. The α grain (⟨$10\bar{1}1$⟩ (point A2)) contributed to another center of α texture, which nucleated in the grain boundaries and grains according to the Burgers relationship {0001}//{110} and ⟨111⟩//⟨$11\bar{2}0$⟩ during β → α process [23]. Moreover, ⟨101⟩ β texture was maximum center in Figure 3a,b, which was commonly found in the fully recrystallized β grains of the deformed β phase due to the orientated nucleation mechanism [24,25].

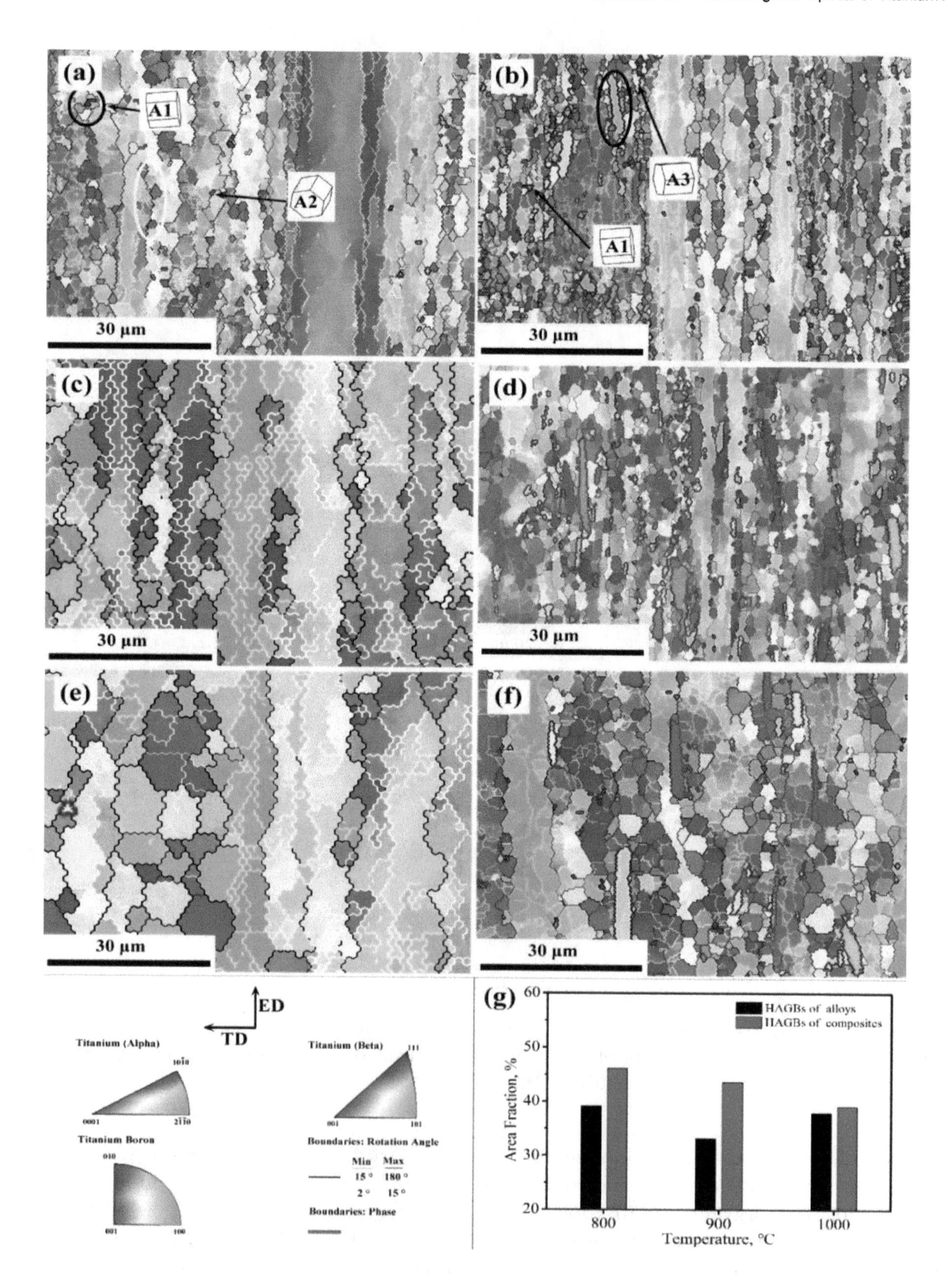

Figure 2. IPF mapping of alloys extruded at (**a**) 800 °C, (**c**) 900 °C, (**e**) 1000 °C and composites extruded at (**b**) 800 °C, (**d**) 900 °C, (**f**) 1000 °C; (**g**) histogram high-angle grain boundaries (HAGBs) counts of as-extruded alloys and composites.

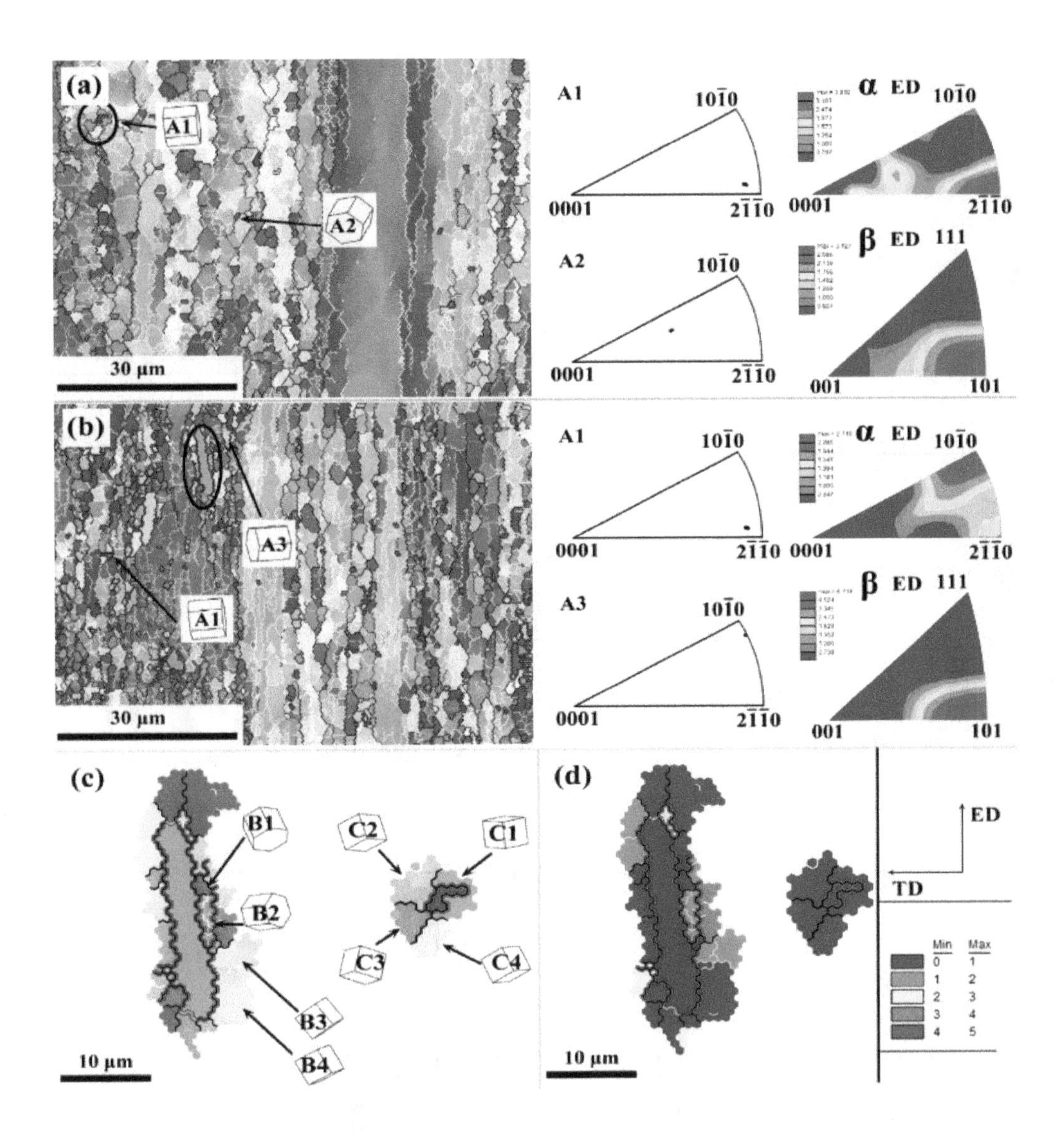

Figure 3. (**a**) IPF mapping, discrete plot and texture of alloy extruded at 800 °C; (**b**) IPF mapping, discrete plot and texture along extrusion direction (ED) of composite extruded at 800 °C; (**c**) highlighted mapping of (**a**,**b**); (**d**) grain orientation spread (GOS) mapping of (**c**).

The IPF and grain orientation spread (GOS) mapping of marked TiBw and α_p in Figure 3a,b are shown in Figure 3c,d. From Figure 3c, TiBw was surrounded by α grains and equiaxed β grains. Meanwhile, the GOS values of them are shown in Figure 4d. Basu et al. [26] revealed that the GOS values of the recrystallized grains were less than 1°, indicating β grains with blue color occurred DRX in Figure 3d. TiBw was a brittle phase with high elastic modulus, resulting in inharmonious deformation [27]. During extrusion, TiBw rotated to parallel to ED with the metal flow, and high extrusion stress concentrated around them by inharmonious deformation [24]. TiBw hindered dislocations motion, resulting in the piled up of dislocation around TiBw, which had been revealed by Park [28]. LAGBs emerged in the adjacent grains (point B2) and transformed into subgrains (point B3) with the accumulation of driving force. Subgrains rotated and formed equiaxed grains as further deformation, resulting in continuous dynamic recrystallization (CDRX) [25]. Meanwhile, the LAGBs evolved into HAGBs, which could be proved by the HAGBs volume fraction of composites increased by TiBw in Figure 2g. It should be pointed out that equiaxed β subgrains (point B4) occurred partial dynamic recrystallization without sufficient driving force [25]. A similar phenomenon was observed in Figure 2b,d,f, which may be attributed to a small rotation angle of TiBw.

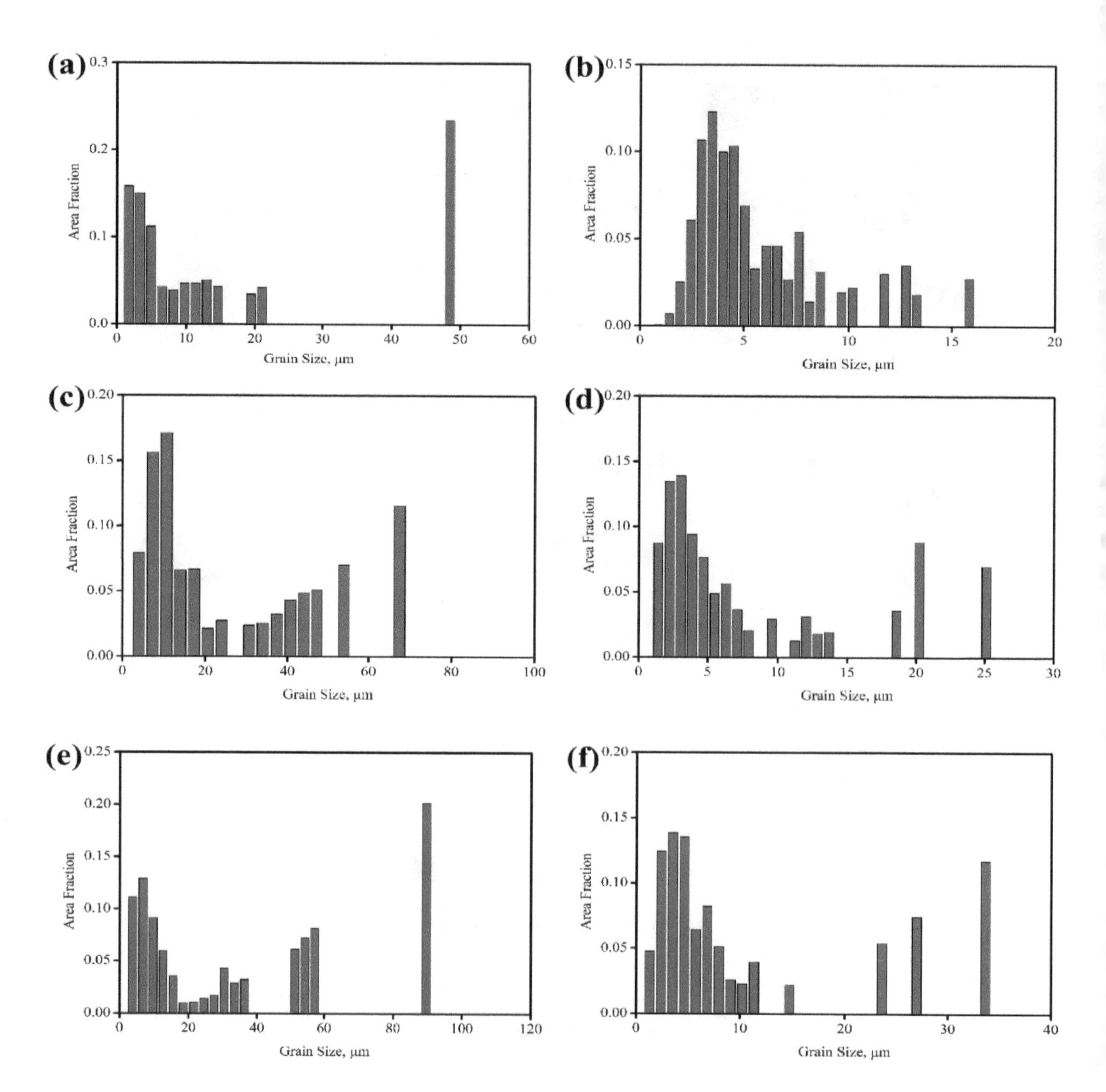

Figure 4. Histogram size counts of (**a**) alloys extruded at 800 °C, (**b**) composites extruded at 800 °C, (**c**) alloys extruded at 900 °C, (**d**) composites extruded at 900 °C, (**e**) alloys extruded at 1000 °C, (**f**) composites extruded at 1000 °C.

During extrusion at 800 °C, the basal planes {0001} of α_p would rotate to parallel to ED with the metal flow [24]. Meanwhile, the slip systems of α were limited with a hexagonal crystal structure (HCP), which led to inharmonious deformation. The relative high dislocation density would be accumulated in the grain boundaries between adjacent β grains and α_p [20], and the dislocation substructure formed [29]. The LAGBs in the adjacent β grains were contributed from the dislocation substructure slip and transformed into subgrains with the accumulation of deformation. Finally, the subgrains rotated and formed fine equiaxed β grains as further deformation, resulting in equiaxed β grains with random orientation (point C1, C2, C3, C4) [30]. Meanwhile, the LAGBs evolved into HAGBs, which also could be proved by that the HAGBs volume fraction of alloys reduced as α_p disappeared by extrusion temperature increased from 800 °C to 900 °C in Figure 2g.

The grain sizes distribution of alloys and composites extruded at different temperatures were shown by a histogram, as shown in Figure 4. The average sizes of β grains reduced obviously with the precipitation of TiBw, particularly the volume fraction of β grains less than 10 μm increased obviously. Moreover, the average sizes of β grains increased with the increasing of extrusion temperature, resulting from more energy was obtained to promote the migration of grain boundaries for DDRX and growth with higher extrusion temperature [25]. During deformation, the rotation of TiBw promoted CDRX of β grains in inharmonious deformation, and the growth of CDRXed β grains was strongly inhibited by them, resulting in the higher volume fraction of fine β grains. Noting that coarse elongated β grains were shown in Figure 2. Figure 4 shows the sizes of coarse grains increased with the increasing extrusion temperature, resulting from the migration of grain boundaries promoted by more energy. The DDRXed and CDRXed β grains and coarse elongated β grains were attributed to bimodal grain size distribution in Figure 4 [31].

The room temperature tensile curves of alloys and composites extruded at 800, 900 and 1000 °C are shown in Figure 5a,b. The tensile strength of alloys decreases from 1086 MPa to 925 MPa with the increasing of extrusion temperature. During a tensile test, α_p distributed along grain boundary trapped dislocation for strengthening, which had been revealed by Liu [32]. Meanwhile, small grains were beneficial for strengthening effect [19]. In a word, when the alloy was extruded at 800 °C, α_p and refined grains contributed mainly for strengthening, with a strength of 1086 MPa. Meanwhile, α_p triggered massive stress concentration and served a nucleation of crack [32], which led to fracture, with 3.5% elongation.

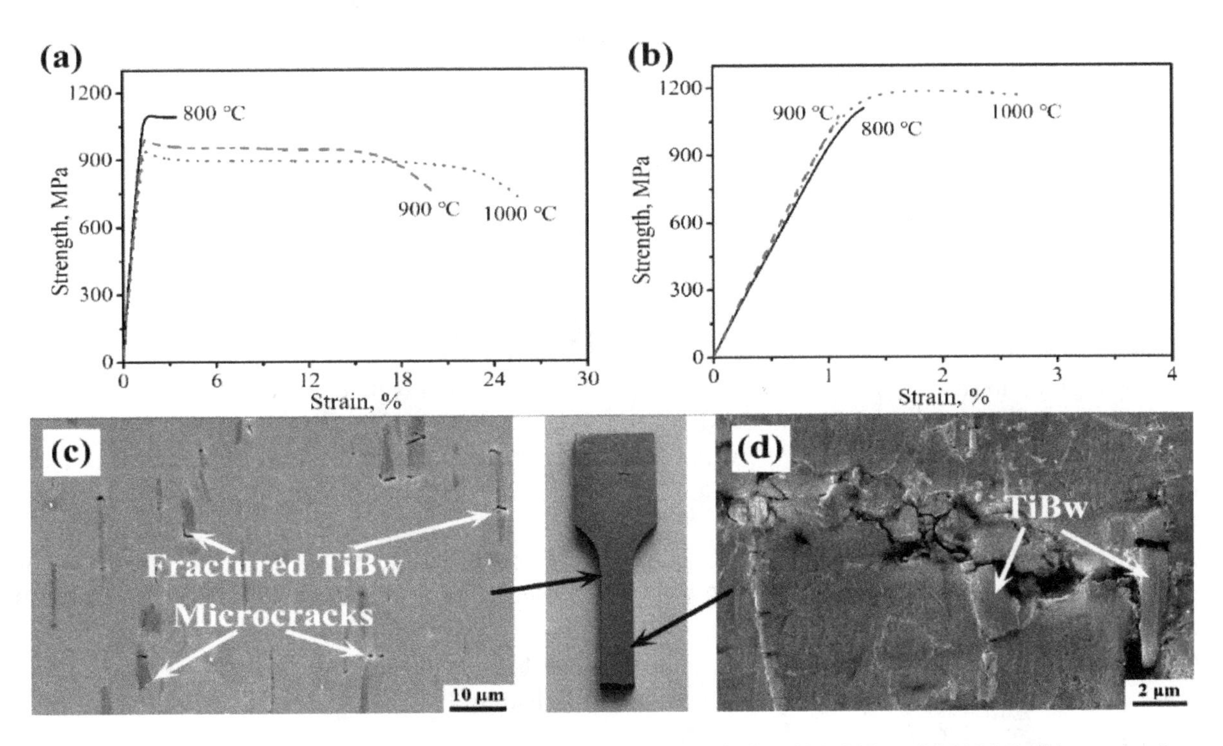

Figure 5. Room temperature tensile curves of (**a**) alloys extruded at 800, 900 and 1000 °C, (**b**) composites extruded at 800, 900 and 1000 °C; SEM of the longitudinal sections of the tensile test specimen of composites extruded at 1000 °C far away from the fracture surface (**c**), near the fracture surface (**d**).

During a tensile test for composites, the stress transfers from the matrix to TiBw due to inharmonious deformation, resulting in strengthening [27]. The findings for the composites extruded at 1000 °C will be discussed in the following in more detail and in Figure 5c,d [5]. Massive fractured TiBw and microcracks around TiBw were shown in the SEM image (Figure 5c) far away from the fracture surface., and the crack path included lots of cracks along TiBw and β grains boundaries in the SEM image (Figure 5d) near the fracture surface. These observations indicate that TiBw fractured with the accumulation of load transfer due to the inherent brittleness of TiBw [17], and then, microcracks emerged in the interface of TiBw and matrix. With further deformation, the cracks extended from microcracks to the β grains boundaries of Ti matrix, leading to the final fracture. When the composite was extruded at 1000 °C, TiBw and grain refinement contributed for strengthening, with a strength of 1200 MPa, raising 27.9%.

4. Conclusions

In this work, the grain refinement and strengthening of TiBw/Ti-15Mo-3Al-2.7Nb-0.2Si alloy fabricated by powder metallurgy and canned hot extrusion were studied. The following conclusions were drawn:

(1) The β grains of composites were refined with the rotation and inhibition of TiBw. In inharmonious deformation, the dislocation motion was inhibited by TiBw, and CDRX of β grains was promoted with the rotation of TiBw. Meanwhile, the growth of CDRXed β grains was strongly inhibited.

(2) During extrusion below β phase region, the basal planes {0001} of α_p rotated to parallel to ED, resulting in grain refinement. The high dislocation density was accumulated in α_p grain boundaries, and inharmonious deformation supplied driving force to promote CDRX of adjacent β grains. Meanwhile, α_p slipped along the prismatic glide of $\{10\bar{1}0\}\langle 11\bar{2}0\rangle$, resulting in $\langle 2\bar{1}\bar{1}0\rangle$ and $\langle 10\bar{1}0\rangle$ α texture.

(3) The strength of composites extruded at 1000 °C was improved. TiBw loaded the stress transmitted from matrix until fracture, and grain refinement contributed to strengthening. Meanwhile, the microcracks initiated from the fractured TiBw.

Author Contributions: Conceptualization, J.H. and G.C.; methodology, L.G.; software, W.C.; validation, G.C., W.Z. and W.T.; investigation, L.G.; resources, W.Z.; data curation, L.G.; writing—original draft preparation, J.H. and L.G.; writing—review and editing, G.C., W.Z. and W.T.; project administration, W.Z.; funding acquisition, G.C. All authors have read and agree to the published version of the manuscript.

References

1. Zheng, Y.; Williams, R.E.; Wang, D.; Shi, R.; Nag, S.; Kami, P.; Sosa, J.M.; Banerjee, R.; Wang, Y.; Fraser, H.L. Role of ω phase in the formation of extremely refined intragranular α precipitates in metastable β-titanium alloys. *Acta Mater.* **2016**, *103*, 850–858. [CrossRef]
2. Yao, T.; Du, K.; Wang, H.; Huang, Z.; Li, C.; Li, L.; Hao, Y.; Yang, R.; Ye, H. In situ scanning and transmission electron microscopy investigation on plastic deformation in a metastable β titanium alloy. *Acta Mater.* **2017**, *133*, 21–29. [CrossRef]
3. Xiao, J.; Nie, Z.; Tan, C.; Zhou, G.; Chen, R.; Li, M.; Yu, X.; Zhao, X.; Hui, S.; Ye, W. The dynamic response of the metastable β titanium alloy Ti-2Al-9.2 Mo-2Fe at ambient temperature. *Mater. Sci. Eng. A* **2019**, *751*, 191–200. [CrossRef]
4. Cherukuri, B.; Srinivasan, R.; Tamirisakandala, S.; Miracle, D.B. The influence of trace boron addition on grain growth kinetics of the beta phase in the beta titanium alloy Ti–15Mo–2.6Nb–3Al–0.2Si. *Scr. Mater.* **2009**, *60*, 496–499. [CrossRef]
5. Okulov, I.; Bönisch, M.; Okulov, A.; Volegov, A.; Attar, H.; Ehtemam-Haghighi, S.; Calin, M.; Wang, Z.; Hohenwarter, A.; Kaban, I. Phase formation, microstructure and deformation behavior of heavily alloyed TiNb-and TiV-based titanium alloys. *Mater. Sci. Eng. A* **2018**, *733*, 80–86. [CrossRef]
6. Okulov, I.; Kühn, U.; Marr, T.; Freudenberger, J.; Schultz, L.; Oertel, C.-G.; Skrotzki, W.; Eckert, J. Deformation and fracture behavior of composite structured Ti-Nb-Al-Co (-Ni) alloys. *Appl. Phys. Lett.* **2014**, *104*, 071905. [CrossRef]
7. Okulov, I.; Kühn, U.; Marr, T.; Freudenberger, J.; Soldatov, I.; Schultz, L.; Oertel, C.-G.; Skrotzki, W.; Eckert, J. Microstructure and mechanical properties of new composite structured Ti–V–Al–Cu–Ni alloys for spring applications. *Mater. Sci. Eng. A* **2014**, *603*, 76–83. [CrossRef]
8. Okulov, I.; Okulov, A.; Soldatov, I.; Luthringer, B.; Willumeit-Römer, R.; Wada, T.; Kato, H.; Weissmüller, J.; Markmann, J. Open porous dealloying-based biomaterials as a novel biomaterial platform. *Mater. Sci. Eng. C* **2018**, *88*, 95–103. [CrossRef]

9. Okulov, I.; Okulov, A.; Volegov, A.; Markmann, J. Tuning microstructure and mechanical properties of open porous TiNb and TiFe alloys by optimization of dealloying parameters. *Scr. Mater.* **2018**, *154*, 68–72. [CrossRef]
10. Okulov, I.; Soldatov, I.; Sarmanova, M.; Kaban, I.; Gemming, T.; Edström, K.; Eckert, J. Flash Joule heating for ductilization of metallic glasses. *Nat. Commun.* **2015**, *6*, 7932. [CrossRef]
11. Okulov, I.; Volegov, A.; Attar, H.; Bönisch, M.; Ehtemam-Haghighi, S.; Calin, M.; Eckert, J. Composition optimization of low modulus and high-strength TiNb-based alloys for biomedical applications. *J. Mech. Behav. Biomed. Mater.* **2017**, *65*, 866–871. [CrossRef] [PubMed]
12. Okulov, I.V.; Wendrock, H.; Volegov, A.S.; Attar, H.; Kühn, U.; Skrotzki, W.; Eckert, J. High strength beta titanium alloys: New design approach. *Mater. Sci. Eng. A* **2015**, *628*, 297–302. [CrossRef]
13. Okulov, I.; Bönisch, M.; Volegov, A.; Shahabi, H.S.; Wendrock, H.; Gemming, T.; Eckert, J. Micro-to-nano-scale deformation mechanism of a Ti-based dendritic-ultrafine eutectic alloy exhibiting large tensile ductility. *Mater. Sci. Eng. A* **2017**, *682*, 673–678. [CrossRef]
14. Zhang, W.; Wang, M.; Chen, W.; Feng, Y.; Yu, Y. Evolution of inhomogeneous reinforced structure in TiBw/Ti-6AL-4V composite prepared by pre-sintering and canned β extrusion. *Mater. Des.* **2015**, *88*, 471–477. [CrossRef]
15. Wang, B.; Huang, L.J.; Liu, B.X.; Geng, L.; Hu, H.T. Effects of deformation conditions on the microstructure and substructure evolution of TiBw/Ti60 composite with network structure. *Mater. Sci. Eng. A* **2015**, *627*, 316–325. [CrossRef]
16. Feng, Y.; Zhang, W.; Cui, G.; Wu, J.; Chen, W. Effects of the extrusion temperature on the microstructure and mechanical properties of TiBw/Ti6Al4V composites fabricated by pre-sintering and canned extrusion. *J. Alloy. Compd.* **2017**, *721*, 383–391. [CrossRef]
17. Okulov, I.; Sarmanova, M.; Volegov, A.; Okulov, A.; Kühn, U.; Skrotzki, W.; Eckert, J. Effect of boron on microstructure and mechanical properties of multicomponent titanium alloys. *Mater. Lett.* **2015**, *158*, 111–114. [CrossRef]
18. Ma, X.; Li, C.; Du, Z.; Zhang, W. Thermodynamic assessment of the Ti–B system. *J. Alloy. Compd.* **2004**, *370*, 149–158. [CrossRef]
19. Dyakonov, G.; Mironov, S.; Semenova, I.; Valiev, R.; Semiatin, S. EBSD analysis of grain-refinement mechanisms operating during equal-channel angular pressing of commercial-purity titanium. *Acta Mater.* **2019**, *173*, 174–183. [CrossRef]
20. Lin, P.; Sun, Y.; Zhang, S.; Zhang, C.; Wang, C.; Chi, C. Microstructure and texture heterogeneity of a hot-rolled near-α titanium alloy sheet. *Mater. Charact.* **2015**, *104*, 10–15. [CrossRef]
21. Su, J.; Sanjari, M.; Kabir, A.S.H.; Jung, I.-H.; Yue, S. Dynamic recrystallization mechanisms during high speed rolling of Mg–3Al–1Zn alloy sheets. *Scr. Mater.* **2016**, *113*, 198–201. [CrossRef]
22. Belyakov, A.; Sakai, T.; Miura, H.; Kaibyshev, R.; Tsuzaki, K. Continuous recrystallization in austenitic stainless steel after large strain deformation. *Acta Mater.* **2002**, *50*, 1547–1557. [CrossRef]
23. Wang, S.C.; Aindow, M.; Starink, M.J. Effect of self-accommodation on α/ α boundary populations in pure titanium. *Acta Mater.* **2003**, *51*, 2485–2503. [CrossRef]
24. Huang, G.; Han, Y.; Guo, X.; Qiu, D.; Wang, L.; Lu, W.; Zhang, D. Effects of extrusion ratio on microstructural evolution and mechanical behavior of in situ synthesized Ti-6Al-4V composites. *Mater. Sci. Eng. A* **2017**, *688*, 155–163. [CrossRef]
25. Doherty, R.D.; Hughes, D.A.; Humphreys, F.J.; Jonas, J.J.; Rollett, A.D. Current issues in recrystallization: A review. *Mater. Sci. Eng. A* **1997**, *238*, 219–274. [CrossRef]
26. Basu, I.; Al-Samman, T. Twin recrystallization mechanisms in magnesium-rare earth alloys. *Acta Mater.* **2015**, *96*, 111–132. [CrossRef]
27. Panda, K.B.; Chandran, K.S.R. First principles determination of elastic constants and chemical bonding of titanium boride (TiB) on the basis of density functional theory. *Acta Mater.* **2006**, *54*, 1641–1657. [CrossRef]
28. Park, J.G.; Keum, D.H.; Lee, Y.H. Strengthening mechanisms in carbon nanotube-reinforced aluminum composites. *Carbon* **2015**, *95*, 690–698. [CrossRef]
29. Murty, S.V.S.N.; Nayan, N.; Kumar, P.; Narayanan, P.R.; Sharma, S.C.; George, K.M. Microstructure–texture–mechanical properties relationship in multi-pass warm rolled Ti–6Al–4V Alloy. *Mater. Sci. Eng. A* **2014**, *589*, 174–181. [CrossRef]

30. Dong, R.; Li, J.; Kou, H.; Tang, B.; Hua, K.; Liu, S. Characteristics of a hot-rolled near β titanium alloy Ti-7333 *Mater. Charact.* **2017**, *129*, 135–142. [CrossRef]
31. He, J.; Jin, L.; Wang, F.; Dong, S.; Dong, J. Mechanical properties of Mg-8Gd-3Y-0.5 Zr alloy with bimodal grain size distributions. *J. Magnes. Alloy.* **2017**, *5*, 423–429. [CrossRef]
32. Liu, C.; Lu, Y.; Tian, X.; Liu, D. Influence of continuous grain boundary α on ductility of laser melting deposited titanium alloys. *Mater. Sci. Eng. A* **2016**, *661*, 145–151. [CrossRef]

2

Mechanical Behavior and Microstructure Evolution of a Ti-15Mo/TiB Titanium–Matrix Composite during Hot Deformation

Sergey Zherebtsov [1], Maxim Ozerov [1,*], Margarita Klimova [1], Dmitry Moskovskikh [2], Nikita Stepanov [1] and Gennady Salishchev [1]

[1] Laboratory of Bulk Nanostructured Materials, Belgorod State University, Belgorod 308015, Russia; zherebtsov@bsu.edu.ru (S.Z.); klimova_mv@bsu.edu.ru (M.K.); stepanov@bsu.edu.ru (N.S.); salishchev@bsu.edu.ru (G.S.)

[2] Centre of Functional Nanoceramics, National University of Science and Technology, Moscow 119049, Russia; mos@misis.ru

* Correspondence: ozerov@bsu.edu.ru

Abstract: A Ti-15Mo/TiB titanium–matrix composite (TMC) was produced by spark plasma sintering at 1400 °C under a load of 40 MPa for 15 min using a Ti-14.25(wt.)%Mo-5(wt.)%TiB_2 powder mixture. Microstructure evolution and mechanical behavior of the composite were studied during uniaxial compression at room temperature and in a temperature range of 500–1000 °C. At room temperature, the composite showed a combination of high strength (the yield strength was ~1500 MPa) and good ductility (~22%). The microstructure evolution of the Ti-15Mo matrix was associated with the development of dynamic recovery at 500–700 °C and dynamic recrystallization at higher temperatures (≥800 °C). The apparent activation energy of the plastic deformation was calculated and a processing map for the TMC was constructed using the obtained results.

Keywords: titanium–matrix composite; deformation; microstructure evolution; mechanical properties

---

1. Introduction

Due to a combination of high specific strength, excellent corrosion properties, and remarkable biocompatibility, titanium alloys are widely used in industry (e.g., shipbuilding, aircraft building, chemistry, food industry, etc.) and medicine (e.g., orthopedic and dental implants, surgery instruments) [1,2]. However, the absolute strength and hardness of titanium alloys are rather low and that limits their utilization for certain applications, especially in the case of most biocompatible pure titanium and some low-alloyed titanium alloys. A considerable increase in strength can be attained by complex alloying; however, the most accepted alloying elements (e.g., Al, V) can substantially deteriorate biocompatibility [3]. Searching for new titanium alloy compositions with satisfactory properties is still a challenging problem attracting a great deal of interest from materials scientists.

Another promising way to attain high strength and hardness without loss of biocompatibility and corrosion properties is associated with the production of titanium-based composites [4,5]. Among a variety of reinforcements, TiB seems to be the most suitable option due to very similar properties (density, thermal expansion coefficient, good crystallographic matching) with the Ti matrix [5,6]. Titanium–matrix composites (TMCs) can be produced using traditional metallurgical methods through the addition of B into melted Ti [6]. However, powder metallurgy is more preferable in order to obtain a finer microstructure. In this case, TiB reinforcements form in the Ti matrix during the in-situ $3Ti + TiB_2 = 2Ti + 2TiB$ reaction [6–8]. Due to high heating rate and high pressures, the spark plasma sintering (SPS) process can be used for powder consolidation at relatively low temperatures and for short time intervals, thereby preserving a very fine microstructure in the specimen [8]. Hard fibers of TiB increase the strength and hardness of TMC, however the ductility of the composite can decrease to nearly zero [9,10].

Some improvement in both ductility and technological properties was attained in thermomechanically treated TMCs [11–17] due to the shortening and redistribution of TiB whiskers along with the development of dynamic recrystallization in the titanium matrix. Another promising way to improve ductility of the composite can be associated with an increase in plasticity of the titanium matrix by adding a β-stabilizing element. In this case, a hard-to-deform hexagonal close-packed (hcp) lattice should become more ductile due to the greater number of slip systems in a body-centered cubic (bcc) lattice [1]. Meanwhile, there is a lack of information on both the mechanical properties of β-Ti-based TMCs and an influence of thermomechanical treatment on structure and properties of these materials.

In this work a bcc Ti-15 (wt.%)Mo matrix (which is widely used for bio-medical applications) was reinforced by 8.5 vol.% of TiB. This amount of TiB was selected using literature data for hcp Ti/TiB composites to attain a combination of high strength and satisfactory ductility [6]. It was selected to investigate the mechanical properties of the TBC at room temperature, to study the effect of hot deformation at 500–1000 °C and strain rates in the interval 5×10^{-4}–10^{-2} s^{-1} on microstructure evolution and mechanical behavior. The obtained results will be used for the development of a proper thermomechanical treatment of the TMC.

2. Materials and Procedure

Powders of Ti (99.1% purity), Mo (99.95% purity), and TiB_2 (99.9% purity) were used for the sintering of the Ti-15Mo/TiB composites with the bcc Ti matrix. The average sizes of the Ti, Mo, and TiB_2 particles were 25, 3, and 4 μm, respectively (Figure 1). A mixture of the powders containing 80.75 wt.% Ti, 14.25 wt.% of Mo, and 5 wt.% of TiB_2 (to obtain a Ti-15wt.% Mo alloy with 8.5 vol.% of TiB) were produced using a Retsch RS200 vibrating cup mill (RETSCH, Haan, Germany) in ethanol at a milling rotation speed of 700 rpm. The mixing duration was 1 h.

Figure 1. SEM images of (**a**) Ti, (**b**) Mo, and (**c**) TiB_2 powders.

Cylindrical Ti-15Mo/TiB TMC specimens measured 15 mm in height and 19 mm in diameter and were produced using the SPS process under vacuum on a Thermal Technology SPS 10-3 machine (Thermal Technology, LLC, Santa Rosa, CA, USA) at 1400 °C and 40 MPa for 15 min. Then, the sintered samples of the composite were homogenized at 1200 °C for 24 h and cooled in air. To prevent oxidation during the homogenization annealing, the specimens were sealed in vacuumed quartz tubes filled with a titanium getter. The homogenized condition of the TMC is referred to as the initial one hereafter.

Specimens Ø7 × 10 mm in high were cut from the homogenized composite using a Sodick AQ300L electro-discharge machine (Sodick Inc., Schaumburg, IL, USA). Then, the specimens were isothermally strained by compression in air at 20, 500, 600, 700, 800, 900, or 1000 °C on an Instron mechanical testing machine (INSTRON, Horwood, MA, USA). This was done at a nominal strain rate of 10^{-3} s^{-1} to 70% height reduction (a corresponding true strain was $\varepsilon \approx 1.2$) or to the specimens' fracture. The holding time at the required temperatures before the onset of deformation was 15 min. After deformation, the specimens were air cooled. Strain-rate jump compressive tests were conducted at strain rates of 10^{-2}, 5×10^{-3}, 10^{-3}, and 5×10^{-4} s^{-1} in the temperature interval 400–1000 °C. The obtained data were used to determine the activation energy of deformation for the Ti-15Mo/TiB TMC.

Microstructure of the TMC in the initial condition and after deformation was studied using an FEI Quanta 600 FEG scanning electron microscope (Thermo Fisher Scientific, Hillsboro, OR, USA) and a JEOL JEM 2100 transmission electron microscope (JEOL, Tokyo, Japan). Mechanically polished specimens for SEM were etched with Kroll's reagent (95% H_2O, 3% HNO_3, 2% HF). Samples for TEM analysis were prepared by twin-jet electro-polishing in a mixture of 60 mL perchloric acid, 600 mL methanol, and 360 mL butanol at −35 °C and 29 V. The spacing between the TiB whiskers was evaluated based on seven TEM images (more than 100 measurements in total) using the linear-intercept method. Phase composition of the TMC was evaluated using X-ray diffraction (XRD) on an ARL-Xtra diffractometer (Thermo Fisher Scientific, Portland, OR, USA) with Cu Kα radiation. The volume fraction of the phases was determined using the Rietveld method [18].

To determine the optimal processing window, a processing map was constructed using the obtained results for a true strain of 1.2. Toward this end, power dissipation efficiency (η) was evaluated in comparison with an ideal linear dissipator (i.e., with $m = 1$): $\eta = 2m/(m + 1)$ [19], with the strain rate sensitivity of flow stress $m = \frac{\Delta log\sigma}{\Delta log\dot{\varepsilon}}$ (here σ and $\dot{\varepsilon}$ are the flow stress and strain rate, respectively). The processing map shows a three-dimensional area projection describing the η isolines, depending on the strain rate and temperature on the $T - \dot{\varepsilon}$ surface.

3. Results

Microstructure of the Ti-15Mo/TiB composite in the initial condition (i.e., after the homogenization annealing at 1200 °C for 24 h) consisted of the β-Ti matrix reinforced by TiB whiskers (Figure 2a–c). Due to rather high sintering and/or homogenization temperatures, the cross-section of some TiB whiskers attained ~3–4 μm (Figure 2b,c). Nevertheless, the average diameter was almost an order of magnitude smaller (~200 nm). Dislocation density in the titanium matrix was higher nearby or between the TiB whiskers (Figure 2d). Grain boundaries in the titanium matrix were not detected on TEM images and the spacing between the TiB whiskers (which can be accepted in a first approximation as a free dislocation path) was ~0.5–0.7 μm.

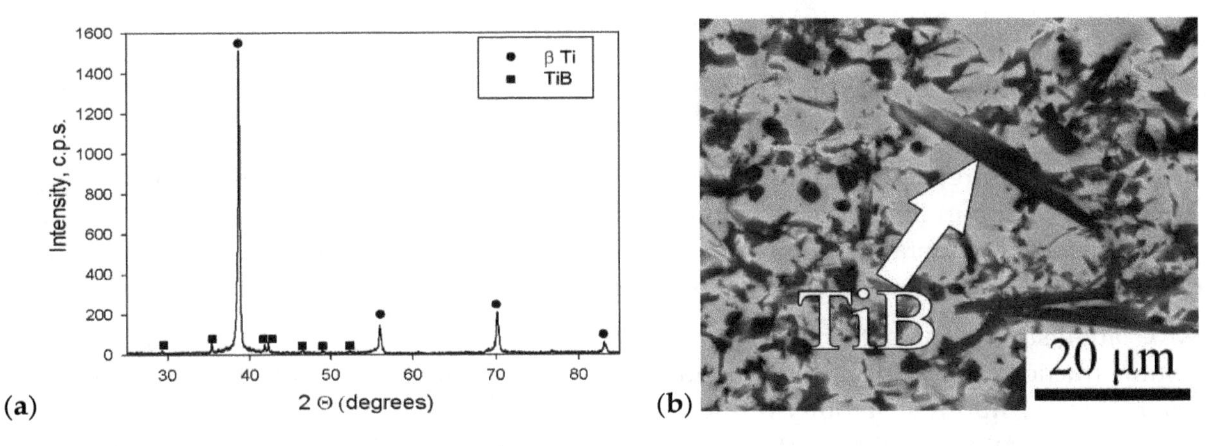

Figure 2. *Cont.*

Figure 2. Microstructure of the Ti-15Mo/TiB titanium–matrix composite (TMC) in the initial condition: (**a**) XRD pattern, (**b**) SEM image of unetched surface; (**c**) SEM image of etched surface; (**d**) TEM bright field image.

During compression at room temperature, the composite showed an attractive combination of strength (the yield strength was ~1480 MPa and the ultimate strength was almost 2 GPa) and ductility (~22% height reduction) (Figure 3a). A short plateau just before the fracture can most likely be associated with strain localization.

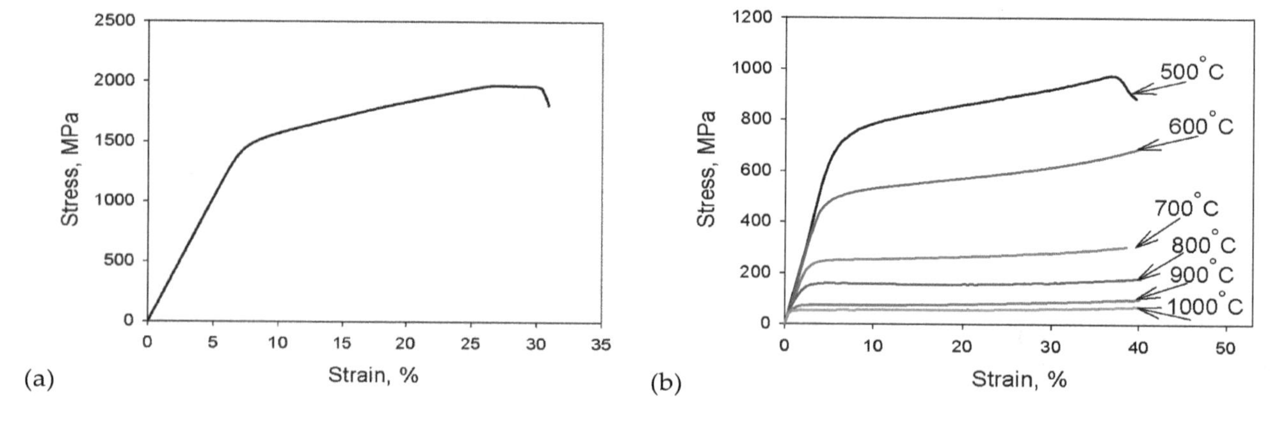

Figure 3. Flow curves obtained during compression at (**a**) 20 °C or at (**b**) 500–1000 °C of the Ti-15Mo/TiB TMC at a nominal strain rate 10^{-3} s^{-1}.

Deformation of the Ti-15Mo/TiB TMC in compression resulted—after initial hardening transient—in continuous strengthening at 500–700 °C (this interval is below the $\alpha + \beta \leftrightarrow \beta$ transition, which occurs at 727 °C for the Ti-15Mo alloy [20]) or steady-state flow at higher temperatures, that is, at 800–1000 °C (Figure 3b). The latter can be ascribed to dynamic recrystallization/recovery processes typical of hot deformation. The yield strength gradually decreased from 630 MPa at 500 °C to 45 MPa at 1000 °C. At 500 °C, the TMC fractured at ~30% reduction.

The microstructure evolution during deformation was associated with some shortening of TiB whiskers (from ~6 μm in the initial condition to ~3.5 μm in all specimens deformed at 500–1000 °C) and their preferred orientation along the metal flow direction (shown by arrows in Figure 4). The redistribution process appeared to be more pronounced with an increase in temperature.

XRD analysis indicated the formation of the α phase in the β matrix during deformation. The volume fraction of the α phase was found to be ~17%–22% in the interval 500–800 °C (this interval corresponds to the minimal stability of the β phase [21]) and decreased to ~7% while increasing in deformation temperature to 1000 °C (Figure 5). The volume fraction of TiB was found to be ~9% in the whole temperature interval.

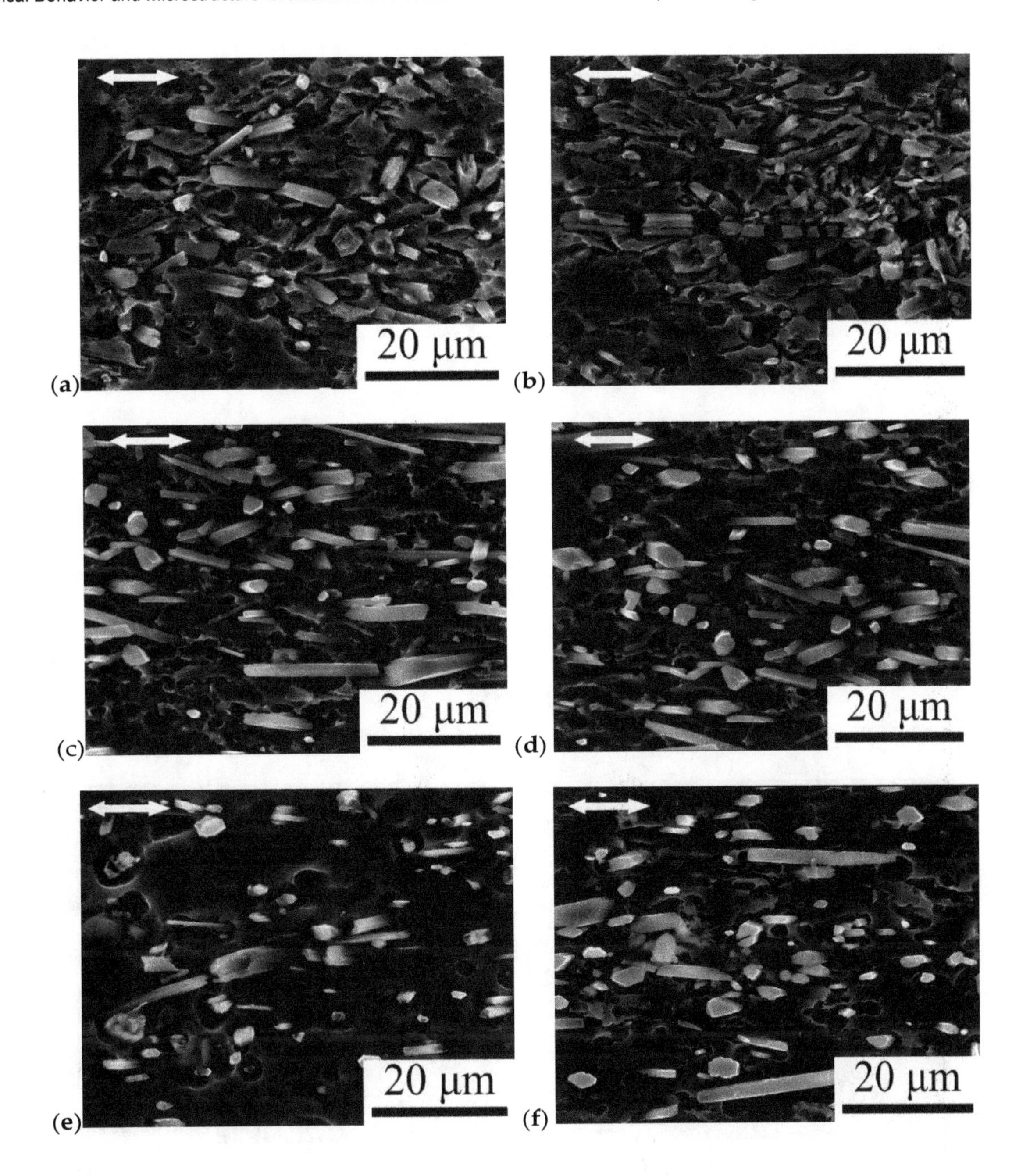

Figure 4. SEM microstructure of the Ti-15Mo/TiB TMC produced using spark plasma sintering (SPS) at 1400 °C; uniaxial compression at a nominal strain rate of 10^{-3} s^{-1} and temperatures (**a**) 500 °C, (**b**) 600 °C, (**c**) 700 °C, (**d**) 800 °C, (**e**) 900 °C, and (**f**) 1000 °C. The compression axis is vertical in all cases and the metal flow direction is shown by arrows.

The TEM investigation showed the formation of a typical deformed microstructure with a rather high dislocation density and dislocation cells/pile-ups after deformation at low temperatures (500–600 °C) (Figure 6a,b). However, areas with a decreased dislocation density were already observed at 600 °C (Figure 6b). In addition, lens-shaped α-phase lamellae (up to 0.1 µm width) were found (Figure 6b). An increase in deformation temperature resulted in the development of dynamic recrystallization, associated with an overall decrease in dislocation density, formation of new recrystallized grains, and straightening of grain boundaries (Figure 6c,d). The size of the recrystallized grains increased with temperature from ~1 µm at 800 °C (Figure 6c) to 2–3 µm at 1000 °C (Figure 6d). At all temperatures, the Ti/TiB interfaces were quite clean and did not contain any cracks or pores.

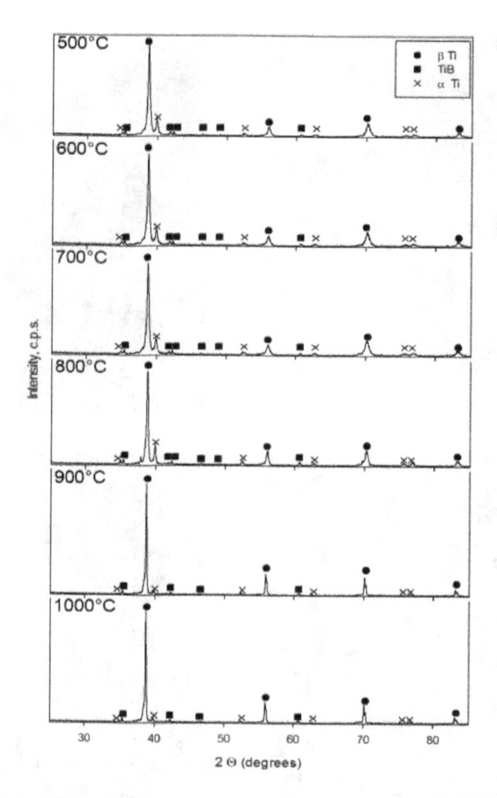

Figure 5. XRD patterns of the Ti-15Mo/TiB MMC after compression at different temperatures.

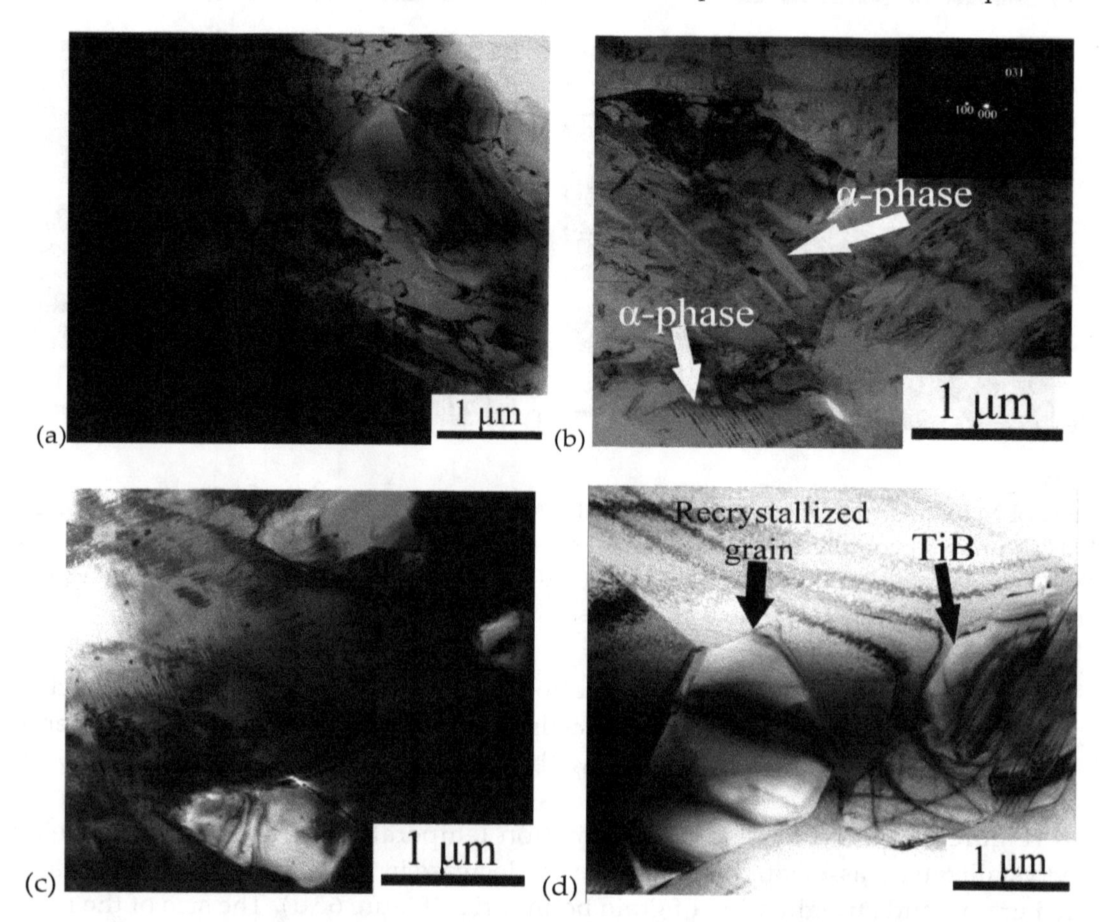

Figure 6. TEM microstructure of the Ti-15Mo/TiB TMC after uniaxial compression at (**a**,**b**) 600 °C, (**c**) 800 °C and (**d**) 1000 °C. Selected area diffraction pattern for the α-phase is inserted in (**b**).

4. Discussion

The obtained results demonstrate a definite beneficial effect of the matrix transition from the hcp alpha titanium to the bcc beta titanium. The latter was obtained through the addition of 15 wt.% of Mo into Ti. The composite produced by SPS showed a very attractive combination of strength and ductility (1.5–2 GPa and ~22%, respectively) (Figure 3a) at room temperature, which is superior to most of the β-rich alloys in the heat-strengthened condition [22] or metal–matrix composites with the hcp α-Ti matrix [4,23,24]. The observed increase in ductility is obviously associated with a greater number of slip systems in the bcc lattice in comparison with that of the hcp one [25]. Lower compression ductility of the hcp Ti/TiB (4%–17%, depending on sintering methods) was also reported in [23,24]. The very high strength of the Ti-15Mo/TiB composite can be ascribed to the superposition of several strengthening mechanisms with Orowan strengthening providing the main contribution [26,27]. Some additional strengthening effects in comparison with the hcp Ti/TiB composites [28] is most likely the result of solid solution strengthening.

An increase in testing temperature led to a decrease in both yield and flow stresses and an improvement in ductility (Figure 3b) due to the activation of additional deformation mechanisms. For example, the pronounced softening at T ≥ 700 °C can be associated with the development of dynamic recrystallization (Figure 6c,d). Meanwhile, precipitations of the α-phase particles (Figures 5 and 6b) should result in some increase in the overall strength. Deformation-induced precipitation of the α-phase particles could be the reason for the hardening transient at the stress–strain curves obtained at 500–700 °C (Figure 3b). Note that the α phase was found in the composite after compression in the full temperature range of 500–1000 °C (Figure 5). However, the α phase formation at temperatures above the α + β ↔ β transition (727 °C for the Ti-15Mo alloy [18]) obviously occurred during cooling from the deformation temperatures. That is why the fraction of the α phase was lower after deformation at the high temperatures.

Since some improvement in mechanical properties can be attained in the composites through hot/warm working [14,29], the obtained results on mechanical behavior during hot deformation were used to determine the optimal processing window. The temperature–strain rate map in Figure 7 shows domains where deformation capacity of the composite was high enough for hot/warm working (i.e., areas associated, for example, with superplasticity or dynamic recrystallization). The map also predicts domains of unstable plastic flow where deformation of the composite can be associated with early necking and/or crack formation [30]. The maximum values of dissipation are expectably related to the highest temperature and the lowest strain rate (the bottom-right corner in Figure 7). However, even at relatively low temperatures (700–800 °C) and strain rates 5×10^{-4}–10^{-3} s^{-1} the value of η fell in the interval 0.35–04, thereby suggesting sufficiently favorable hot working conditions. The obtained results (Figure 7) suggest the occurrence of dynamic recrystallization/recovery in these domains, warranting satisfactory deformability. At higher temperatures, good ductility of the composite was accompanied by intensive grain growth, which is not advantageous for the performance properties.

To gain insight into the operative deformation mechanisms, the apparent activation energy of plastic deformation was calculated using the obtained results (Figure 8). The calculated values of the apparent activation energy were found to be Q = 241 kJ/mol for the high-temperature interval (700–1000 °C) with $n \approx 5$ and Q = 135 kJ/mol with n = 10 for lower temperatures. The former value of Q is very similar to that observed in [20] for the Ti/TiB composite with the hcp α-Ti matrix during deformation in the β phase field region (i.e., at temperatures above 900 °C), where Q was found to be 250 kJ/mol. This value being higher than that reported for self-diffusion in β titanium (153 kJ/mol) [31] can be ascribed to dislocation slip inhibition due to the presence of the TiB whiskers [32]. In addition, similar values of Q = 250–330 kJ/mol were reported in [33] for glide along the prism planes with the thermally activated overcoming of solute atoms as the main rate-controlling mechanism. An increase in the value of Q can also be related to the occurrence of discontinuous dynamic recrystallization during hot deformation [34].

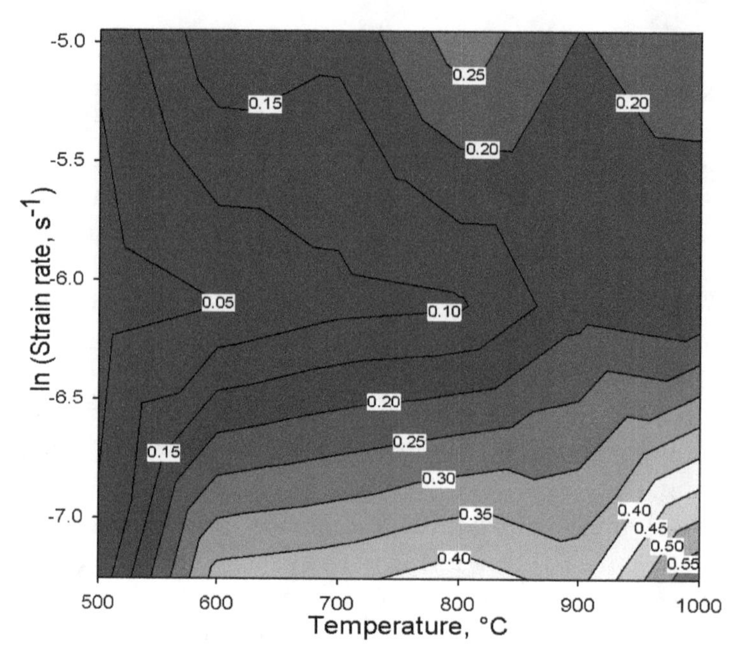

Figure 7. Processing map for the Ti-15Mo/TiB composite.

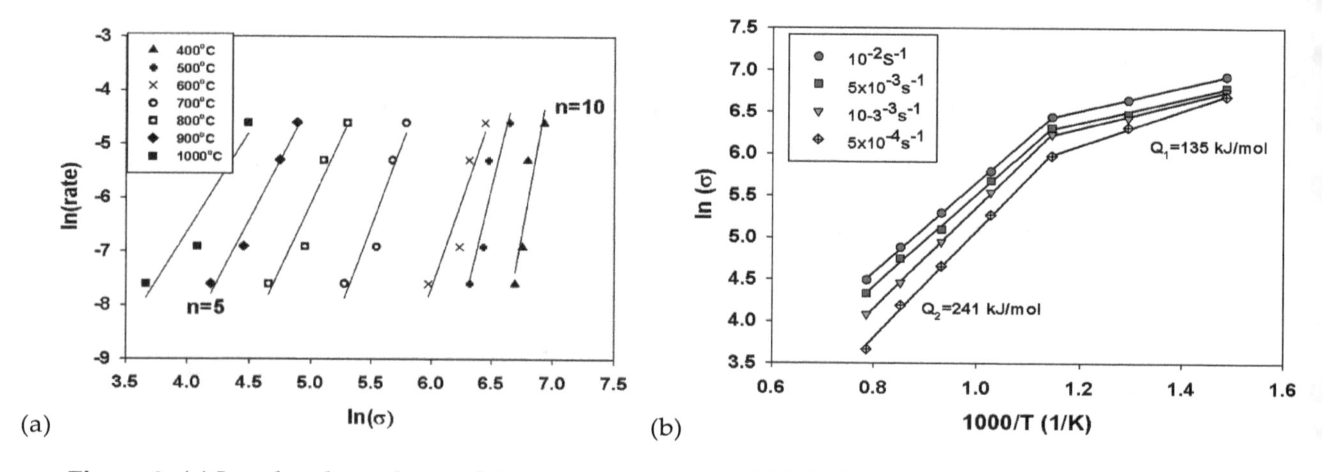

Figure 8. (**a**) Log–log dependence of strain rate on stress and (**b**) Arrhenius semi-log plot of steady-state flow stress vs. inversed temperature for the Ti-15Mo/TiB TMC strained in the interval T = 400–1000 °C.

In the lower temperature range (400–600 °C) the value of Q = 135 kJ/mol suggests that deformation is controlled by volume or pipe diffusion [31,35] associated with the development of dynamic recovery or continuous dynamic recrystallization. Microstructural observations confirmed the development of dynamic recovery in the TMC during compression at 600 °C (Figure 6b).

Quite low values of strain rate sensitivity $m = 1/n \approx 0.1$–0.2 (Figure 8a) along the obtained values of activation energy suggest the operation of dislocation-related mechanisms, mainly (dislocation glide/climb) during deformation in the studied temperature interval [33,36].

5. Summary

In this work, the structure and compression mechanical behavior of a Ti-15Mo/TiB titanium–matrix composite (TMC) produced by spark plasma sintering were studied. The following conclusions were drawn:

(1) The Ti-15Mo/TiB TMC fabricated by spark plasma sintering at 1400 °C under a load of 40 MPa for 15 min using a Ti-14.25(wt.)%Mo-5(wt.)%TiB_2 powder mixture was composed of β-Ti matrix reinforced with TiB whiskers with the average diameter of 200 nm. The as-sintered TMC had

attractive compression mechanical properties at room temperature: yield strength ~1480 MPa and ductility ~22%.

(2) The uniaxial compression in the temperature range of 500–1000 °C demonstrated gradual softening with an increase in deformation temperature. Microstructure evolution of the Ti-15Mo matrix was associated with the development of dynamic recovery at 500–700 °C and dynamic recrystallization at temperatures ≥800 °C. Reorientation of TiB whiskers toward the metal flow direction and some shortening of the whiskers also occurred. In addition, the precipitation of the α particles was found after deformation at 500–1000 °C.

(3) The optimal parameters of thermomechanical processing for the Ti-15Mo/TiB TMC were associated with deformation temperatures of 700–800 °C and strain rates of 5×10^{-4}–10^{-3} s^{-1}.

Author Contributions: S.Z. conceived and designed the experiments. M.O. performed the experiments. D.M. prepared the TEM specimens. M.K. carried out microstructure analysis. S.Z., N.S., M.O., and G.S. analyzed the data and wrote the paper.

Acknowledgments: The authors gratefully acknowledge the financial support from the Russian Science Foundation (Grant Number 15-19-00165). The authors are grateful to the personnel of the Joint Research Centre at Belgorod State University for their assistance with the instrumental analysis.

References

1. Leyens, C.; Peters, M. *Titanium and Titanium Alloys: Fundamentals and Applications*; Wiley-VCH: Weinheim, Germany, 2003; pp. 1–499.
2. Khorasani, A.M.; Goldberg, M.; Doeven, E.H.; Littlefair, G. Titanium in biomedical applications—Properties and fabrication: A review. *J. Biomater. Tissue Eng.* **2015**, *5*, 593–619. [CrossRef]
3. Chen, Q.; Thouas, G.A. Metallic implant biomaterials. *Mater. Sci. Eng. R Rep.* **2015**, *87*, 1–57. [CrossRef]
4. Saito, T.; Furuta, T.; Yamaguchi, T. Development of low cost titanium matrix composite. In *Advances in Titanium Metal Matrix Composites, the Minerals, Metals and Materials Society*; Froes, F.H., Storer, J., Eds.; TMS: Warrendale, PA, USA, 1995; pp. 33–44.
5. Godfrey, T.M.T.; Goodwin, P.S.; Ward-Close, C.M. Titanium Particulate Metal Matrix Composites—Reinforcement, Production Methods, and Mechanical Properties. *Adv. Eng. Mater.* **2000**, *2*, 85–91. [CrossRef]
6. Morsi, K.; Patel, V.V. Processing and properties of titanium-titanium boride (TiBw) matrix composites—A review. *J. Mater. Sci.* **2007**, *42*, 2037–2047. [CrossRef]
7. Ravi Chandran, K.S.; Panda, K.B.; Sahay, S.S. TiBw-reinforced Ti composites: Processing, properties, application, prospects, and research needs. *JOM* **2004**, *56*, 42–48. [CrossRef]
8. Feng, H.; Zhou, Y.; Jia, D.; Meng, Q.; Rao, J. Growth mechanism of in situ TiB whiskers in spark plasma sintered TiB/Ti metal matrix composites. *Cryst. Growth Des.* **2006**, *6*, 1626–1630. [CrossRef]
9. Ozerov, M.; Stepanov, N.; Kolesnikov, A.; Sokolovsky, V.; Zherebtsov, S. Brittle-to-ductile transition in a Ti–TiB metal-matrix composite. *Mater. Lett.* **2017**, *187*, 28–31. [CrossRef]
10. Ozerov, M.; Klimova, M.; Vyazmin, A.; Stepanov, N.; Zherebtsov, S. Orientation relationship in a Ti/TiB metal-matrix composite. *Mater. Lett.* **2017**, *186*, 168–170. [CrossRef]
11. Gaisin, R.A.; Imayev, V.M.; Imayev, R.M. Effect of hot forging on microstructure and mechanical properties of near α titanium alloy/TiB composites produced by casting. *J. Alloys Compd.* **2017**, *723*, 385–394. [CrossRef]
12. Ozerov, M.S.; Gazizova, M.Y.; Klimova, M.V.; Stepanov, N.D.; Zherebtsov, S.V. Effect of Plastic Deformation on the Structure and Properties of the Ti/TiB Composite Produced by Spark Plasma Sintering. *Russ. Metall. (Met.)* **2018**, *7*, 638–644. [CrossRef]
13. Zherebtsov, S.; Ozerov, M.; Stepanov, N.; Klimova, M.; Ivanisenko, Y. Effect of high-pressure torsion on structure and microhardness of Ti/TiB metal-matrix composite. *Metals* **2017**, *7*, 507. [CrossRef]
14. Ozerov, M.; Klimova, M.; Sokolovsky, V.; Stepanov, N.; Popov, A.; Boldin, M.; Zherebtsov, S. Evolution of microstructure and mechanical properties of Ti/TiB metal-matrix composite during isothermal multiaxial forging. *J. Alloys Compd.* **2019**, *770*, 840–848. [CrossRef]

15. Imayev, V.; Gaisin, R.; Gaisina, E.; Imayev, R.; Fecht, H.-J.; Pyczak, F. Effect of hot forging on microstructure and tensile properties of Ti–TiB. *Mater. Sci. Eng. A* **2014**, *609*, 34–41. [CrossRef]
16. Ozerov, M.S.; Klimova, M.V.; Stepanov, N.D.; Zherebtsov, S.V. Microstructure evolution of a Ti/TiB metal-matrix composite during high-temperature deformation. *Mater. Phys. Mech.* **2018**, *38*, 54–63. [CrossRef]
17. Zherebtsov, S.; Ozerov, M.; Stepanov, N.; Klimova, M. Structure and properties of Ti/TiB metal–matrix composite after isothermal multiaxial forging. *Acta Phys. Pol. A* **2018**, *134*, 695–698. [CrossRef]
18. Will, G. *Powder Diffraction: The Rietveld Method and the Two-Stage Method to Determine and Refine Crystal Structures from Powder Diffraction Data*; Springer: Berlin, Germany, 2005.
19. Prasad, Y.V.R.K.; Rao, K.P.; Sasidhara, S. *Hot Working Guide: A Compendium of Processing Maps*; ASM International: Materials Park, OH, USA, 2015; pp. 1–625.
20. Weiss, I.; Semiatin, S.L. Thermomechanical processing of beta titanium alloys—An overview. *Mater. Sci. Eng. A* **1998**, *243*, 46–65. [CrossRef]
21. Jiang, B.; Tsuchiya, K.; Emura, S.; Min, X. Effect of high-pressure torsion process on precipitation behavior of α phase in β-type Ti-15Mo alloy. *Mater. Trans.* **2014**, *55*, 877–884. [CrossRef]
22. Ilyin, A.A.; Kolachev, B.A.; Polkin, I.S. *Titanium Alloys. Composition, Structure, Properties*; VILS-MATI Publishing: Moscow, Russia, 2009.
23. Jeong, H.W.; Kim, S.J.; Hyun, Y.T.; Lee, Y.T. Densification and Compressive Strength of In-situ Processed Ti/TiB Composites by Powder Metallurgy. *Metals Mater. Int.* **2002**, *8*, 25–35. [CrossRef]
24. Attar, H.; Bönisch, M.; Calin, M.; Zhang, L.-C.; Scudino, S.; Eckert, J. Selective laser melting of in situ titanium–titanium boride composites: Processing, microstructure and mechanical properties. *Acta Mater.* **2014**, *76*, 13–22. [CrossRef]
25. Meyers, M.A.; Chawla, K.K. *Mechanical Behavior of Materials*; Cambridge University Press: New York, NY, USA, 2009.
26. Casati, R.; Vedani, M. Metal Matrix Composites Reinforced by Nano-Particles—A review. *Metals* **2004**, *4*, 65–83. [CrossRef]
27. Zherebtsov, S.; Ozerov, M.; Klimova, M.; Stepanov, N.; Vershinina, T.; Ivanisenko, Y.; Salishchev, G. Effect of High-Pressure Torsion on Structure and Properties of Ti-15Mo/TiB Metal-Matrix Composite. *Materials* **2018**, *11*, 2426. [CrossRef] [PubMed]
28. Morsi, K. Review: Titanium–titanium boride composites. *J. Mater. Sci.* **2019**, *54*, 6753–6771. [CrossRef]
29. Ozerov, M.; Klimova, M.; Kolesnikov, A.; Stepanov, N.; Zherebtsov, S. Deformation behavior and microstructure evolution of a Ti/TiB metal-matrix composite during high-temperature compression tests. *Mater. Des.* **2016**, *112*, 17–26. [CrossRef]
30. Rao, K.P.; Prasad, Y.V.R.K. Advanced Techniques to Evaluate Hot Workability of Materials. *Compr. Mater. Process.* **2014**, *3*, 397–426. [CrossRef]
31. Frost, H.J.; Ashby, M.F. *Deformation-Mechanism Maps*; Pergamon Press: Oxford, UK, 1982; pp. 1–166.
32. Zhang, Y.; Huang, L.; Liu, B.; Geng, L. Hot deformation behavior of in-situ TiBw/Ti6Al4V composite with novel network reinforcement distribution, Trans. *Nonferrous Metals Soc.* **2012**, *22*, 465–471. [CrossRef]
33. Conrad, H. Effect of interstitial solutes on the strength and ductility of titanium. *Prog. Mater. Sci.* **1981**, *26*, 123–403. [CrossRef]
34. Raj, S.V.; Langdon, T.G. Creep behavior of copper at intermediate temperatures—I. Mechanical characteristics. *Acta Metall.* **1989**, *37*, 843–852. [CrossRef]
35. Walsöe De Reca, N.E.; Libanati, C.M. Autodifusion de titanio beta y hafnio beta. *Acta Metall.* **1968**, *16*, 1297–1305. [CrossRef]
36. Kumari, S.; Prasad, N.E.; Chandran, K.S.R.; Malakondaiah, G. High-temperature deformation behavior of Ti-TiBw in-situ metal-matrix composites. *JOM* **2004**, *56*, 51–55. [CrossRef]

3

Incremental Forming of Titanium Ti6Al4V Alloy for Cranioplasty Plates—Decision-Making Process and Technological Approaches

Sever Gabriel Racz, Radu Eugen Breaz *, Melania Tera, Claudia Gîrjob, Cristina Biriş, Anca Lucia Chicea and Octavian Bologa

Department of Industrial Machines and Equipment, Engineering Faculty, "Lucian Blaga" University of Sibiu, Victoriei 10, 550024 Sibiu, Romania; gabriel.racz@ulbsibiu.ro (S.G.R.); melania.tera@ulbsibiu.ro (M.T.); claudia.girjob@ulbsibiu.ro (C.G.); cristina.biris@ulbsibiu.ro (C.B.); anca.chicea@ulbsibiu.ro (A.L.C.); octavian.bologa@ulbsibiu.ro (O.B.);

* Correspondence: radu.breaz@ulbsibiu.ro

Abstract: Ti6Al4V titanium alloy is considered a biocompatible material, suitable to be used for manufacturing medical devices, particularly cranioplasty plates. Several methods for processing titanium alloys are reported in the literature, each one presenting both advantages and drawbacks. A decision-making method based upon AHP (analytic hierarchy process) was used in this paper for choosing the most recommended manufacturing process among some alternatives. The result of AHP indicated that single-point incremental forming (SPIF) at room temperature could be considered the best approach when manufacturing medical devices. However, Ti6Al4V titanium alloy is known as a low-plasticity material when subjected to plastic deformation at room temperature, so special measures had to be taken. The experimental results of processing parts from Ti6Al4V titanium alloy by means of SPIF and technological aspects are considered.

Keywords: single-point incremental forming; AHP; cranioplasty plates; decision-making; titanium alloys; medical devices

---

1. Introduction

Titanium alloys are considered eligible materials for biomedical applications (implants and prosthetic devices) due to their biocompatibility. The work presented in [1] provides a comprehensive analysis regarding the main types of titanium alloys used in biomedical applications, as well as their advantages and main drawbacks. A review regarding the titanium alloys seen as the best solution for orthopedic implants is presented in [2], where also the main requirements for a material to be considered a biomaterial are introduced. One of the requirements for this is biocompatibility, which according to [3] is measured by how the human body reacts to the device made of this material when it is implanted. In this work, hip and knee implants are defined as the main orthopedic implants. Both studies presented in [1] and [2] mention Ti6Al4V alloy as one of the titanium alloys; it was initially developed for the aeronautical industry, but can be successfully used for biomedical applications.

Cranioplasty is another main field where titanium alloys may be used due to their biocompatibility. A detailed review about the techniques and materials used in cranioplasty is presented in [4]. Titanium is considered one of the suitable materials, being biocompatible but hard to shape. Another review [5] also indicated titanium alloys as one the materials of choice for cranioplasty plates.

The work presented in [6] reported the successful use of 300 plates of titanium for cranioplasty. The requested shape of the cranioplasty plate was determined either by a traditional technique or by means of computer tomographic scans. Finally, the plates were shaped by pressing them against a counter-die, which could be considered a plastic deformation process. A comprehensive study presented in [7] confirmed titanium alloys as one of the recommended materials for cranioplasty plates, but also highlighted the fact that complications occurred in 29% out of 127 cases. However, the recommendations were to strengthen the prophylaxis measures against infections, rather than replacing titanium as the material for cranioplasty plates. The research was not focused upon the method of manufacturing titanium cranioplasty plates.

Another study reporting the use of titanium plates for cranioplasty was presented in [8]. The work was towards using CAD (computer aided design)/CAM (computer aided manufacturing) methods for manufacturing the plates. A rapid prototyping method based upon fine casting was used and it was reported as having several advantages compared with a traditional milling process.

Titanium alloys were considered the best choice for cranioplasty of large skull defects, according to the results presented in [9]. The study was based upon long-term observations of 26 patients and emphasized the fact that none of the titanium plates implanted had to be removed. Even if the study mentioned CAD/CAM techniques for manufacturing cranioplasty plates, these techniques were not described in detail, and were mostly oriented on the generation of the requested shape of the plate using computer tomography, rather than presenting how the plates were manufactured.

The approaches regarding the methods of manufacturing the titanium plates for cranioplasty are very diversified. The study presented in [10] emphasized the advantages of a manual approach (the shape of the plate was obtained by pressing the titanium sheet against a template model using a manual press), while in [11] the shape of the plate was obtained by means of multiforming, a method which requires very complex technological equipment with a high degree of automatization.

Thus, it can be concluded that titanium alloys are suited for manufacturing cranioplasty plates and there is no consecrated technological approach, either manual or automated, for that. Finding a suitable method for manufacturing the plates was one of the objectives of this work and it involves, in the first stage, a review of the main methods of manufacturing parts from titanium alloys.

The work presented in [12] emphasizes the effects of machining Ti6Al4V alloy by means of cutting, using tools made of straight-grade cemented carbide. Microstructure alterations were reported and, moreover, the reported surface roughness falls into the rough machining category. Consequently, cutting processes may not be the recommended solution when machining biomedical devices from titanium alloys.

Usually, the titanium alloys are also low-plasticity materials, so processing them by means of plastic deformation is also difficult. Single-point incremental forming (SPIF) is one of the manufacturing processes used for processing titanium alloys to overcome the drawback introduced by the low plasticity of these materials, which prevents their processing by other plastic deformation processes. A schematic diagram of the SPIF process is presented in Figure 1, where the sheet metal workpiece (2) is fixed by means of the retaining plate (3) and active plate (backing plate) (1). The punch (4) is moving in the vertical direction, along the Z axis, while the assembly formed by (1)–(3) executes a movement in the XY plane. By combining these movements, various trajectories can be achieved and, subsequently, different shapes of the sheet metal final part.

Literature reviews regarding the SPIF process are presented in [13], which covers the results obtained before 2005, and most recently in [14], which synthesizes the research results reported between 2005 and 2015. The influence of various SPIF process parameters upon the results is synthesized in [15]. It is here noticeable the fact that the maximum achievable angle for a truncated cone part made of Ti6Al4V alloy processed by SPIF using a 10 mm diameter punch reported in [14] was 32°, the lowest one compared with all materials processed by means of SPIF. For comparison, for similar geometry and tool but for DC04 steel and AA 5754 (AlMg3) aluminum alloy, the reported maximum wall angles were 64° and 71°, respectively. These results stress the fact that special measures must be taken when unfolding the SPIF process using titanium alloys as workpieces.

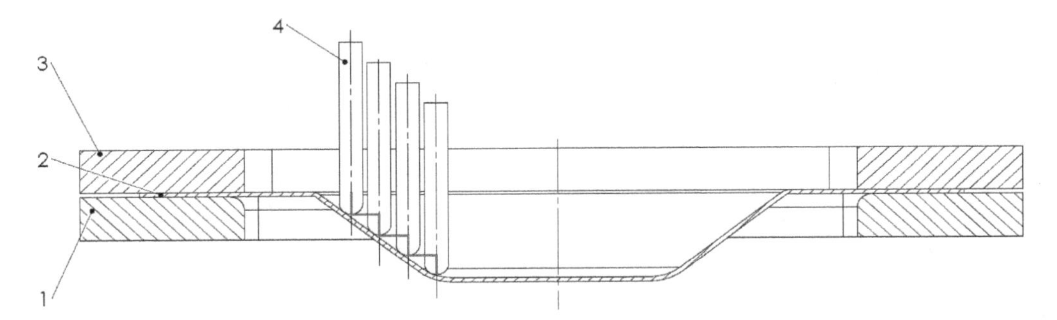

Figure 1. Schematic diagram of asymmetric single-point incremental forming (SPIF).

A comprehensive study about machining commercially pure titanium (CP Ti) by means of incremental forming is presented in [16]. The experiments have proven that by a proper selection of tool (diameter, material) and lubricant (type and lubrication method), wall angles up to 65° can be achieved for CP Ti. Certain values for the ratio between tool diameter (d) and processing pitch (p) were recommended ($d/p \leq 40$) for better results with regards to surface quality.

High feed rates and rotation speeds were tested for parts processed by SPIF and promising results were presented in [17], but the experiments took place on aluminum alloys. Moreover, high feed rates were used for reducing the processing time, in contrast with the results shown in [18], which indicated that formability is inversely proportional with the feed rate. Another experimental study using aluminum alloys for the test parts presented in [19] indicated that the use of high rotation speeds for the forming tools can improve the formability by lowering the forming force. On the other hand, the surface roughness is improved by using the punch rotation, while the rotating speed does not influence it. In [20], Titanium grade 2 and Ti6Al4V were machined on a CNC (computer numerically controlled) lathe using high feed rates. The results have shown that high-speed SPIF does not adversely affect the microstructure of the materials. However, the study was not focused upon the formability, and the geometrical shape of the part was a cone frustum with a wall angle of 25° (for the Ti6Al4V), which was considered by the authors as safe from this point of view.

A new technique of SPIF which involves the heating of the machined workpiece was proposed in [21]. Sheet metal workpieces made of AZ31 magnesium and TiAl2Mn1.5, both with low formability at room temperature, were processed. The parts were heated using direct current (DC) with values between 300 and 600 A, and good results were reported for machining symmetrical parts (cone frustum) from a formability point of view. However, because the method is subject to a patent it was not clearly described how the temperature was controlled and, moreover, how the heat did affect the microstructure of the materials, which is very important when considering the biocompatibility. Another work, presented in [22], used also heat supplied by means of DC to machine several materials including Ti6Al4V by means of SPIF. The studies have reported an increase in the formability (a maximum wall angle of 35° for the cone frustum part made of Ti6Al4V), but also microstructure alterations in the form of different grain distributions were observed. The roughness of the machined part also increased with the increase of the wall angle. Another approach, presented in [23], combined the local heating with high tool speed to machine a car body element made of Ti6Al4V alloy by means of SPIF. The studies have indicated that at 400 °C, the formability of the parts increases, while the normal anisotropy was not influenced by the temperature.

Local heating of the workpiece by a laser beam system, coaxial with the punch, integrated in the main spindle system of the CNC machine-tool used for the SPIF process, was presented in [24]. The path for the laser beam was calculated by taking into consideration the part geometry. Improvements with regards to formability were reported on parts made of Ti6Al4V alloy, where the maximum machining depths were higher than those obtained at room temperature. No references were made with regards to temperature measurement and control or the influence of the heat upon the microstructure of the machined materials. Another method for applying heat on parts machined by SPIF was presented in [25], where friction obtained by tool rotation was used. At speeds between 2000 and 7000 rpm, the heat generated by friction improved the formability of the parts, but as pointed out by the authors it is not only due to material softening, but also to recrystallization. However the material used for this research was AA5052-H32 aluminum alloy, so no information about using frictional heat for machining titanium alloys by means of SPIF was available.

A master–slave tool layout for double-side incremental forming (DSIF) combined with electrically assisted heating was presented in [26]. Good results were reported in processing lightweight materials (AZ31B magnesium alloy) with regards to surface quality and maximum wall angle for truncated cone parts and a new type of hybrid toolpath lead to better geometric accuracy. However, the machining layout for DSIF, which must be custom-built, and the necessity to control and synchronize the toolpaths of the master and slave tools could lead to very high machining costs in contrast to the real value of the machined parts.

With regards to the influence of high temperatures applied during manufacture processes upon the biocompatibility of titanium alloys, there are still many opinions. The biocompatibility of these materials is related to the spontaneous formation of a passive oxide layer at room temperature, which is reported to reduce oxygen diffusion and further oxidation at lower temperatures [27–29] However, at higher temperatures the situation is changing. The works presented in [30–32] consider that oxidation at high temperatures limits the applications of these kind of alloys. In [30] is stated the fact that diffusion of oxygen at temperatures above 400 °C leads to the development of a hard and brittle oxygen diffusion zone which leads to a loss of tensile ductility and of fatigue resistance, reducing the life expectancy of titanium alloys. Above 600 °C, a thick and defective oxide layer is formed, facilitating the penetration of oxygen into the material [30]. Aluminum was found to diffuse outwards through the oxide layer in the later oxidation stage [33]. A study reported in [29] stated that the presence of aluminum in outer layers of Ti6Al4V alloy may hinder osteointegration (bone bonding to the implant) when used as an implant material. Also, aluminum is known to cause neurological disorders [34].

However, a controlled oxidation process called "thermal oxidation" at high temperature is also seen as a promising technique to improve protection against friction and wear [30,32,35–37].

There are also arguments which support the fact that forming titanium alloys at high temperature does not affect their compatibility. Promising results regarding the manufacturing of cranioplasty plates by SPIF with material heating were presented in [38]. The workpiece was heated at 650 °C during the SPIF process and impact tests were performed to measure the maximum force and energy absorbed by the plate. Furthermore, a cytotoxicity test was also performed to assess if the manufacturing process affected the biocompatibility of the plate. The test showed no differences between the processed surfaces and the control ones with regards to biocompatibility. The work presented in [39] had shown that the influence of oxygen enrichment during the process of manufacturing cranial prostheses by means of superplastic forming did not affect the biocompatibility of the Ti6Al4V alloy. A cytotoxicity test was performed, and the viability of the cells was not affected.

Consequently, it is not yet fully demonstrated if heating the material during the process does or does not affect its biocompatibility. However, processing at high temperatures significantly improves the plasticity of the titanium alloys. On the other hand, the complexity of the equipment, the costs associated with that complexity (and with higher energy consumption), and the difficulty of controlling the process favor processing at room temperature, if plasticity requirements can be fulfilled.

From an environmental and sustainability point of view, recent works have pointed out that SPIF is a process with higher amounts of energy consumption compared with other forming processes, such as stamping [40]. The studies presented in [41] have indicated tool speed, type of material, and vertical incremental step in this order as the main influence factors upon the amount of energy consumption during the SPIF process. A similar study which also took into consideration the technological equipment (CNC machine, six-axes industrial robot, and dedicated Amino machine) was presented in [42] and emphasized the fact that forming time is the most influential factor upon the electric energy consumption. Based upon this assumption, the study presented in [43] compared the energy efficiency of performing SPIF machining on a CNC milling machine and on a high-speed CNC lathe. The results have demonstrated that high-speed processing significantly reduces the processing time and, consequently, the energy consumption. Environmental aspects must be considered every time machining is involved, but medical devices such as implants and prosthetic devices are usually machined as prototypes; thus, the environmental impact manufacturing them should be considered quite low. However, reducing the overall machining time is one of the most recommended approaches from this point of view.

The authors of this work have performed some previous studies with regards to using complex trajectories and computer-assisted techniques in SPIF processes [44,45] and some preliminary work in the field of using titanium alloys in cranial implants [46].

As presented above, there are many techniques in use for manufacturing cranioplasty plates from Ti6Al4V alloy, each one with advantages and drawbacks. However, as presented in the next section, single-point incremental forming at room temperature could be considered the best choice, if some criteria are considered. Of course, as reported in the literature, the low plasticity of the Ti6Al4V alloy makes it difficult to process it in this condition. Thus, the approach presented in this paper was oriented towards finding some technological approaches which could allow for better processing of Ti6Al4V alloy at room temperature. As results, some incremental findings, linked mainly with the types of toolpaths and the values of processing steps, with regards to the proposed objective were synthesized.

2. Materials and Methods

2.1. Decision-Making Process

Cranioplasty plates may be manufactured using either a manual or a digital approach (Figure 2). The input data come either from a physical template (a bone fragment taken form the patient) or from a computer tomography (CT) scan. If the manual manufacturing approach is chosen, by means of the physical template, a negative cast is made using laboratory putty. The following steps involve several manual operations, which may differ from one method to another. A comprehensive description of a manual method for manufacturing cranioplasty plates is presented in [10].

The digital approach relies heavily on CAD (computer-aided design)/CAM (computer-aided manufacturing) techniques. The data from the CT scan is processed and converted from a point cloud model to a 3D STL (stereolithography) file format by means of any CAD software package. The 3D model is imported into any CAM software where processing technology and machining code are generated and sent to the technological equipment (CNC machine tool, industrial robot, or even a specialized machine). The format of the machining code and the type of the technological equipment depend on the technological process used for the actual manufacturing of the cranioplasty plate. No matter the chosen manufacturing method, in the final stage, the cranioplasty plate is subjected to some specific operations to prepare it for implantation (i.e., sterilization).

According to the literature review presented above, the most recommended manufacturing processes for cranioplasty plates are cutting (CUT), single-point incremental forming (SPIF), single-point incremental forming with heating (SPIFH), and double-side incremental forming (DSIF). A special mention should be made with regards to additive manufacturing methods, which were recently reported as effective methods for manufacturing cranioplasty plates. An approach presented in [47] presented a two-stage method for manufacturing a cranioplasty plate. In the first stage, by means of 3D printing, a mold was manufactured which was further used for casting a polymethyl-methacrylate (PMMA) cranioplasty plate in the second stage. More recently, an industrial company presented a case study [48] in which a cranioplasty plate was manufactured using a metal 3D printing machine, using Ti MG1 (ISO 10993) as the material.

However, the current research is oriented upon using Ti6Al4V as the material for manufacturing a cranioplasty plate; thus, only CUT, SPIF, SPIFH, and DSIF will be considered for the analysis.

A decision-making method for selecting between these processes based upon AHP (analytic hierarchy process) was developed during this research.

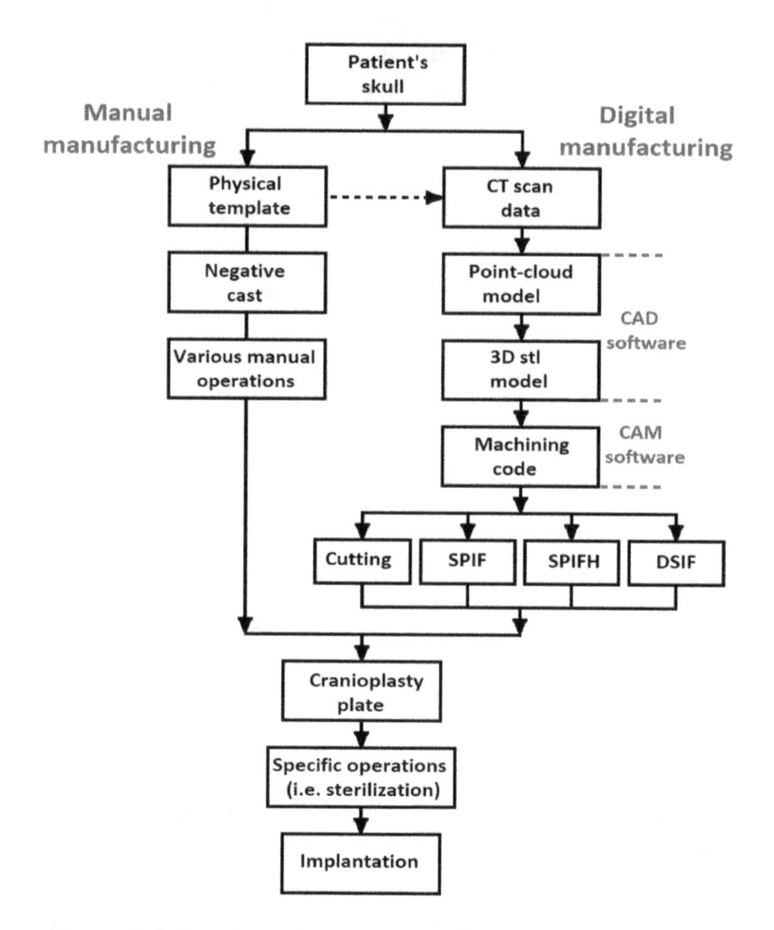

Figure 2. Manufacturing processes for cranioplasty plates.

Comparing the four considered manufacturing processes is a multiattribute decision-making problem due to the factors involved. One of the methods which can be used for this purpose is the Analytic Hierarchy Process (AHP), a method introduced by Saaty [49,50]. The method is based upon pairwise comparison. Elements i and j are compared, and the result is expressed by value a_{ij}. A given hierarchization criteria is used for the comparison:

$$\begin{array}{c} a_{ij} = 1 \quad for\ \ i = j, \quad where\ i, j = 1, 2, \ldots n \\ a_{ij} = \frac{1}{a_{ji}} \quad for\ i \neq j \end{array} \tag{1}$$

The judgement scale, used for AHP, was proposed by Saaty: 1, equally important; 3, weakly more important; 5, strongly more important; 7, demonstrably more important; 9, absolutely more important. The values in between (2, 4, 6, and 8) represent compromise judgements.

To use the AHP process for comparing the four manufacturing processes, a set of seven criteria were proposed and compared pairwise against each other. The preference matrix from Table 1 is used to store the results. The six proposed criteria are presented below:

- C1—Formability: seen here as the ability of the manufacturing process to modify the shape of the workpiece by redistributing the material (plastic deformation). It is noticeable here that three of the analyzed processes are plastic deformation processes, while one of them (CUT) is based upon shaping the part by removing material. However, it was considered that this criterion could be also applied to the CUT process;

- C2—Microstructure: seen here as a measure of how the microstructure of the material is affected by the manufacturing process and, consequently, how the biocompatibility of the processed part could be affected;
- C3—Degree of control: seen here as a measure of how the parameters of the process and the shape and dimensional parameters of the parts (cranioplasty plates) can be controlled;
- C4—Roughness: the meaning of this criterion is quite straightforward, as it expresses the surface quality achievable for the processed parts;
- C5—Energy consumption: it is related with the amount of energy required by each manufacturing process;
- C6—Accuracy: seen here as the maximum achievable accuracy for the parts processed by each of the analyzed manufacturing processes;
- C7—Production time: seen here as the total amount of time to produce a cranioplasty plate.

Table 1. Preference matrix A.

Criterion	C1	C2	C3	C4	C5	C6	C7
C1	1	1/3	5	3	7	3	5
C2	3	1	9	3	9	5	3
C3	1/5	1/9	1	1/5	3	1/5	1/5
C4	1/3	1/3	5	1	7	5	1/3
C5	1/7	1/9	1/3	1/7	1	1/7	1/7
C6	1/3	1/5	1/5	1/5	7	1	1/3
C7	1/5	1/3	5	3	7	3	1

As an example, the way in which the first line of Table 1 was filled is presented below:

- Microstructure (C2) and formability (C1) are very important characteristics of a cranioplasty plate; however, for a device in contact with the human tissue, the state of the microstructure should be considered weakly more important that the ability of shaping the plate;
- The degree of control (C3) is a measure of the quality and repeatability of the process. A higher degree of control will allow the process to be automated, but, finally, for a prosthetic device (which can also be manufactured manually), the ability to shape the plate exactly as required (C1) is strongly more important;
- Roughness (C4) of the part is also important for a prosthetic device, but while the microstructure cannot be repaired if affected by the manufacturing process, roughness could be improved (even by manual operations); thus, the formability of the plate (C1) should be considered weakly more important;
- Energy consumption (C5) should be reduced as possible for any manufacturing process; however, when it comes to cranioplasty plates (which usually are manufactured as prototypes), the ability of shaping the part should (C1) be considered demonstrably more important than saving energy (C5);
- Manufacturing accuracy of the cranioplasty plate (C6) is important, but from the point of view of its functional role (prosthetic device, which is not moving or being in contact with other moving parts), the formability (C1) should be considered weakly more important;
- Production time (C7) is a measure of the efficiency of a production process, but taking into consideration of the fact that, as stated for the (C5) criterion, the cranioplasty plates are manufactured as prototypes, the (C1) criterion should be considered strongly more important.

The next step of the AHP process involves the normalization of the preference matrix by transforming it into matrix B, where

$$B = [b_{ij}]$$
$$b_{ij} = \frac{a_{ij}}{\sum_{i=1}^{n} a_{ij}} \quad (2)$$

It is now required to calculate the eigenvector $w = [w_i]$, which expresses the preference between the elements, by using the following relationship:

$$w_{ij} = \frac{\sum_{i=1}^{n} b_{ij}}{n} \quad (3)$$

The normalized B is presented in Table 2. The eigenvector w was placed on the last column of matrix B, calculated using Equation (3).

Table 2. Normalized matrix B.

Criterion	C1	C2	C3	C4	C5	C6	C7	w
C1	0.1996	0.1458	0.3024	0.3842	0.2188	0.2901	0.2239	0.2256
C2	0.5989	0.4375	0.4234	0.3842	0.2188	0.2901	0.3134	0.3905
C3	0.0399	0.0625	0.0605	0.0427	0.0938	0.0193	0.1343	0.0558
C4	0.0665	0.1458	0.1815	0.1281	0.2188	0.2901	0.1343	0.1777
C5	0.0285	0.0625	0.0202	0.0183	0.0313	0.0138	0.0149	0.0233
C6	0.0665	0.1458	0.0121	0.0427	0.2188	0.0967	0.1343	0.0829
C7	0.0384	0.0640	0.0160	0.0423	0.0811	0.0227	0.0448	0.0442

The comparisons must be checked from the point of view of consistency, according to [48–51] The check is made by calculating the maximal eigenvalue according to

$$\lambda_{max} = \frac{1}{n} \sum_{i=1}^{n} \frac{(Aw)_i}{w_i} = 7.4469 \quad (4)$$

where λ_{max} is the matrix's largest eigenvalue [34].

Using the random consistency index table (Table 3) from [50], the consistency ratio CR may be determined (for a 6-dimensional matrix, the r coefficient is 1.32).

Table 3. Values for CI indices.

Size of Matrix (n)	1	2	3	4	5	6	7	8	9	10
Random average CI (r)	0	0	0.58	0.90	1.12	1.24	1.32	1.41	1.45	1.51

According to Equation (5), the value of CR is smaller than 10%, showing that the comparisons made during the building of matrices A and B are consistent [36,37].

$$CR = \frac{\lambda_{max} - n}{r(n-1)} = 5.64\% \quad (5)$$

The evaluation of the four manufacturing strategies with respect to the seven criteria will be unfolded below. The evaluation for each criterion is presented in Tables 4–10, together with the eigenvectors (introduced in the last column of each table). For exemplification, the way in which the second line of Table 4 was filled is presented below:

- Cranioplasty plates are manufactured starting from a sheet metal workpiece; thus, a plastic deformation process (SPIF) should be considered as an intermediate between equally important and weakly more important than a cutting process (CUT) from the point of view of formability (C1). Even the workpiece is different for cutting, and cutting also allows the user to machine complex shapes; thus an intermediate value has been considered;

Ti6Al4V alloy is known as a low-formability material, and heating it leads to an increase in the formability. However, applying heat could lead to some problems described above. Thus, SPIFH should be considered weakly more important than SPIF, from the (C1) point of view;

- Using a master–slave tools layout with punch and counter-punch will significantly improve the formability of the part, but will also lead to the use of very complex layouts and equipment; this is why DSIF should be considered weakly more important than SPIF, from the (C1) point of view.

Table 4. Comparison of the processing strategies with regards to C1 (formability).

C1	CUT	SPIF	SPIFH	DSIF	w
CUT	1	1/2	1/2	1/2	0.1386
ASPIF	2	1	1/3	1/3	0.1622
ASPIFH	2	3	1	1/2	0.2902
DSPIF	2	2	2	1	0.4090

Table 5. Comparison of the processing strategies with regards to C2 (microstructure).

C2	CUT	SPIF	SPIFH	DSIF	w
CUT	1	1/9	1/5	1/7	0.0399
ASPIF	9	1	7	7	0.6440
ASPIFH	5	1/7	1	1/3	0.1145
DSPIF	7	1/7	3	1	0.2016

Table 6. Comparison of the processing strategies with regards to C3 (degree of control).

C3	CUT	SPIF	SPIFH	DSIF	w
CUT	1	3	5	5	0.5143
ASPIF	1/3	1	5	5	0.3045
ASPIFH	1/5	1/5	1	3	0.1158
DSPIF	1/5	1/5	1/3	1	0.0654

Table 7. Comparison of the processing strategies with regards to C4 (roughness).

C4	CUT	SPIF	SPIFH	DSIF	w
CUT	1	1/7	1/5	1/7	0.0328
ASPIF	7	1	5	1/3	0.3520
ASPIFH	5	1/5	1	7	0.3199
DSPIF	7	3	1/7	1	0.2953

Table 8. Comparison of the processing strategies with regards to C5 (energy consumption).

C5	CUT	SPIF	SPIFH	DSIF	w
CUT	1	3	5	5	0.4941
ASPIF	1/3	1	9	7	0.3713
ASPIFH	1/9	1/5	1	1/2	0.0528
DSPIF	1/7	1/5	2	1	0.0818

Table 9. Comparison of the processing strategies with regards to C6 (accuracy).

C6	CUT	SPIF	SPIFH	DSIF	w
CUT	1	7	5	3	0.5761
ASPIF	1/7	1	1/2	1/3	0.0715
ASPIFH	1/5	2	1	1/3	0.1125
DSPIF	1/3	3	3	1	0.2399

Table 10. Comparison of the processing strategies with regards to C7 (production time).

C7	CUT	SPIF	SPIFH	DSIF	w
CUT	1	5	5	7	0.5430
ASPIF	1/5	1	3	5	0.2445
ASPIFH	1/5	1/3	1	3	0.0765
DSPIF	1/7	1/5	1/3	1	0.1360

The matrix C will be built using the results from Tables 4–10. The columns of matrix C represent the eigenvectors resulting by comparing the four processes pairwise. The order of the columns within matrix C takes into consideration the order of the criteria determined in Table 2: C2, C1, C4, C6, C3, C7, and C5. Performing the multiplication of matrix C and the vector w, the preference vector **x** for the four manufacturing strategies may be obtained, according to the following relation:

$$x = Cw = \begin{bmatrix} 0.0399 & 0.1386 & 0.0328 & 0.5761 & 0.5143 & 0.5967 & 0.4941 \\ 0.6440 & 0.1622 & 0.3520 & 0.0715 & 0.3045 & 0.2292 & 0.3713 \\ 0.1145 & 0.2902 & 0.3199 & 0.1125 & 0.1158 & 0.0188 & 0.0528 \\ 0.2016 & 0.4090 & 0.2953 & 0.2399 & 0.0654 & 0.0553 & 0.0818 \end{bmatrix} \times \begin{bmatrix} 0.2256 \\ 0.3905 \\ 0.0558 \\ 0.1777 \\ 0.0233 \\ 0.0829 \\ 0.0442 \end{bmatrix} = \begin{bmatrix} 0.2506 \\ 0.2835 \\ 0.1919 \\ 0.2740 \end{bmatrix} \quad (6)$$

As can be noticed from Equation (6), the resulting column matrix has the highest value on the second line, 0.2835, a position which corresponds to the second analyzed manufacturing strategy, (ASPIF). According to this result, the AHP process has returned ASPIF as the most recommended approach if the seven proposed criteria are considered. Consequently, the experimental program was oriented to this process as the preferred solution for manufacturing cranioplasty plates. To assess the robustness and the reliability of the AHP process results, a sensitivity analysis was introduced according to the method proposed in [52,53]. According to this, the weights were changed while maintaining the ranking order previously determined. According to the proposed method, a coefficient $\alpha \geq 0$ is introduced and the matrix A is transformed into $\left[a_{ij}^{\alpha}\right]$. If $\alpha > 1$, more dispersed weights are obtained and if $\alpha < 1$, the weights become more concentrated, without any change in the ranking order.

Table 11 shows the weights obtained for α = 0.5, 0.7, 0.9, 1.0, 1.1, 1.3, 1.5 (values proposed in [52]). Table 12 presents the simulation results of calculating the preference vector x for the weights from Table 11. A graphical synthesis of the sensitivity analysis is presented in Figure 3. It can be noticed that the changes in the weights do not affect the hierarchy of the preference vectors x; consequently, SPIF is the most recommended process for the entire range of the analysis.

Table 11. Sensitivity analysis for the weights.

	Coefficient α						
Criterion	**0.5**	**0.7**	**0.9**	**1.0**	**1.1**	**1.3**	**1.5**
C1	0.1128	0.15792	0.20304	0.2256	0.24816	0.29328	0.3384
C2	0.19525	0.27335	0.35145	0.3905	0.42955	0.50765	0.58575
C3	0.0279	0.03906	0.05022	0.0558	0.06138	0.07254	0.0837
C4	0.08885	0.12439	0.15993	0.1777	0.19547	0.23101	0.26655
C5	0.01165	0.01631	0.02097	0.0233	0.02563	0.03029	0.03495
C6	0.04145	0.05803	0.07461	0.0829	0.09119	0.10777	0.12435
C7	0.0221	0.03094	0.03978	0.0442	0.04862	0.05746	0.0663

Table 12. Results of the sensitivity analysis simulations for the preference vector x.

	Coefficient α/Preference Vector x						
Strategy	**0.5**	**0.7**	**0.9**	**1.0**	**1.1**	**1.3**	**1.5**
CUT	0.1253	0.1754	0.2255	0.2506	0.2757	0.3258	0.3759
SPIF	0.1417	0.1984	0.2551	0.2835	0.3118	0.3685	0.4252
SPIFH	0.0959	0.1343	0.1727	0.1919	0.2111	0.2495	0.2878
DSPIF	0.1370	0.1918	0.2466	0.2740	0.3014	0.3562	0.4110

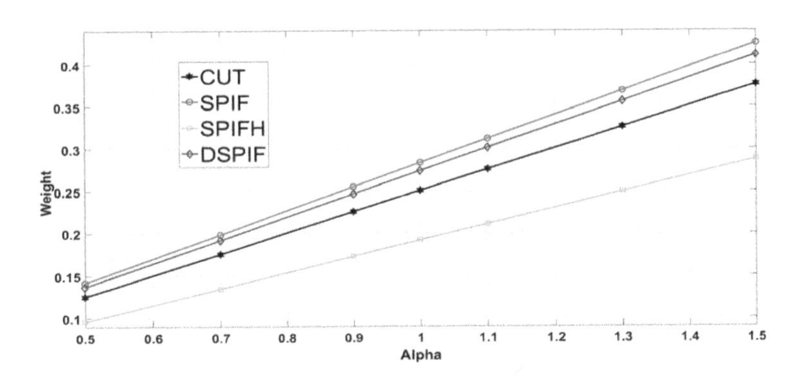

Figure 3. Graphical synthesis of the sensitivity analysis.

2.2. Experimental Layout

As stated in the literature review, one of the most used types of technological equipment for ASPIF processing are CNC milling machines. For this research, a Haas MiniMill CNC machining center was used. The milling machine and the experimental layout mounted on the machine are presented in Figure 4.

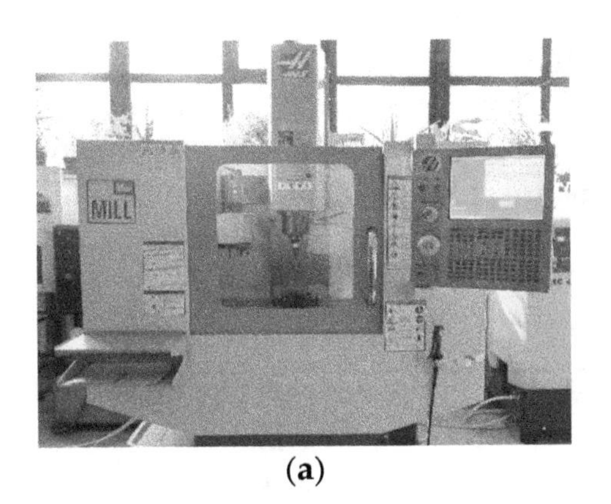

(a)

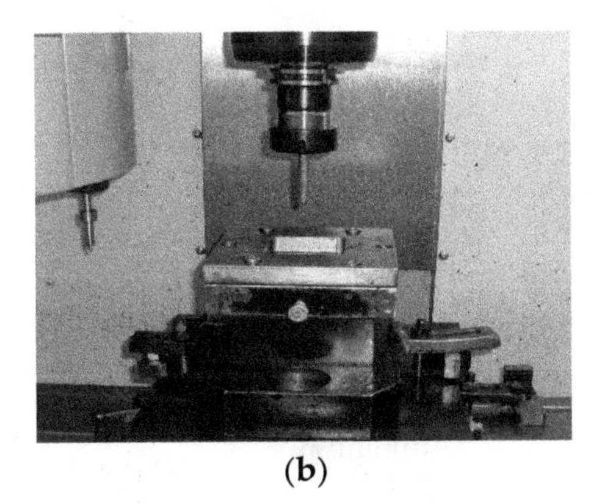

(b)

Figure 4. Technological equipment: (**a**) Haas MiniMill computer numerically controlled (CNC) machining center; (**b**) Forming equipment (active die and retaining plate).

Figure 5 presents a 3D model of the forming equipment, where 1, punch; 2, active die; 3, retaining plate; 4, active die support; 5, baseplate; 6, fixing screws; 7, centering screw; and 8, sheet metal workpiece.

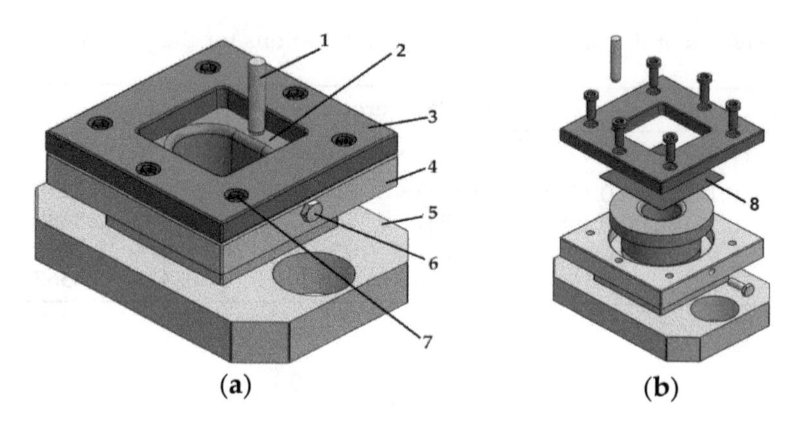

Figure 5. Forming equipment: (**a**) 3D model of the forming assembly—without the sheet metal workpiece; (**b**) Exploded view—with the sheet metal workpiece.

The active plate used for processing the parts is presented in Figure 6. Both the active die and the punch were made from 20Cr115 SR EN ISO 4957:2002 alloyed steel, heat treated.

Figure 6. Active die used for manufacturing the test parts.

2.3. Material

The chemical composition of Ti6Al4V titanium alloy in mass percentage is shown in Table 13.

Titanium is an allotropic substance consisting of a cubic structure (α-Ti) and a compact hexagonal structure up to a temperature of 882 ° C (β-Ti). As can be seen in Table 13, the main alloying elements are aluminum and vanadium, but besides these there are also other minor alloying elements such as iron, oxygen, nitrogen, hydrogen, silicon, and so on.

Table 13. Alloying elements of the Ti6Al4V material.

Alloy Element	Chemical Symbol	Mass Percentage (%)
Aluminum	Al	5.5–6.75
Vanadium	V	3.5–4.5
Carbon	C	0.10
Iron	Fe	0.3
Oxygen	O	0.02
Nitrogen	N	0.05
Hydrogen	H	0.015
Silicon	Si	0.15
Remainder	-	0.4

The mechanical characteristics of the titanium alloy Ti6Al4V are given in Table 14.

Table 14. Mechanical characteristics of the titanium alloy Ti6Al4V.

Characteristic	Measurement Unit	Value
Yield Strength	[MPa]	965–1103
Tensile Strength	[MPa]	896–1034
Density	[g/cm^3]	4.5
Modulus of Elasticity (Young modulus)	[GPa]	116

To determine the other mechanical characteristics of the material needed for finite element method (FEM) analysis, the tensile test was used. The tests were carried out for Ti6Al4V titanium alloy with a thickness of 0.5 mm, using the following laboratory equipment:

- tensile testing machine Instron 5587;
- optical strain measurement system GOM Aramis.

One of the methods of testing the deformation is the uniaxial traction test. On this machine, the specimen is fixed at both ends and deformed at a constant speed until cracking occurs.

Test specimens used for tensile testing are specimens with a calibrated length of 75 mm, a width of 12.5 mm, and a rectangular cross section (Figure 7) in accordance with the standard for the traction testing of metallic materials, SR EN 10002-1: 2002.

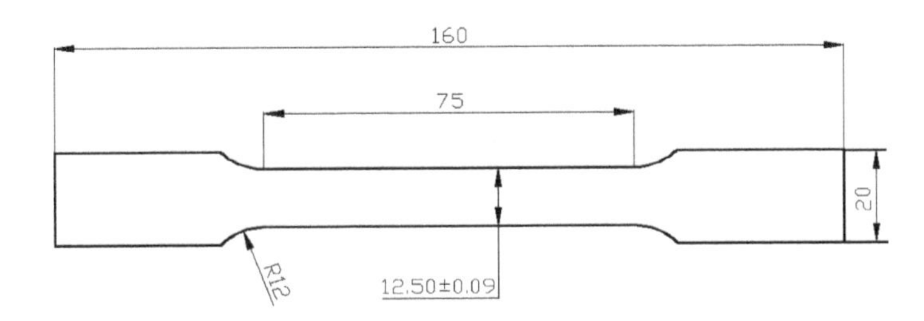

Figure 7. Specimens used for tensile testing.

To study the material anisotropy, sets of specimens were cut (by waterjet cutting) at 0°, 45°, and 90° angles to the sheet rolling direction; these are shown in Figure 8.

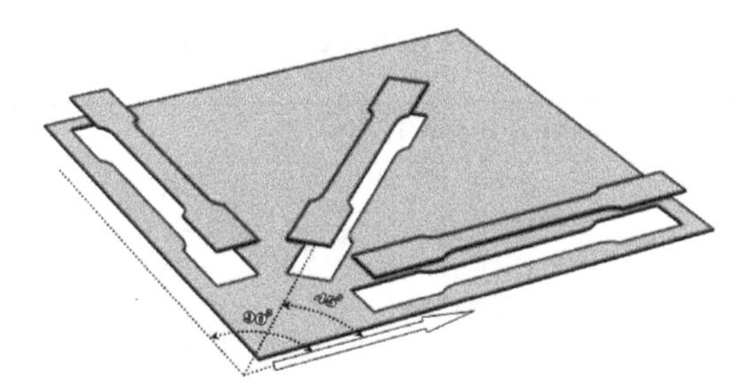

Figure 8. The different angles of the specimens.

The parameters related to the intrinsic properties of the material measured by traction test are hardening coefficient, coefficient of resistance, and coefficients of plastic anisotropy. The values will be used to define the elastoplastic behavior of the material in the FEM simulation. The tensile tests were performed on 3 sets of samples at room temperature, according to Table 15.

Table 15. Tensile tests.

No.	No. of Specimens/Set	Direction of Lamination (°)	Temperature (°C)
1.	3	90°	25 °C
2.	3	0°	25 °C
3.	3	45°	25 °C

Using the BlueHill version 2.0 software (produced by Instron company, Norwood, MA, USA) to control the Instron 5587 Traction Testing Machine (produced also by Instron company), the following were set as input data: type of test, initial dimensions of the specimen, and deformation speed. Both BlueHill software and Instron machine are in the laboratories of Lucian Blaga University of Sibiu.

Following the data processing, the conventional strain curves (σ) versus elongation (ε_{max}) were obtained for the titanium alloy Ti6Al4V at room temperature, which are shown in Figure 9.

The mechanical characteristics of the material that were determined by the traction test are

- modulus of elasticity E [MPa],
- flow limit $R_{p0.2}$ [MPa],
- tensile strength R_m [MPa],
- hardening coefficient n [-],
- resistance coefficient K [Pa],
- elongation ε_{max} [%].

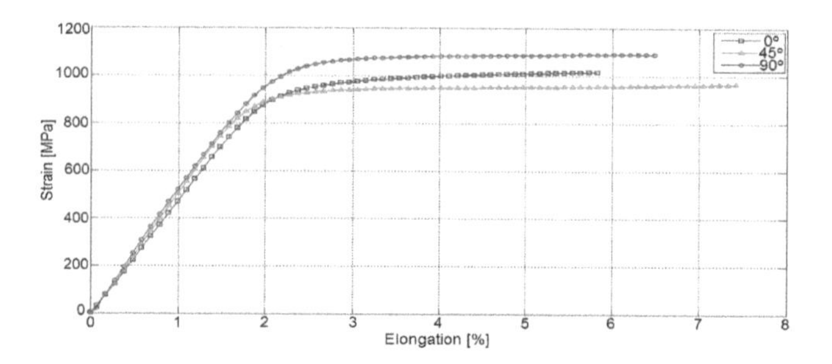

Figure 9. Conventional strain curves (σ) versus elongation (ε_{max}).

In Table 16, a synthesis of the data obtained for the tensile testing for the three types of samples is presented.

Table 16. Synthesis of the data obtained from tensile tests.

Characteristic	Measurement Unit	Value		
Specimen Cutting Angle	[°]	0	45	90
The modulus of elasticity E	[MPa]	49,645.24	49,779.71	52,587.8
Flow Limit $Rp_{0.2}$	[MPa]	881.9	863.11	922.51
Tensile Strength R_m	[MPa]	960.76	992.76	1001.35
Coefficient of hardening n	[-]	0.16	0.1	0.13
Resistance coefficient K	[Pa]	1618.9	1190.88	1392.81
Elongation ε_{max}	[%]	5.5	7.8	6.4

2.4. *Shape of Test Parts*

To test the proposed technological approach, a truncated cone shape of the part was chosen. The geometry of the test part was defined by the cone angle (α°), the height of the part (h), and the diameter of the upper base (d). The shape and dimensions of the parts are presented in Figure 10.

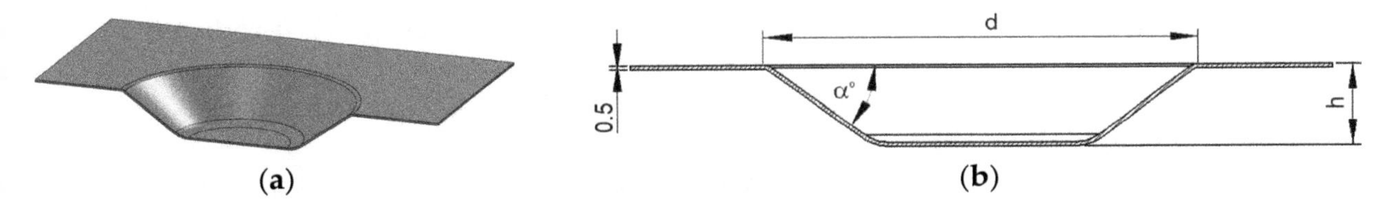

Figure 10. Test parts: (**a**) 3D model; (**b**) Characteristic dimensions.

2.5. Processing Trajectories

The processing trajectories were selected by taking into consideration the third objective stated above. According to the literature review, two main solutions have been imposed lately:

- contour-curves-based trajectory (a contour curve is obtained by intersecting the 3D shape by an XY plane—for the truncated cone, the contour curve is a circle);
- spatial spiral trajectory.

The trajectories used during the experimental test are synthesized in Table 17 and Figure 11.

Table 17. Trajectories.

Trajectory Type	Geometrical Primitive	Code	Observations
Circular trajectories	Contour curve (circle)	CT	The lead-in/lead-out points are lying on the same line (cone generatrix)
Circular trajectories with special entry points	Contour curve (circle)	CTSEP	The lead-in/lead-out points are distributed on the part surface
Spiral trajectories	Spatial spiral	ST	Only one lead-in and one lead-out point

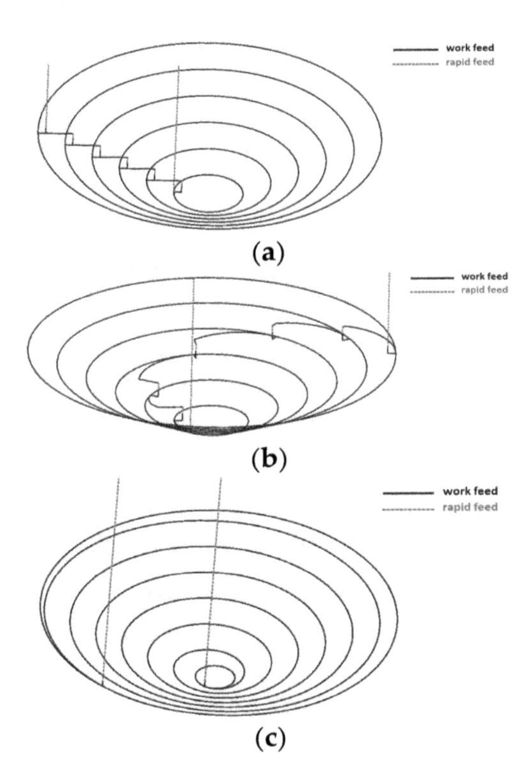

Figure 11. Processing trajectories: (**a**) Circular trajectories with the lead-in/lead-out points lying on the same line; (**b**) Circular trajectories with the lead-in/lead-out points distributed on the surface; (**c**) Spiral trajectories with only one lead-in and one lead-out point.

A computer-aided manufacturing software package, SprutCAM v. 11 (produced by Sprut Technology Ltd., Naberezhnye Chelny, Russia) dedicated for milling operations, was used for generating the trajectories. By using a commercially available CAM solution, the goal regarding ease of generation has been reached.

The circular trajectories (CT) (Figure 11a) have the drawback that all the lead-in points are situated on the same line, a cone generatrix. Lead-in points are the points where the tool (punch) enters in contact with the part. For the circular trajectories, the punch approaches the part with rapid feed changes it in work feed in the near vicinity of the part, and enters in contact with the workpiece all these movements being unfolded on the Z axis. After that, the relative movement between the punch and workpiece is unfolded in the XY plane, until a full circle is completed. After completing the circle, the punch performs a new lead-in movement combined on the XY plane and Z axis, and engages the part on a new circle, situated at distance p from the first one, where p is the vertical step of the ASPIF process. In Figure 11a, all the lead-in points are situated on the same line, a situation which can lead to stress accumulation and, finally, to cracks. It is here noticeable that for a CAM solution for milling, aligning the lead-in points is a default procedure.

To avoid this drawback, the second approach was used. In the circular trajectories with special entry points (CTSEP) situation (Figure 11b), the lead-in points were distributed on the lateral surface of the truncated cone. To achieve this distribution, approach and retraction paths in the XY plane had to be defined (Figure 12) where 1, workpiece; 2, approach path in the XY plane; 3, start of the contour curve (circle); 4, retraction path in the XY plane; and 5, finish of the contour curve (circle).

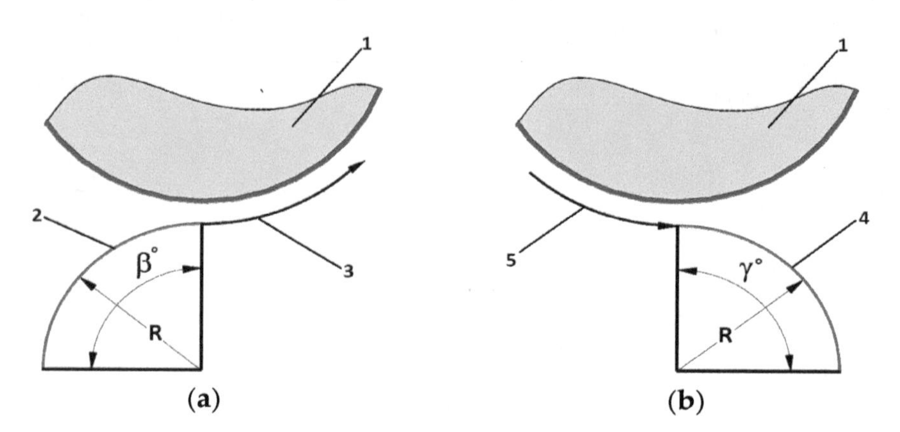

Figure 12. Approach and retraction: (**a**) Approach path in the XY plane defined by radius R and angle β; (**b**) Retraction path in the XY plane defined by radius R and angle γ.

The following relations for the values R, β, and γ were used:

$$\beta = \gamma = 45^\circ \tag{6}$$

By combining the approach and retraction paths in the XY plane with the approach and retraction paths on the Z axis, the movement cycle of the punch may be divided as follows (Figure 13):

- the punch approaches on the Z axis with rapid feed (a);
- the punch continues the approach on the Z axis with work feed, until it reaches the contour curve level (b);
- the punch follows the approach path, in an XY plane at the Z level of the contour curve, until it is positioned on the contour curve (c);
- the punch follows the contour curve (d);
- the punch follows the retraction path, in the XY plane situated at the Z level of the contour curve (e);
- the punch approaches on the Z axis with rapid feed, travelling to the next contour curve (f);

- the punch continues the approach on the Z axis with work feed, until it reaches the next contour curve level—a new XY plane (g);
- after phase (g), the movements are repeating in a cycle, from d to g, until the last contour curve is processed.

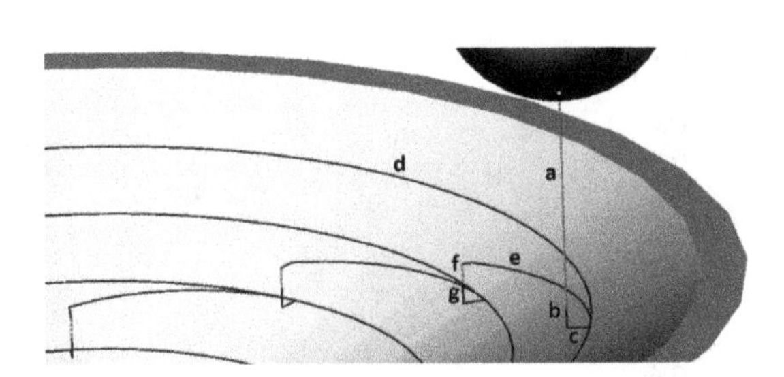

Figure 13. Movement phases for circular trajectories with special entry points (CTSEP).

By performing the movement phases described in Figure 12, the lead-in and lead-out points are distributed on the lateral surface of the cone, thus avoiding the accumulation of stresses and consequently avoiding the occurrence of cracks.

The third approach uses a spatial spiral trajectory (Figure 11c). In this case the trajectory is a continuous one, with only one entry point (lead-in) and one exit point (lead-out).

2.6. *Processed Parts*

The experimental tests were conducted according to the following parameters:

- The working feedrate was fixed to 150 mm/min;
- The punch was fixed in the main spindle of the machine and driven with a rotational speed of 150 rev/min around its own axis. According to the literature review, this rotation reduces the friction and has a favorable influence upon the formability of the material. However, the rotational speed was limited to avoid the heating of the material, which could affect its formability. The temperature limit in this case is 400 °C. At 150 rev/min, the temperature (measured during the process with an FLIR TermoVision A320 thermal imaging camera (manufactured by FLIR Systems, Inc., Wilsonville, OR, USA) was found to be lower than 100 °C;
- The starting angle of the truncated cone was set to 30°, the next one was 35°, and afterwards the angle was incremented by 1°;
- Three vertical steps were considered: 0.2, 0.4, and 0.6 mm. Smaller steps, i.e., 0.1 mm were considered too small to be considered from a technological point of view, while steps greater than 0.6 mm lead to crack occurrence events at an angle of 30°;
- Two punch diameters were considered: 8 and 10 mm;
- Mineral oil was used as lubricant;
- At each angle, the first approach involved the use of the simplest trajectory (CT). If for a given angle this trajectory failed (crack occurrence), the CTSE was used instead. If the latter failed also, the ST trajectory was considered;
- The experimental tests are synthesized in Table 18. The lines in Table 18 only present the parts which were processed without cracks (successful tests). Each successful test was confirmed by performing it three times.

Table 18. Synthesis of the experimental tests.

Crt. No.	Base Diameter d [mm]	Vertical Step p [mm]	Height h [mm]	Cone Angle α [°]	Punch Diameter d_p [mm]	Trajectory Type
1.	55	0.4	12	30	8	CT
2.		0.4		30	10	CT
3.		0.6		30	8	CT
4.		0.6		30	10	CT
5.		0.2		35	8	CT
6.		0.2		35	10	CT
7.		0.4		35	8	CTSEP
8.		0.4		35	10	CTSEP
9.		0.6		35	8	CTSEP
10.		0.6		35	10	CTSEP
11.		0.2		36	8	CT
12.		0.2		36	10	CT
13.		0.4		36	8	CTSEP
14.		0.4		36	10	CTSEP
15.		0.6		36	8	ST
16.		0.6		36	10	CTSEP
17.		0.2		37	8	CTSEP
18.		0.2		37	10	CTSEP
19.		0.4		37	8	ST
20.		0.4		37	10	ST
21.		0.6		37	10	ST
22.		0.2		38	8	ST
23.		0.2		38	10	ST
24.		0.4		38	10	ST

Some of the successful tests are presented in Figures 14–16.

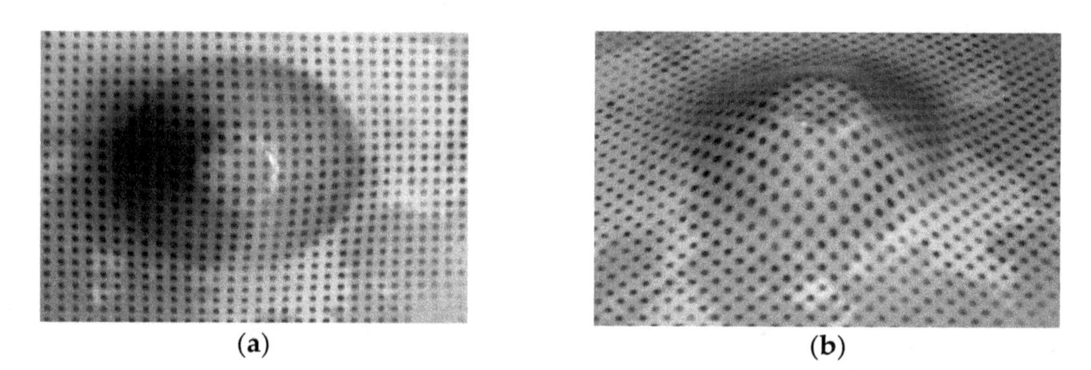

Figure 14. Part with $p = 0.2$ mm, $\alpha = 30°$, $d_p = 8$ mm, and CT. (**a**) view from above; (**b**) side view.

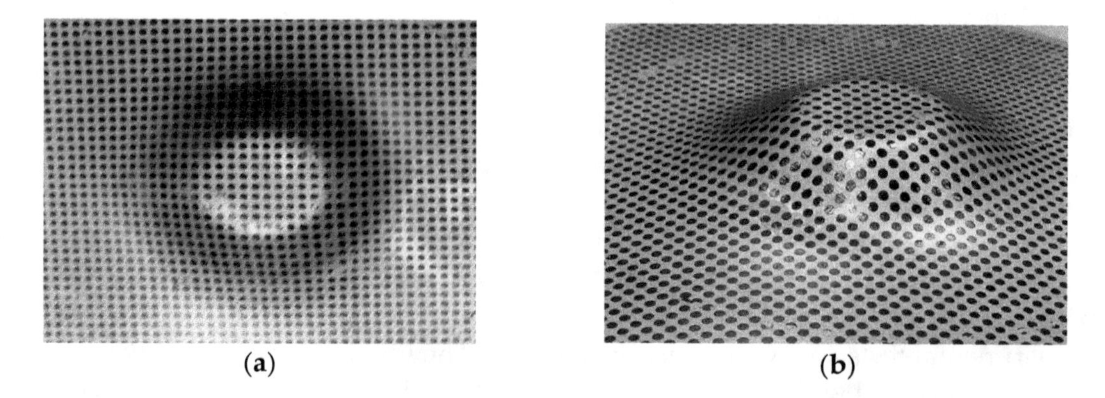

Figure 15. Part with $p = 0.6$ mm, $\alpha = 35°$, $d_p = 10$ mm, and CTSEP. (**a**) view form above; (**b**) side view.

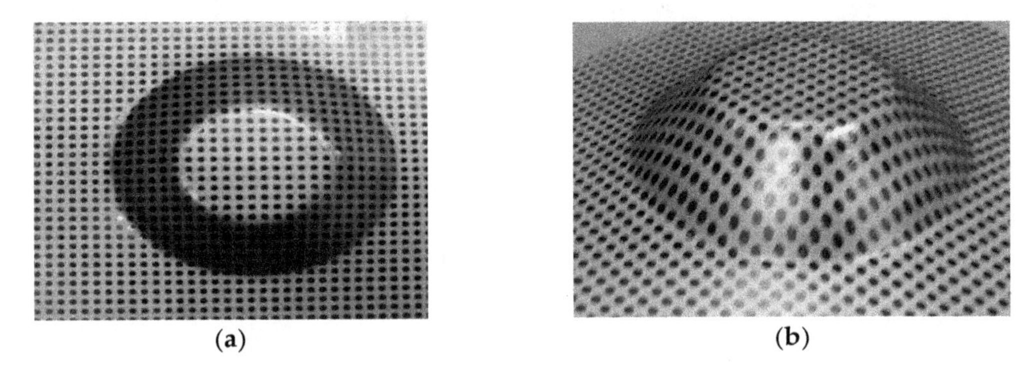

Figure 16. Part with $p = 0.4$ mm, $\alpha = 38°$, $d_p = 10$ mm, and ST. (a) view from above; (b) side view.

Some examples of processed parts which cracked during the ASPIF process are presented in Figures 17–19.

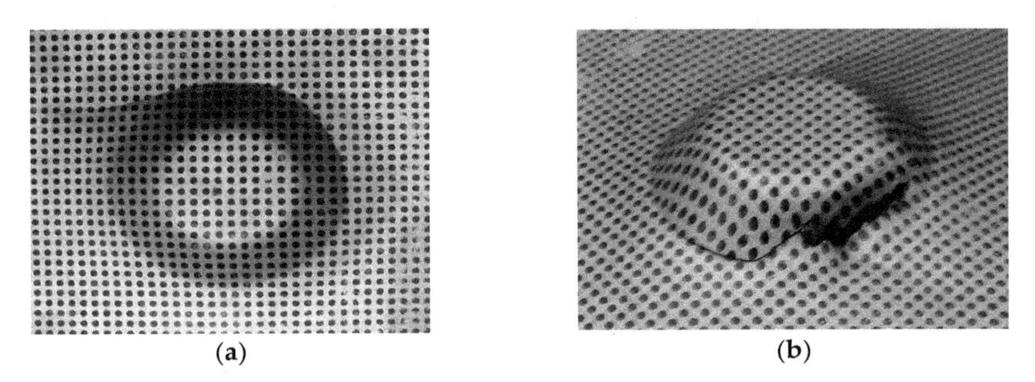

Figure 17. Part with $p = 0.6$ mm, $\alpha = 35°$, $d_p = 8$ mm, and CT. (a) view from above; (b) side view.

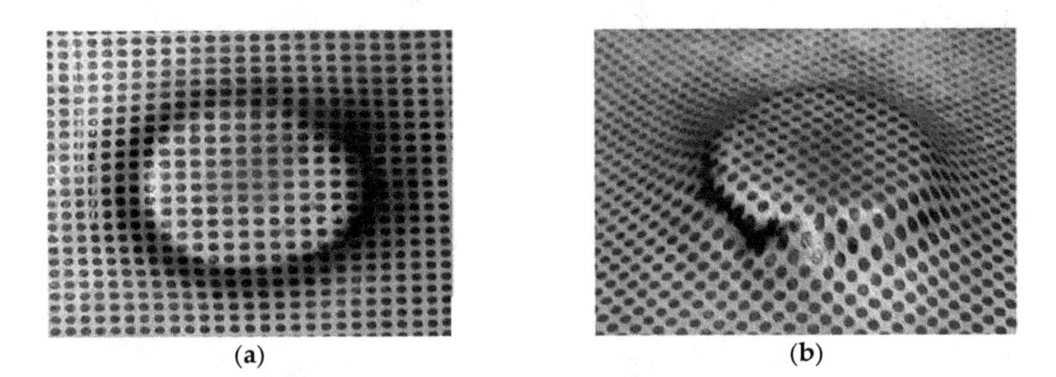

Figure 18. Part with $p = 0.4$ mm, $\alpha = 38°$, $d_p = 8$ mm, and ST. (a) view from above; (b) side view.

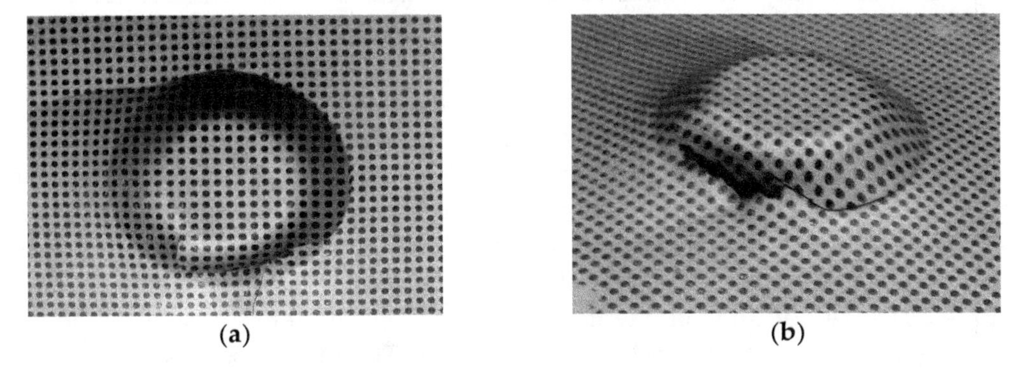

Figure 19. Part with $p = 0.4$ mm, $\alpha = 40°$, $d_p = 10$ mm, and ST. (a) view from above; (b) side view.

To test the accuracy of the processed parts, some measurements were performed using a Mahr profilometer (from Mahr Gmbh, Göttingen, Germany). Figure 20 presents a graphical display of the measurement results for Part 24, while Table 19 presents a synthesis of the results for the measured parts. Even if more dimensional characteristics were measured, the results were focused on the angle α° and on the surface roughness (expressed by Ra and Rz). It is here noticeable the fact that taking into consideration the functional role intended for the parts (cranioplasty plates), their requested accuracy lies in a very different range compared with parts from the manufacturing industry, for example.

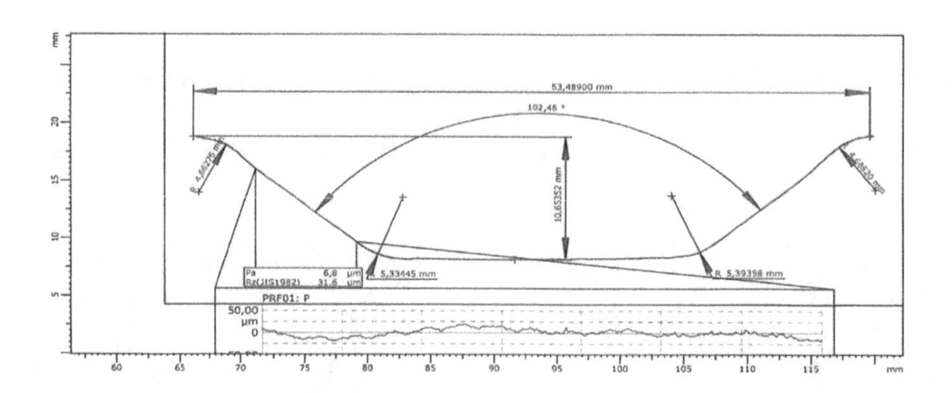

Figure 20. Measurement results for Part 24.

Table 19. Synthesis of results for the measured parts.

Part No.	Characteristics	Measured Values
1.	$\alpha^\circ = 30/p = 0.4/d_p = 8$ mm/CT	$\alpha^\circ = 31.485$ Ra = 15.4 µm/Rz = 23.1 µm
3.	$\alpha^\circ = 30/p = 0.6$ mm$/d_p = 8$ mm/CT	$\alpha^\circ = 31.125$ Ra = 19 µm/Rz = 52.4 µm
6.	$\alpha^\circ = 35/p = 0.2$ mm$/d_p = 10$ mm/CT	$\alpha^\circ = 36.125$ Ra = 6 µm/Rz = 35.3 µm
8.	$\alpha^\circ = 35/p = 0.4$ mm$/d_p = 10$ mm/CTSEP	$\alpha^\circ = 35.86$ Ra = 16.5 µm/Rz = 58.2 µm
11.	$\alpha^\circ = 36/p = 0.2$ mm$/d_p = 8$ mm/CT	$\alpha^\circ = 35.865$ Ra = 10.3 µm/Rz = 37.5 µm
18.	$\alpha^\circ = 37/p = 0.2$ mm$/d_p = 10$ mm/CTSEP	$\alpha^\circ = 38.8$ Ra = 8.3 µm/Rz = 21 µm
23.	$\alpha^\circ = 38/p = 0.2$ mm$/d_p = 10$ mm/ST	$\alpha^\circ = 38.56$ Ra = 3.5 µm/Rz = 18.3 µm
24.	$\alpha^\circ = 38/p = 0.4$ mm$/d_p = 10$ mm/ST	$\alpha^\circ = 38.77$ Ra = 6.8 µm/Rz = 31.6 µm

2.7. FEM Analysis

The Abaqus/Explicit software package, v.14 (produced by Dassault Systèmes®, Vélizy-Villacoublay, France) as used for the FEM analysis. A parameterized model based upon the play between the punch and the active plate, the retention pressure, the diameter of the blank, and the radius of the active plate was developed. The geometric model included the sheet metal workpiece (considered as a deformable body), the active plate, the retention plate, and the punch (all being considered as rigid bodies). For the finite element mesh, four-node shell elements were used. The modeling was done on medium fiber, with five integration points per thickness being considered. The FEM model is presented in Figure 21.

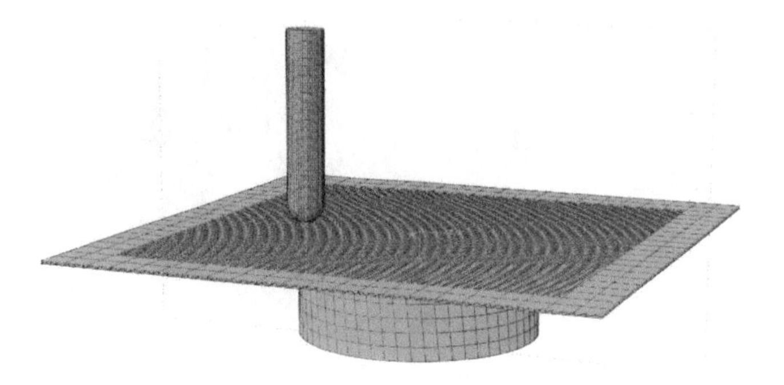

Figure 21. FEM model.

The parameters targeted by the FEM simulations were:

- major strains (ε_1);
- minor strains (ε_2);
- thickness reduction (s_{max});
- evolution of the forming force on Z axis.

Figures 22–24 present the variations in ε_1, ε_2, and s_{max} for a truncated cone with diameter of the upper base d = 55 mm, cone angle α = 30°, vertical step p = 0.4 mm, punch diameter d_p = 10 mm, and spatial spiral trajectories.

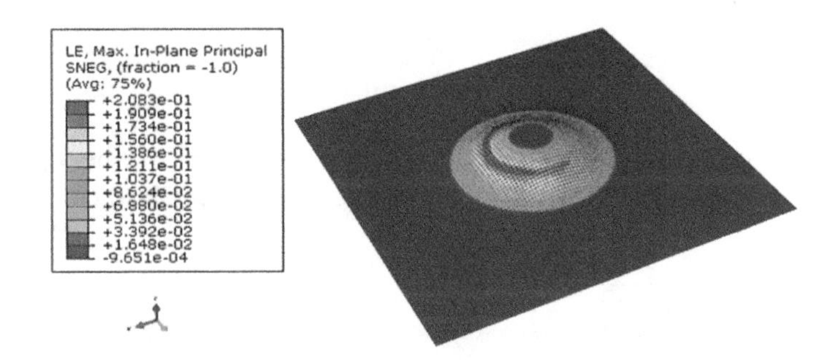

Figure 22. Distribution of major strains (ε_1)—diameter of the upper base d = 55 mm, cone angle α = 30°, vertical step p = 0.4 mm, punch diameter d_p = 10 mm, and spatial spiral trajectories (ST).

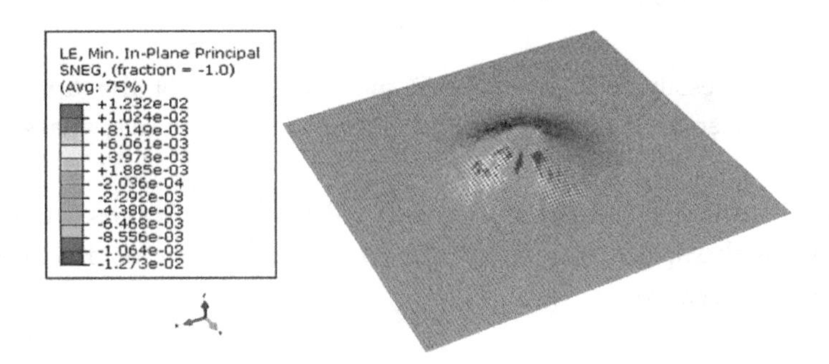

Figure 23. Distribution of minor strains (ε_2)—diameter of the upper base d = 55 mm, cone angle α = 30°, vertical step p = 0.4 mm, punch diameter d_p = 10 mm, and spatial spiral trajectories (ST).

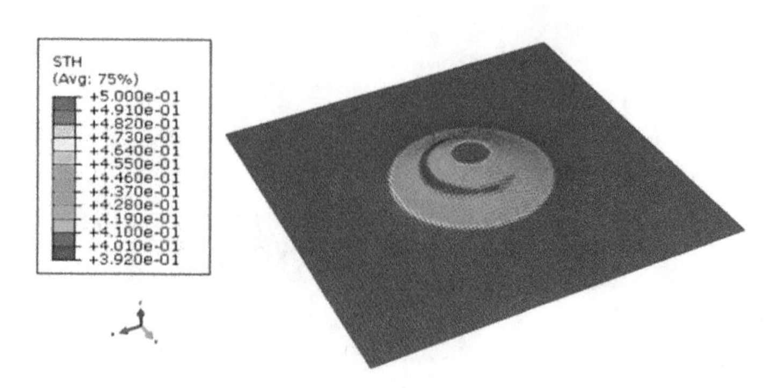

Figure 24. Thickness reduction (s_{max})—diameter of the upper base d = 55 mm, cone angle α = 30°, vertical step p = 0.4 mm, punch diameter d_p = 10 mm, and spatial spiral trajectories (ST).

Figures 25 and 26 present the simulated processing force on the Z axis, for the same part, diameter of the upper base d = 55 mm, cone angle α = 30°, vertical step p = 0.4 mm, and punch diameter d_p = 10 mm, but for different types of trajectories—circles (Figure 22) and spatial spiral (Figure 23). It is here noticeable the fact that the simulation speed was increased by a magnification factor of 100; thus, the time scale covers the whole process.

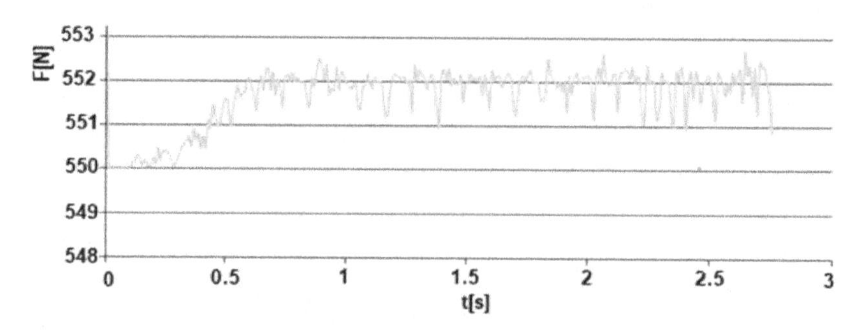

Figure 25. Simulated processing forces on the Z axis—diameter of the upper base d = 55 mm, cone angle α = 30°, vertical step p = 0.4 mm, punch diameter d_p = 10mm, and circular trajectories (CT).

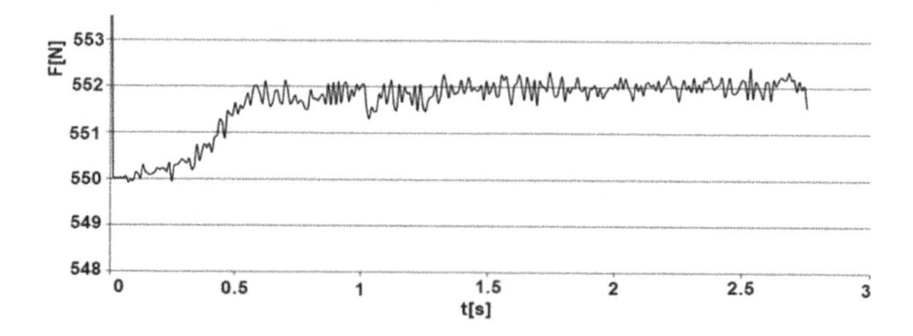

Figure 26. Simulated processing forces on the Z axis—diameter of the upper base d = 55 mm, cone angle α = 30°, vertical step p = 0.4 mm, punch diameter d_p = 10mm, and spatial spiral trajectories (ST).

A preliminary analysis reveals that the maximum values of the forces are quite similar, oscillating around 552 N. However, for the spiral trajectory, the amplitude of the oscillations is higher, a fact that could favor the occurrence of cracks.

2.8. Experimental Measurements

A GOM Argus optical system (produced by GOM company, Braunschweig, Germany) was used for measuring the parts. Figure 27 presents the experimental results for major strain (ε_1) distribution for the part with diameter of the upper base d = 55 mm, cone angle α = 30°, vertical step p = 0.4 mm, punch diameter d_p = 8 mm, and circular trajectories (CT). The maximum value of the major strain is 37.6%. Figure 28 presents the results for major strain (ε_1) distribution for the part with d = 55 mm, α = 35°, p = 0.4 mm, d_p = 8 mm, and circular trajectories with separate entry points (CTSE). The maximum value of the major strain is 34.1%. Figure 29 presents the results for major strain (ε_1) distribution for the part with d = 55 mm, α = 38°, p = 0.4 mm, d_p = 10 mm, and spatial spiral trajectories (ST). The maximum value of the major strain is 23.7%.

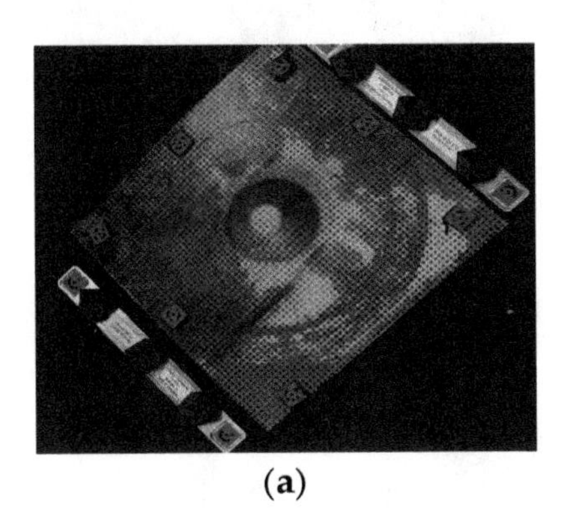

(**a**)

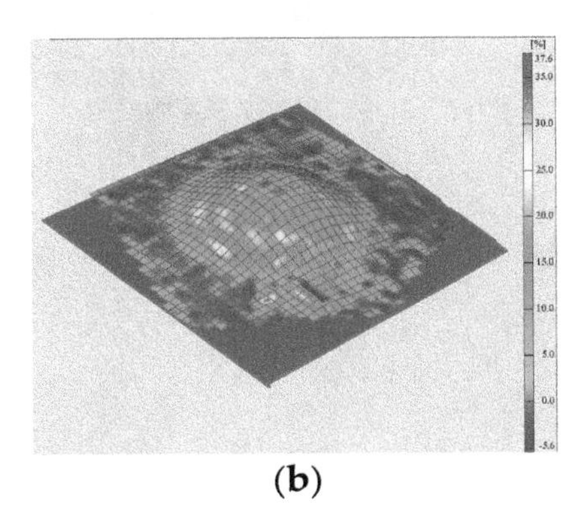

(**b**)

Figure 27. Measured distribution of major strains (ε_1)—diameter of the upper base d = 55 mm, cone angle α = 30°, vertical step p = 0.4 mm, punch diameter d_p = 8 mm, and circular trajectories (CT), (**a**) measured part; (**b**) measurement results.

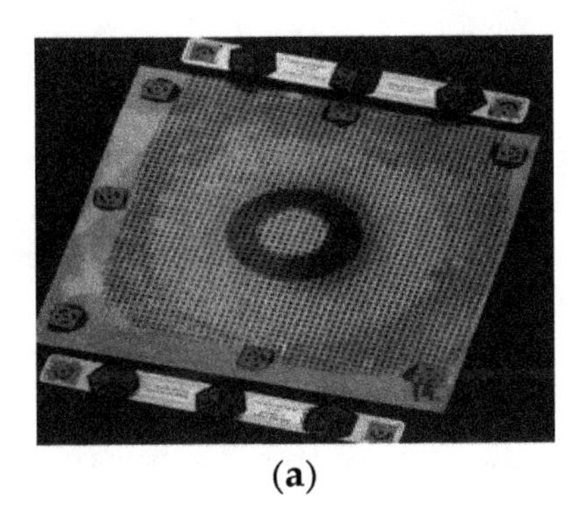

(**a**)

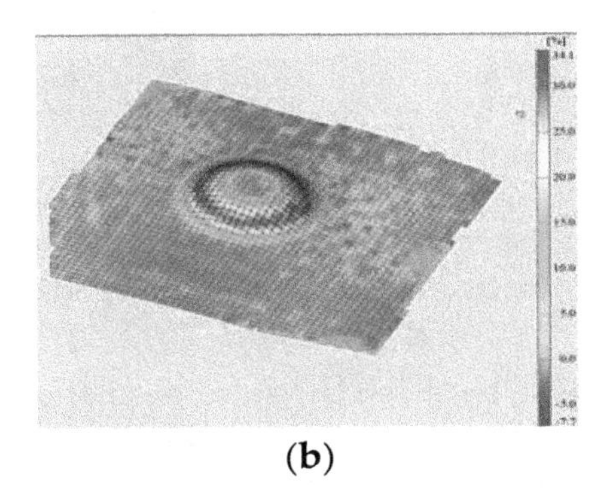

(**b**)

Figure 28. Measured distribution of major strains (ε_1)—diameter of the upper base d = 55 mm, cone angle α = 35°, vertical step p = 0.4 mm, punch diameter d_p = 8 mm, and circular trajectories with separate entry points (CTSEP), (**a**) measured part; (**b**) measured results.

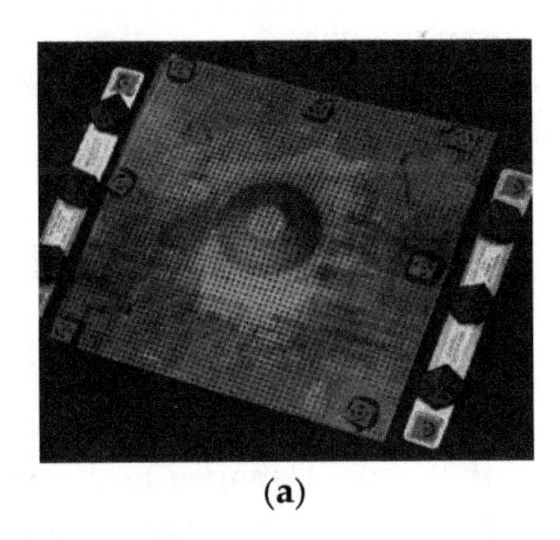

(**a**)

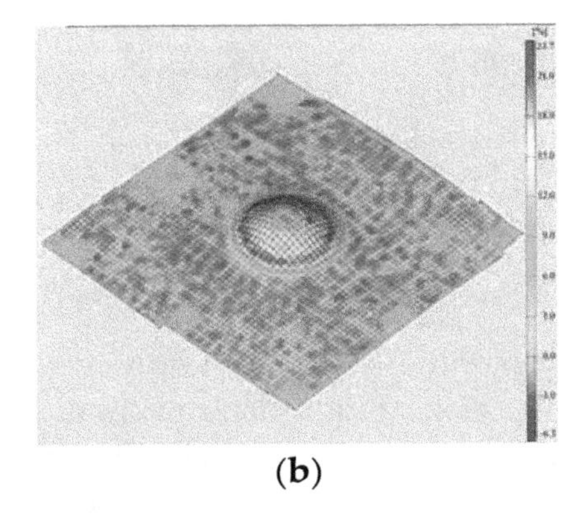

(**b**)

Figure 29. Measured distribution of major strains (ε_1)—diameter of the upper base d = 55 mm, cone angle α = 38°, vertical step p = 0.4 mm, punch diameter d_p = 10 mm, and spatial spiral trajectories (ST), (**a**) measured part; (**b**) measurement results.

A complete set of measured values of major strains (ε_1), minor strains (ε_2), and thickness reduction (s_{max}) for the part with diameter of the upper base d = 55 mm, cone angle α = 30°, vertical step p = 0.4 mm, punch diameter d_p = 10 mm, and spatial spiral trajectories (ST) is presented in Figure 30.

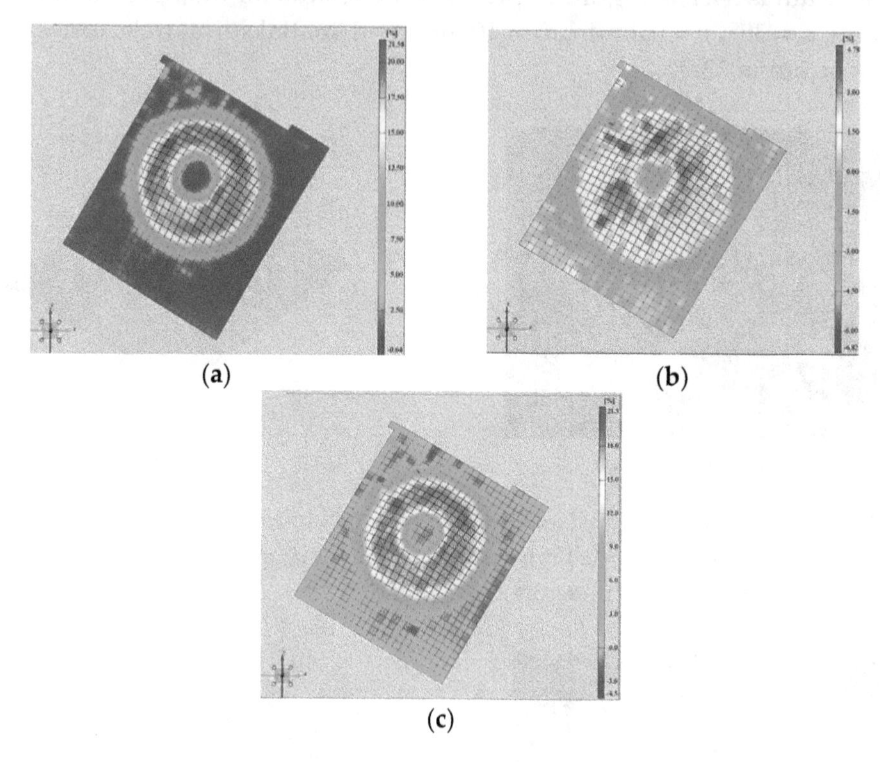

Figure 30. Measured values for major strains (ε_1) (**a**), minor strains (ε_2) (**b**), and thickness reduction (s_{max}) (**c**) for the parts with diameter of the upper base d = 55 mm, cone angle α = 30°, vertical step p = 0.4 mm, punch diameter d_p = 10 mm, and spatial spiral trajectories (ST).

A comparison between the simulated and the experimentally measured values for major strains (ε_1), minor strains (ε_2), and thickness reduction (s_{max}) for the part with diameter of the upper base d = 55 mm, cone angle α = 30°, vertical step p = 0.4 mm, punch diameter d_p = 10 mm, and spatial spiral trajectories (ST) is presented in Table 20.

Table 20. Comparison between simulated and measured values.

Part d = 55 mm, α = 30°, p = 0.4 mm, d_p = 10 mm, ST	Characteristic Input					
	Major Strains ε_1		Minor Strains ε_2		Thickness Reduction s_{max}	
	%	log	%	log	%	log
Experimental	21.52	0.1951	4.78	0.0467	21.5	0.242
Simulated	-	0.2083	-	0.0123	-	0.216

2.9. *Manufacturing a Cranioplasty Plate*

The next step of the work was to process by means of SPIF a part with complex shape, specific for cranioplasty plates, to demonstrate that the proposed technological conditions allow the user to manufacture irregular shapes with rapid variations of the wall shapes and angles at room temperature. A manually made physical model was considered, taking into consideration the following requirements:

- The shape of the model had to be highly irregular, to mimic as close as possibly the human skull;
- The shape of the model had to present rapid variations of the wall shapes and angles;

- Even if the experimental layout had size limitations, the overall area of the model was chosen about 40 cm^2 (exactly 36.5 cm^2). According to the literature [54,55], this size could be considered as a quite common value for a cranial defect surface area.

After scanning the physical model, the 3D model of the part, presented in Figure 31, was stored in an stl file which resulted after processing a CT scan point cloud file.

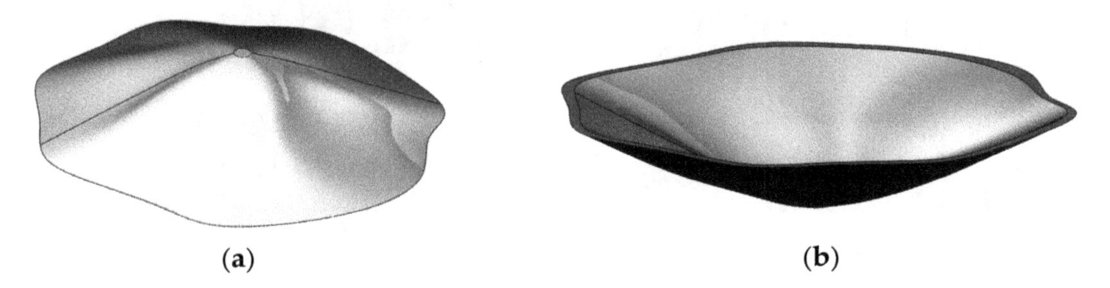

(a) (b)

Figure 31. 3D model of the cranioplasty plate: (**a**) upper side; (**b**) lower side.

The shape of the plate is highly irregular and continuously variable, as can be seen from Figure 32. However, the wall angles were checked to be lower than 38° for any area of the part.

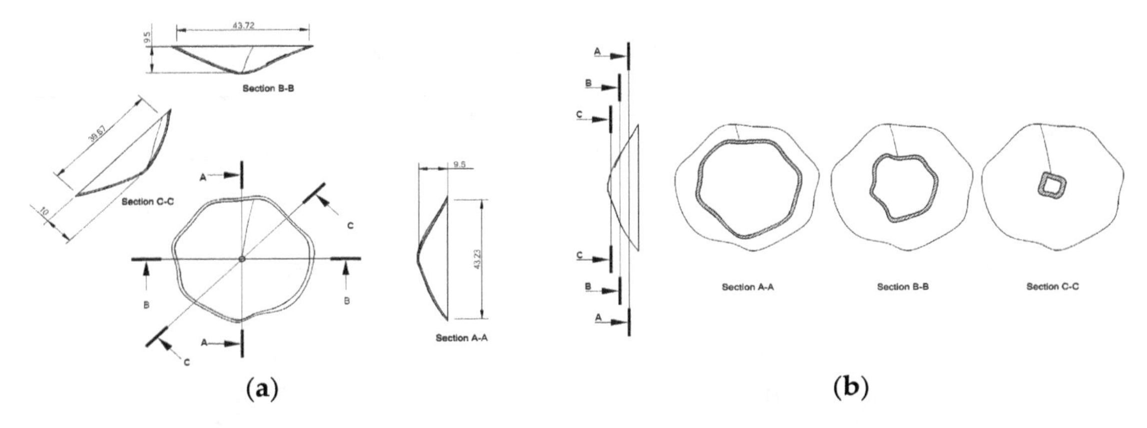

(a) (b)

Figure 32. Geometry of the plate: (**a**) transversal sections; (**b**) horizontal sections.

Spatial spiral trajectories were used with a vertical step of p = 0.2 mm and the punch with d_p = 10 mm was chosen as the processing tool. Figure 33 presents the shape of the processing trajectories (toolpath), but for clarity, the vertical step was enlarged ten times (2 mm).

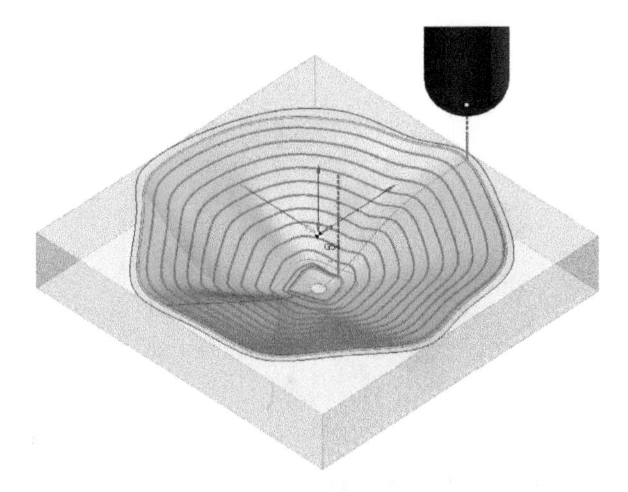

Figure 33. Spiral toolpath (for clarity of presentation, the vertical step was enlarged ten times).

The processed part is presented in Figure 34.

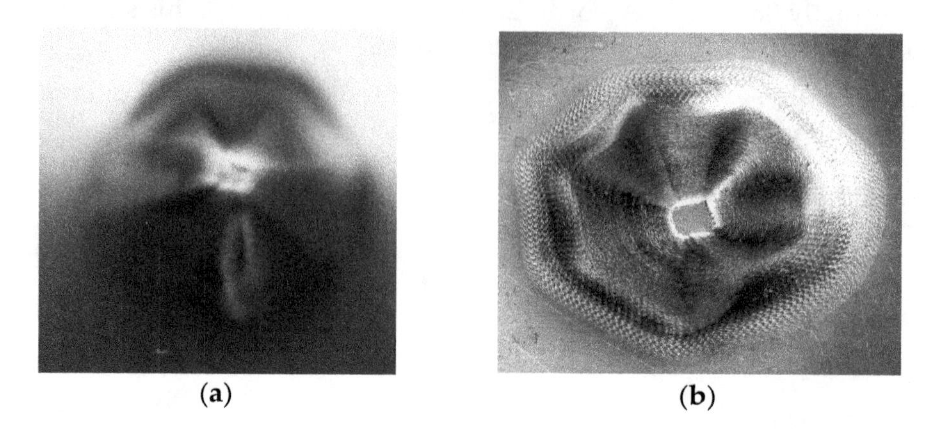

(a) (b)

Figure 34. Processed part: (**a**) upper side; (**b**) lower side.

Using the Argus GOM optical measurement system, major strains (ε_1), minor strains (ε_2), and thickness reduction (s_{max}) for the cranioplasty plates were measured. A synthesis of the values is presented in Table 21.

Table 21. Measured characteristic values for the cranioplasty plate.

Cranioplasty Plate	Characteristic Input		
	Major Strains ε_1 [%]	**Minor strains ε_2 [%]**	**Thickness Reduction s_{max} [%]**
Characteristic	14.5	4.3	17.44

A graphical presentation of the measured values is presented in Figure 35.

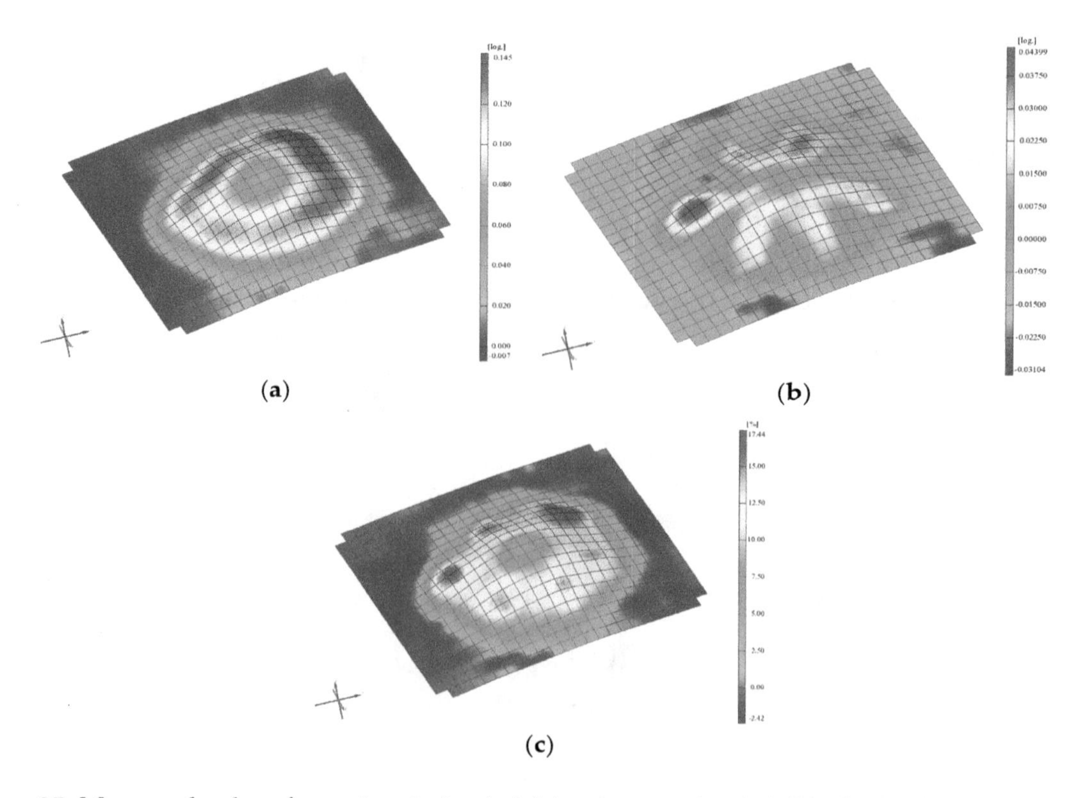

Figure 35. Measured values for major strains (ε_1) (**a**), minor strains (ε_2) (**b**), thickness reduction (s_{max}) (**c**) for the cranioplasty plate.

From Table 21 and Figure 35, it can be noticed that the values of the characteristic inputs are in acceptable ranges for parts manufactured by means of ASPIF. In fact, these values are even smaller than the ones obtained for the test parts (Table 20).

3. Results

The AHP method proposed here has indicated that SPIF is the most recommended manufacturing method if certain criteria are considered. Of course, the method could be affected by subjectivity, and the results could be changed if the analysis is done by other specialists. However, the performed sensitivity analysis guarantees, in some respects, the robustness of the results.

It is here noticeable that AHP indicates the best choice according to the criteria taken into consideration. Thus, if other sets of criteria are chosen, the results may differ significantly. The results presented in this work do not state that SPIF could be considered the best choice in any respect, but it could be considered the best choice if the criteria are those chosen in this approach.

An FEM model was developed which was able to provide simulation results close to values found experimentally for major and minor strains and for thickness reduction.

The experimental program provided some information regarding the technological conditions in which some incremental improvements (mainly an increase in the achievable wall angle) in the results of processing Ti6Al4V alloy at room temperature could be achieved. Also, it was presented that using continuous paths, a part with irregular shapes and with rapid variations of the wall shapes and angles has been processed.

Of course, there are also several shortcomings to this research:

- The processed test parts have limited dimensions (due to the available experimental layout), so it should be demonstrated by further study that the findings are also valid for bigger parts;
- Latest results presented in the literature [38,39] have indicated, by means of a cytotoxicity test, that heating the Ti6Al4V alloy during the SPIF process does not affect its biocompatibility. Corroborated by the superior plastic behavior of the heated material, these results narrow the application range of SPIF at room temperature. However, there are still reasons favoring the approach of SPIF at room temperature, from the points of view of roughness, costs (related to equipment complexity and energy consumption), and degree of control.

4. Conclusions

After conducting the experimental program, from a technological point of view it can be concluded that Ti6Al4V titanium alloy may be used for manufacturing cranioplasty plates, by processing by means of SPIF at room temperature, if the following technological aspects are considered:

- To reduce friction, the punch must rotate around its axis. A rotation speed between 150 and 300 rev/min was found during this experimental work to be appropriate;
- Theoretically, the working feed does not influence the formability of the part; however, it was found that working feeds greater than 200 mm/min may lead to crack occurrence. However, working feed affects the productivity, which is not critically important for this kind of part. Cranioplasty plates are manufactured as prototypes, productivity having less importance from this point of view;
- The vertical step of the punch (no matter if circular or spiral toolpaths are used) must be smaller than 1 mm. Best results were achieved by using vertical steps of 0.6, 0.4, and 0.2 mm. A direct link was noticed between the vertical step and the maximum inclination angle (α°) of the wall: the smaller the vertical step, the larger the achievable angle;
- Continuous toolpaths (which do not use lead-in/lead out entry/exit points) are the best approach, because entry/exit points may become stress concentrators leading to crack occurrence. Spatial

spiral trajectories provide the best results, but for irregular shapes, generating them without the aid of CAM software is difficult. Another solution is to use contour curves spaced on the Z axis as trajectories (circles or another curves) and distribute the lead-in/lead out points on the surface of the part (avoiding placing them close to each other);

- The maximum achievable wall angle was found $\alpha = 38°$;
- The best results, from the point of view of both the accuracy and surface roughness, were obtained using the punch with diameter d_p = 10 mm (compared with the punch with d_p = 8 mm);
- The accuracy of the wall angle was not significantly influenced by the diameter of the punch or by the vertical step. Also, the toolpaths did not influence it. The explanation for this may be found in the fact that the low plasticity of the Ti6Al4V titanium alloy does not lead to significant values of the springback;
- The roughness of the parts was influenced by the vertical step directly, a decrease in the vertical step leading to a decrease in the roughness value.
- Further research will be oriented in the following directions:
- The influence of rotational speed and working feed upon the accuracy of the part will be studied in more detail;
- For the time being, the overall dimensions of the parts were limited by the size of the experimental layout (mainly the size of the active plate and the working space of the CNC machine-tool). A new layout will be designed and implemented to test how greater overall dimensions of the part influence its manufacturability by means of ASPIF at room temperature;
- Industrial robots will be used as technological equipment to test if superior kinematics (more complex processing trajectories) can improve the manufacturability of the parts.

Author Contributions: S.G.R., M.T. and O.B. designed and set up the experimental layouts and the experimental program, R.E.B. set up the machining strategies, performed the AHP analysis, and wrote the paper, S.G.R., C.G., C.B. and A.L.C. performed the formability measurements and data analysis.

References

1. Rack, H.J.; Qazi, J.I. Titanium alloys for biomedical applications. *Mater. Sci. Eng. C* **2006**, *26*, 1269–1277. [CrossRef]
2. Geetha, M.; Singh, A.K.; Asokamani, R.; Gogia, A.K. Ti based biomaterials, the ultimate choice for orthopaedic implants—A review. *Prog. Mater. Sci.* **2009**, *54*, 397–425. [CrossRef]
3. Williams, D.F. On the mechanisms of biocompatibility. *Biomaterials* **2008**, *29*, 2941–2953. [CrossRef] [PubMed]
4. Aydin, S.; Kucukyuruk, B.; Abuzayed, B.; Aydin, S.; Sanus, G.Z. Cranioplasty: Review of materials and techniques. *J. Neurosci. Rural Pract.* **2011**, *2*, 162–167. [PubMed]
5. Servadei, F.; Iaccarino, C. The Therapeutic cranioplasty still needs an ideal material and surgical timing. *World Neurosurg.* **2015**, *83*, 133–135. [CrossRef] [PubMed]
6. Joffe, J.M.; Nicoll, S.R.; Richards, R.; Linney, A.D.; Harris, M. Validation of computer-assisted manufacture of titanium plates for cranioplasty. *Int. J. Oral Maxillofac. Surg.* **1999**, *28*, 309–313. [CrossRef]
7. Wiggins, A.; Austerberry, R.; Morrison, D.; Kwok, H.M.; Honeybul, S. Cranioplasty with custom-made titanium plates—14 years experience. *Neurosurgery* **2013**, *72*, 248–256. [CrossRef] [PubMed]
8. Heissler, E.; Fischer, F.-S.; Boiouri, S.; Lehrnann, T.; Mathar, W.; Gebhardt, A.; Lanksch, W.; Bler, J. Custom-made cast titanium implants produced with CAD/CAM for the reconstruction of cranium defects. *Int. J. Oral Maxillofac. Surg.* **1998**, *27*, 334–338. [CrossRef]
9. Cabraja, M.; Klein, M.; Lehmann, T.-N. Long-term results following titanium cranioplasty of large skull defects. *Neurosurg. Focus* **2009**, *26*, 1–7. [CrossRef] [PubMed]
10. Bhargava, D.; Bartlett, P.; Russell, J.; Liddington, M.; Tyagi, A.; Chumas, P. Construction of titanium cranioplasty plate using craniectomy bone flap as template. *Acta Neurochir.* **2010**, *152*, 173–176. [CrossRef] [PubMed]

11. Chen, J.-J.; Liu, W.; Li, M.-Z.; Wang, C.-T. Digital manufacture of titanium prosthesis for cranioplasty. *Int. J. Adv. Manuf. Technol.* **2006**, *27*, 1148–1152. [CrossRef]
12. Che-Haron, C.H.; Jawaid, A. The effect of machining on surface integrity of titanium alloy Ti–6% Al–4% V. *J. Mater. Process. Technol.* **2005**, *166*, 188–192. [CrossRef]
13. Jeswiet, J.; Micari, F.; Hirt, G.; Bramley, A.; Duflou, J.; Allwood, J. Asymmetric single point incremental forming of sheet metal. *CIRP Ann. Manuf. Technol.* **2005**, *54*, 623–650. [CrossRef]
14. Behera, A.K.; de Sousa, R.A.; Ingarao, G.; Oleksik, V. Single point incremental forming: An assessment of the progress and technology trends from 2005 to 2015. *J. Manuf. Process.* **2017**, *27*, 37–62. [CrossRef]
15. Gatea, S.; Ou, H.; McCartney, G. Review on the influence of process parameters in incremental sheet forming. *Int. J. Adv. Manuf. Technol.* **2016**, *87*, 479–499. [CrossRef]
16. Hussain, G.; Gao, L.; Hayat, N.; Cui, Z.; Pang, Y.C.; Dar, N.U. Tool and lubrication for negative incremental forming of a commercially pure titanium sheet. *J. Mater. Process. Technol.* **2008**, *203*, 193–201. [CrossRef]
17. Hamilton, K.; Jeswiet, J. Single point incremental forming at high feed rates and rotational speeds: Surface and structural consequences. *CIRP Ann. Manuf. Technol.* **2010**, *59*, 311–314. [CrossRef]
18. Kim, Y.H.; Park, J.J. Effect of process parameters on formability in incremental forming of sheet metal. *J. Mater. Process. Technol.* **2002**, *130–131*, 42–46. [CrossRef]
19. Durante, M.; Formisano, A.; Langella, A.; Memola Capece Minutolo, F. The influence of tool rotation on an incremental forming process. *J. Mater. Process. Technol.* **2009**, *209*, 4621–4626. [CrossRef]
20. Ambrogio, G.; Gagliardi, F.; Bruschi, S.; Filice, L. On the high-speed Single Point Incremental Forming of titanium alloys. *CIRP Ann. Manuf. Technol.* **2013**, *62*, 243–246. [CrossRef]
21. Fan, G.; Gao, L.; Hussain, G.; Wu, Z. Electric hot incremental forming: A. novel technique. *Int. J. Mach. Tools Manuf.* **2008**, *48*, 1688–1692. [CrossRef]
22. Ambrogio, G.; Filice, L.; Gagliardi, F. Formability of lightweight alloys by hot incremental sheet forming. *Mater. Des.* **2012**, *34*, 501–508. [CrossRef]
23. Palumbo, G.; Brandizzi, M. Experimental investigations on the single point incremental forming of a titanium alloy component combining static heating with high tool rotation speed. *Mater. Des.* **2012**, *40*, 43–51. [CrossRef]
24. Göttmann, A.; Diettrich, J.; Bergweiler, G.; Bambach, M.; Hirt, G.; Loosen, P.; Poprawe, R. Laser-assisted asymmetric incremental sheet forming of titanium sheet metal parts. *Prod. Eng.* **2011**, *5*, 263–271. [CrossRef]
25. Xu, D.; Wu, W.; Malhotra, R.; Chen, J.; Lu, B.; Cao, J. Mechanism investigation for the influence of tool rotation and laser surface texturing (LST) on formability in single point incremental forming. *Int. J. Mach. Tools Manuf.* **2013**, *73*, 37–46. [CrossRef]
26. Xu, D.K.; Lu, B.; Cao, T.T.; Zhang, H.; Chen, J.; Long, H.; Cao, J. Enhancement of process capabilities in electrically-assisted double sided incremental forming. *Mater. Des.* **2016**, *92*, 268–280. [CrossRef]
27. Feng, B.; Chen, J.Y.; Oi, S.K.; He, L.; Zhao, J.Z.; Zhang, X.D. Characterization of surface oxide films on titanium and bioactivity. *J. Mater. Sci. Mater. Med.* **2002**, *13*, 457–464.
28. Guleryuz, H.; Cimenoglu, H. Surface modification of a Ti–6Al–4V alloy by thermal oxidation. *Surf. Coat. Technol.* **2005**, *192*, 164–170. [CrossRef]
29. Duarte, L.T.; Bolfarini, C.; Biaggio, S.R.; Rocha-Filho, R.C.; Nascente, P.A.P. Growth of aluminum-free porous oxide layers on titanium and its alloys Ti–6Al–4V and Ti–6Al–7Nb by micro-arc oxidation. *Mater. Sci. Eng. C* **2014**, *41*, 343–348. [CrossRef] [PubMed]
30. Guleryuz, H.; Cimenoglu, H. Oxidation of Ti–6Al–4V alloy. *J. Alloys Compd.* **2009**, *472*, 241–246. [CrossRef]
31. Chen, M.; Li, W.; Shen, M.; Zhu, S.; Wang, F. Glass–ceramic coatings on titanium alloys for high temperature oxidation protection: Oxidation kinetics and microstructure. *Corros. Sci.* **2013**, *74*, 178–186. [CrossRef]
32. Zhang, Y.; Maa, G.-R.; Zhang, X.-C.; Li, S.; Tu, S.-T. Thermal oxidation of Ti-6Al–4V alloy and pure titanium under external bending strain: Experiment and modelling. *Corros. Sci.* **2017**, *122*, 61–73. [CrossRef]
33. Du, H.L.; Datta, P.K.; Lewis, D.B.; Burnell-Gray, J.S. Air oxidation behavior ofTi-6Al-4 V alloy between 650 and 850 °C. *Corros. Sci.* **1994**, *36*, 631–642. [CrossRef]
34. McLachlan, D.R.C. Aluminium and the risk for Alzheimer's disease. *Environmetrics* **1995**, *6*, 233–275. [CrossRef]
35. Aniołek, K.; Kupka, M.; Barylski, A.; Dercz, G. Mechanical and tribological properties of oxide layers obtained on titanium in the thermal oxidation process. *Appl. Surf. Sci.* **2015**, *357*, 1419–1426. [CrossRef]

36. Luo, Y.; Chen, W.; Tian, M.; Teng, S. Thermal oxidation of Ti6Al4V alloy and its biotribological properties under serum lubrication. *Tribol. Int.* **2015**, *89*, 67–71. [CrossRef]
37. Aniołek, K. The influence of thermal oxidation parameters on the growth of oxide layers on titanium. *Vacuum* **2017**, *144*, 94–100. [CrossRef]
38. Ambrogio, G.; Sgambitterra, E.; De Napoli, L.; Gagliardi, F.; Fragomeni, G.; Piccininni, A.; Gugleilmi, P.; Palumbo, G.; Sorgente, D.; La Barbera, L.; et al. Performances analysis of titanium prostheses manufactured by superplastic forming and incremental forming. *Procedia Eng.* **2017**, *183*, 168–173. [CrossRef]
39. Palumbo, G.; Sorgente, D.; Vedani, M.; Mostaed, E.; Hamidi, M.; Gastaldi, D.; Villa, T. Effects of superplastic forming on modification of surface properties of Ti alloys for biomedical applications. *J. Manuf. Sci. Eng.* **2018**, *140*, 10. [CrossRef]
40. Ingarao, G.; Ambrogio, G.; Gagliardi, F.; Di Lorenzo, R. A sustainability point of view on sheet metal forming operations: Material wasting and energy consumption in incremental forming and stamping processes. *J. Clean. Prod.* **2012**, *29–30*, 255–268. [CrossRef]
41. Bagudanch, I.; Garcia-Romeu, M.L.; Ferrer, I.; Lupiañez, J. The effect of process parameters on the energy consumption in single point incremental forming. *Procedia Eng.* **2013**, *63*, 346–353. [CrossRef]
42. Ingarao, G.; Vanhove, H.; Kellens, K.; Duflou, J.R. A comprehensive analysis of electric energy consumption of single point incremental forming processes. *J. Clean. Prod.* **2014**, *67*, 173–186. [CrossRef]
43. Ambrogio, G.; Ingarao, G.; Gagliardi, F.; Di Lorenzo, R. Analysis of energy efficiency of different setups able to perform single point incremental forming (SPIF) Processes. *Procedia CIRP* **2014**, *15*, 111–116. [CrossRef]
44. Breaz, R.; Bologa, O.; Tera, M.; Racz, G. Researches Regarding the Use of Complex Trajectories and Two Stages Processing in Single Point Incremental Forming of Two Layers Sheet. *Circles* **2012**, *91*, 128.
45. Breaz, R.; Bologa, O.; Tera, M.; Racz, G. Computer assisted techniques for the incremental forming technology. In Proceedings of the IEEE 18th Conference on Emerging Technologies & Factory Automation (ETFA), Cagliari, Italy, 10–13 September 2013.
46. Cotigă, C.; Bologa, O.; Racz, S.G.; Breaz, R.E. Researches regarding the usage of titanium alloys in cranial implants. *Appl. Mech. Mater.* **2014**, *657*, 173–177. [CrossRef]
47. Kim, B.J.; Hong, K.S.; Park, K.J.; Park, D.H.; Chung, Y.G.; Kang, S.H. Customized cranioplasty implants using three-dimensional printers and polymethyl-methacrylate casting. *J. Korean Neurosurg. Soc.* **2012**, *52*, 541–546. [CrossRef] [PubMed]
48. Digital Evolution of Cranial Surgery. Available online: http://www.renishaw.com/en/digital-evolution-of-cranial-surgery--38602 (accessed on 7 August 2018).
49. Saaty, T.L. *The Analytic Hierarchy Process: Planning, Priority Setting, Resource Allocation*; McGraw-Hill: New York, NY, USA, 1980; p. 287.
50. Saaty, T.L. *Decision Making for Leaders: The Analytic Hierarchy Process for Decisions in a Complex Word*; RWS Publication: Pittsburgh, PA, USA, 1990.
51. Alonso, J.; Lamata, T.M. Consistency in the analytic hierarchy process: A new approach. *Int. J. Uncertain. Fuzziness Knowl. Based Syst.* **2006**, *14*, 445–459. [CrossRef]
52. Cabala, P. Using the analytic hierarchy process in evaluating decision alternatives. *Oper. Res. Decis.* **2010**, *20*, 5–23.
53. Hurley, W.J. The analytic hierarchy process: A note on an approach to sensitivity which preserves rank order. *Comput. Oper. Res.* **2001**, *28*, 185–188. [CrossRef]
54. Williams, L.R.; Fan, K.F.; Bentley, R.P. Custom-made titanium cranioplasty: Early and late complications of 151 cranioplasties and review of the literature. *Int. J. Oral Maxillofac. Surg.* **2015**, *44*, 599–608. [CrossRef] [PubMed]
55. Williams, L.; Fan, K.; Bentley, R. Titanium cranioplasty in children and adolescents. *J. Cranio Maxillofac. Surg.* **2016**, *44*, 789–794. [CrossRef] [PubMed]

Ambivalent Role of Annealing in Tensile Properties of Step-Rolled Ti-6Al-4V with Ultrafine-Grained Structure

Geonhyeong Kim [1], Taekyung Lee [2,*], Yongmoon Lee [3], Jae Nam Kim [1], Seong Woo Choi [4], Jae Keun Hong [4] and Chong Soo Lee [1,*]

[1] Graduate Institute of Ferrous Technology (GIFT), Pohang University of Science and Technology (POSTECH), Pohang 37673, Korea; kkh285@postech.ac.kr (G.K.); jnkims@postech.ac.kr (J.N.K.)
[2] School of Mechanical Engineering, Pusan National University, Busan 46241, Korea
[3] Center for Advanced Aerospace Materials, Pohang University of Science and Technology (POSTECH), Pohang 37673, Korea; ymlee0725@postech.ac.kr
[4] Advanced Metals Division, Korea Institute of Materials Science, Changwon 51508, Korea; rhkdgh@kims.re.kr (S.W.C.); jkhong@kims.re.kr (J.K.H.)
* Correspondence: taeklee@pnu.ac.kr (T.L.); cslee@postech.ac.kr (C.S.L.)

Abstract: Step rolling can be used to mass-produce ultrafine-grained (UFG) Ti-6Al-4V sheets. This study clarified the effect of subsequent annealing on the tensile properties of step-rolled Ti-6Al-4V at room temperature (RT) and elevated temperature. The step-rolled alloy retained its UFG structure after subsequent annealing at 500–600 °C. The RT ductility of the step-rolled alloy increased regardless of annealing temperature, but strengthening was only attained by annealing at 500 °C. In contrast, subsequent annealing rarely improved the elevated-temperature tensile properties. The step-rolled Ti-6Al-4V alloy without the annealing showed the highest elongation to failure of 960% at 700 °C and a strain rate of 10^{-3} s^{-1}. The ambivalent effect of annealing on RT and elevated-temperature tensile properties is a result of microstructural features, such as dislocation tangles, subgrains, phases, and continuous dynamic recrystallization.

Keywords: Ti-6Al-4V; step rolling; grain refinement; superplasticity; continuous dynamic recrystallization

---

1. Introduction

Ti-6Al-4V alloy is widely used in aerospace industries due to its high specific strength, excellent corrosion resistance, and high service temperature [1]. This alloy also has superior superplasticity and consequent high formability. Accordingly, many complicated aircraft structures are manufactured from Ti-6Al-4V alloys, using superplastic forming to reduce the material's weight for cost saving [2]. This process is typically performed at relatively high temperatures of over 850 °C and low strain rates of less than 10^{-3} s^{-1}; these processes are expensive, so they increase production cost. Thus, research has been conducted to find ways to reduce the temperature or increase the strain rate (or both) at which superplastic forming is conducted. Grain refinement may be a solution to this challenge [3,4].

An increased fraction of grain boundaries assists grain-boundary sliding (GBS), which is the main mechanism of superplasticity. GBS by dislocation processes or diffusional flow can accommodate a large amount of deformation. Grain refinement can shift region II (i.e., the region of the highest strain rate sensitivity) to faster strain rates [5]. Therefore, grain refinement can achieve superplasticity at low temperatures or fast strain rates or both. Grain size has been reduced using a variety of severe plastic deformation (SPD) processes, such as equal-channel angular pressing (ECAP) [6], high-pressure torsion (HPT) [7], and multiaxial forging [8]. The grains of Ti-6Al-4V alloy have been reduced to a size of ~0.4 μm, which has enabled superplasticity at 550 °C and a strain rate of 2×10^{-4} s^{-1} [4]. However, SPD cannot easily produce an appreciable size of product, so the method is not useful for industrial production. Multi-pass caliber rolling has been proposed as a solution [9–11] but is only applicable to fabricating metallic rods rather than plates.

Step rolling [12] has been developed as a grain-refining process to fabricate relatively large Ti plates. This method can fabricate ultrafine-grained (UFG) bulk Ti-6Al-4V plate (i.e., 300 mm × 70 mm × 4 mm, with grain size d = 0.5 μm) with only a small cumulative strain of 1.6. Step rolling is performed by decreasing the rolling temperature step by step. Initial deformation is imposed at a high temperature to effectively fragment the martensitic laths. This fragmentation increases the formability of the material so that the plate can be rolled at progressively lower temperatures in subsequent rolling passes. This low-temperature rolling suppresses grain growth and thereby achieves significant grain refinement with a relatively low amount of plastic strain [12].

The superplasticity of Ti-6Al-4V obtained by step rolling and subsequent annealing was recently reported [13], but the study used only one deformation condition (i.e., a deformation temperature of 750 °C and a strain rate of 10^{-3} s^{-1}). However, superplastic behavior drastically varies with deformation temperature and $\dot{\varepsilon}$, so to exploit the advantage of step rolling, the high-temperature deformation behavior must be quantified at a wide range of these parameters. Therefore, the present study is a systematic investigation of the deformation behavior of step-rolled Ti-6Al-4V alloys at room temperature (RT) and elevated temperatures under a range of $\dot{\varepsilon}$.

2. Materials and Methods

The Ti-6Al-4V alloy used in this study (supplied by ATI, USA) was a 20 mm-thick plate with a chemical composition (mass %) of 6.28 Al, 4.00 V, 0.18 Fe, 0.178 O, and 0.012 C, in balance with Ti. Thermomechanical processing was achieved using step rolling (Figure 1). The as-received plate was solution-treated at 1050 °C for 30 min, then quenched in water to induce the formation of a full martensitic structure. The plate was soaked at 800 °C for 1 h, then step-rolled to a thickness of 3 mm. The step rolling was composed of eight passes with decreasing temperatures: 800 °C for the first pass, 725 °C for the second and third passes, and 650 °C for the fourth to eighth passes. The sample was then cooled in air to RT. The step-rolled sample without additional heat treatment is denoted as STEP-0. Sections of the STEP-0 sample were further annealed at 500 °C (sample STEP-5) or 600 °C (STEP-6) for 150 min in an electric furnace (AWF13, Lenton, Hope Valley, UK), then cooled in air to RT.

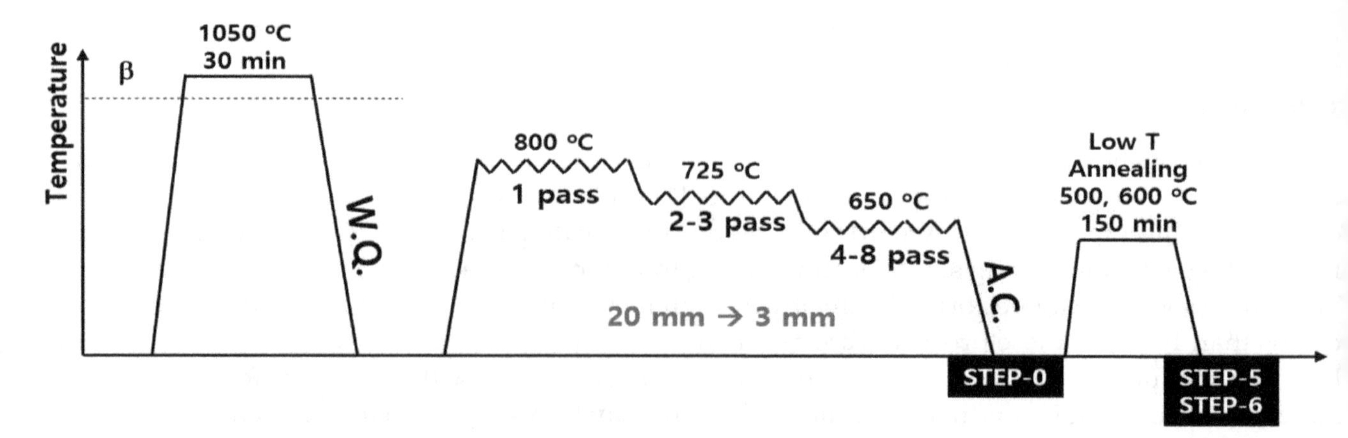

Figure 1. Schematic illustration of the step rolling and subsequent annealing employed in this work.

An RT tensile test was performed at a strain rate of 5×10^{-3} s^{-1} using a universal mechanical testing machine (8801, INSTRON, Norwood, MA, USA) with an extensometer (3542-025M-100-ST, Epsilon, Jackson, WY, USA), based on the procedure of the ASTM-E8 standard. Plate-type tensile specimens with a gage length of 25 mm, width of 6 mm, and thickness of 3 mm were machined along the rolling direction. An elevated-temperature tensile test was performed using a universal mechanical testing machine (8862, INSTRON, Norwood, MA, USA) attached to a halogen furnace (DF-60HG, Dae Heung Scientific Company, Incheon, Korea). For this experiment, specimens were prepared with a gauge length of 5 mm, width of 5 mm, and thickness of 2 mm. The elevated-temperature tensile tests were performed at all combinations of the three temperatures (650 °C, 700 °C, 750 °C) and the three initial strain rates (10^{-2} s^{-1}, 10^{-3} s^{-1}, 2×10^{-4} s^{-1}). Samples were held at a given testing temperature for 10 min before the tensile test was started.

For microstructural observation, samples were electro-polished (LectroPol-5, STRUERS, Denmark) in a solution of 410 mL of methanol, 245 mL of 2-butoxy ethyl alcohol, and 40 mL of $HClO_4$ at 22 V for 30 s. The plane normal to the transverse direction was examined using electron backscatter diffraction (EBSD, FEI QUANTA 3D FEG, Hillsboro, OR, USA) analysis with a step size of 0.1 μm. Quantitative microstructural characterizations were obtained using TSL-OIM Ver. 7.3 software (EDAX, Mahwah, NJ, USA); for data reliability, the scans were only analyzed for points that had a confidence index > 0.07. More than 1000 validated grains per condition were used for the analysis to ensure statistical accuracy. For examination with transmission electron microscope (TEM, JEM-2100F, JEOL, Tokyo, Japan) observation, the samples were mechanically polished to a thickness < 100 μm and then punched to form thin disks that had a diameter of 3 mm. They were electro-polished using a jet polisher (TenuPol-5, STRUERS, Denmark) at 22 V in the same solution used for the preparation of the EBSD samples.

3. Results and Discussion

3.1. Deformation Behavior at Room Temperature

The microstructures of the step-rolled alloy (i.e., STEP-0) and the subsequently annealed alloys (i.e., STEP-5 and STEP-6) were compared by EBSD analysis (Figure 2). The annealing had little effect on grain size: STEP-0 had a UFG structure with α-grain size $d = 0.70$ μm, STEP-5 had $d = 0.72$ μm, and STEP-6 had $d = 0.74$ μm. This small change implies that annealing induced recovery rather than recrystallization. In contrast, annealing significantly decreased the volume fraction of the β phase: STEP-0, 6.2%; STEP-5, 3.0%; STEP-6, 3.6%.

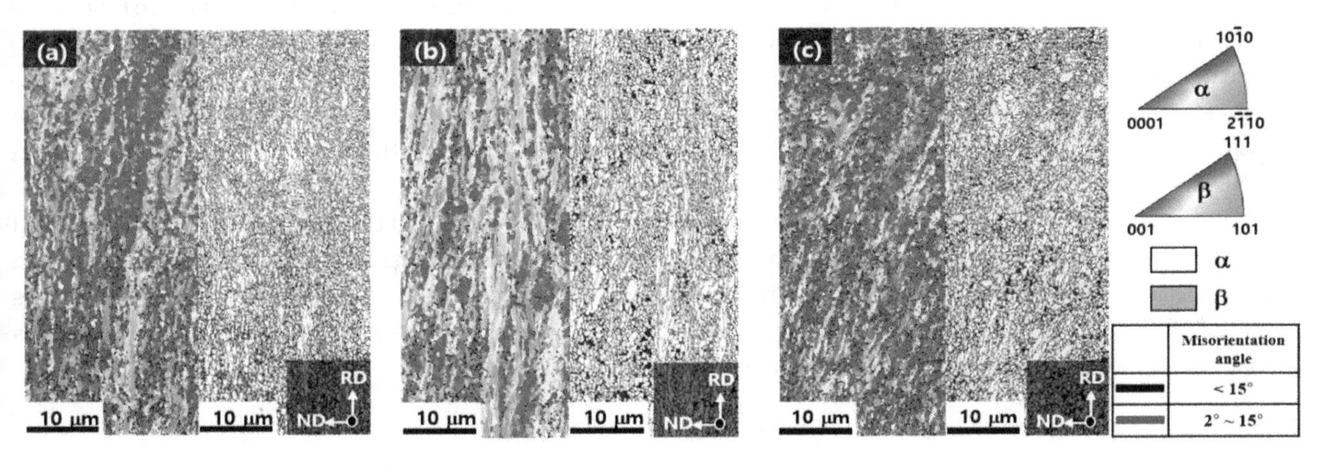

Figure 2. Electron backscatter diffraction (EBSD) inverse pole figure map and corresponding phase map of the samples: (**a**) STEP-0, (**b**) STEP-5, (**c**) STEP-6.

The subsequent annealing caused a marked difference in the RT tensile properties of the samples (Figure 3, Table 1). The STEP-0 sample had an increased yield strength (YS) and ultimate tensile strength (UTS) which were both 20% higher than those of conventional mill-annealed Ti-6Al-4V alloy, which has $d = 10$ μm [14]. However, STEP-0 showed a decrease in elongation to failure (EL) compared to the conventional sample [14]. The subsequent annealing of STEP-0 recovered the low EL without loss of mechanical strength.

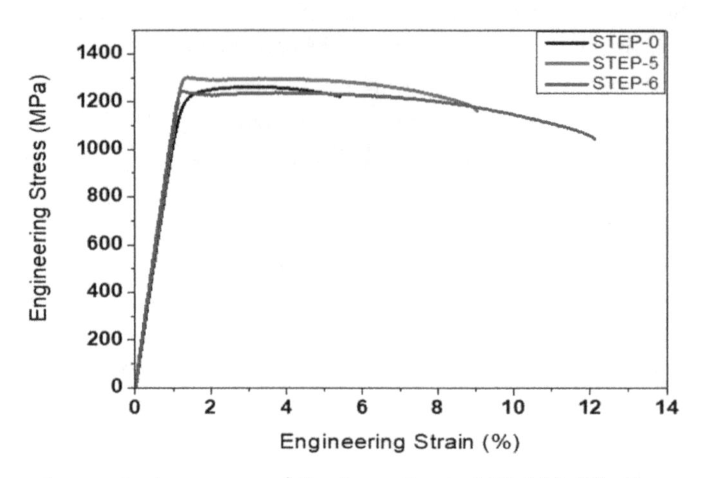

Figure 3. Engineering stress–strain curves of the investigated Ti-6Al-4V alloys at room temperature.

Table 1. Room-temperature tensile properties of the investigated Ti-6Al-4V alloys.

Sample	Yield Strength (YS) (MPa)	Ultimate Tensile Strength (UTS) (MPa)	Elongation to Failure (EL) (%)
STEP-0	1201	1263	5.4
STEP-5	1298	1311	9.0
STEP-6	1243	1243	12.1

Annealing temperature affected the change in strength and ductility differently in each sample. STEP-5 showed an increase in both whereas STEP-6 did not. TEM observation provided a rationalization for the simultaneous improvement in strength and ductility in STEP-5. The TEM micrograph (Figure 4) of STEP-0 showed the formation of dislocation tangles at the grain boundaries (Figure 4a, arrows). Annealing at 500 °C significantly diminished the frequency of dislocation tangles and decreased the dislocation density, resulting in the increased EL. STEP-5 showed fine (i.e., 100–200 nm in diameter) particles; a uniform distribution of fine particles effectively strengthens the matrix [15,16], so they may be the source of the increases in YS and EL in this sample. The selected-area diffraction pattern (Figure 4b, inset) characterizes these particles as α phase. In contrast, the STEP-6 sample showed no evidence of such fine particles. The α particles may have transformed to β phase at this annealing temperature. The absence of fine particles led to the loss of strengthening in STEP-6, compared to STEP-5.

Subsequent annealing affected the substructure of the UFG grains. STEP-0 showed only distributed dislocations (Figure 4c), whereas STEP-5 presented numerous subgrains with diameters of 20–50 nm (Figure 4d). At 500 °C, dislocations at grain boundaries were recovered and organized into subgrains inside a UFG α grain [17]. The consequent increase in the fraction of low-angle grain boundaries (LABs) would also contribute to the increased mechanical strength [18].

3.2. *Deformation Behavior at Elevated Temperatures*

EL at elevated temperatures was affected by strain rate, deformation temperature, and type of sample (Figure 5). ELs measured at a deformation temperature of 750 °C and 2×10^{-4} s^{-1} are not presented, because severe oxidation at the sample surface caused large variability in the measurements. An EL of 400% is regarded as the minimum requirement for superplasticity in metals [19]. The investigated alloys exhibited a significant EL of over 400% under most conditions, except at the lowest deformation temperature (650 °C) and the highest strain rate (10^{-2} s^{-1}). The subsequent annealing had an ambivalent effect on the difference between the mechanical properties at RT and those at elevated temperatures:

t rarely increased EL in the step-rolled alloys at elevated temperatures, and it even degraded the ductility at 650 °C.

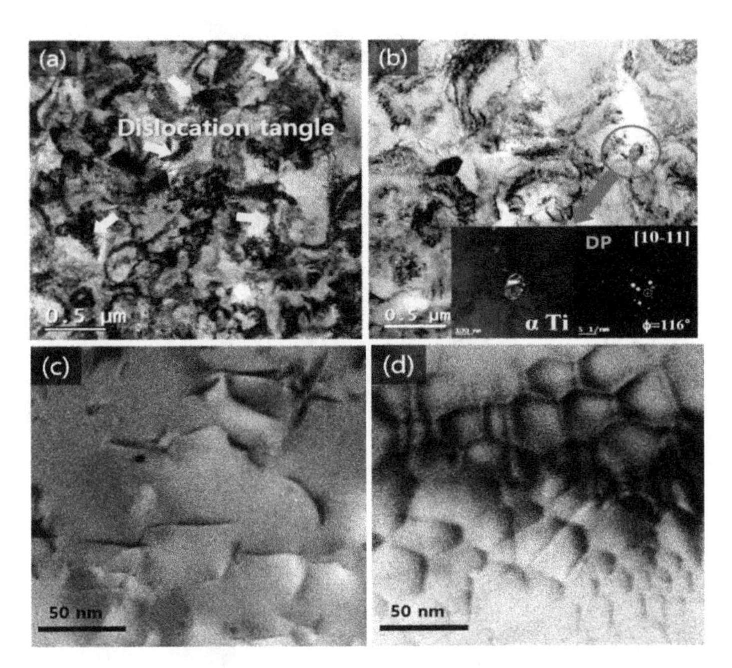

Figure 4. Bright-field transmission electron microscope (TEM) micrographs of (**a**,**c**) STEP-0 and (**b**,**d**) STEP-5. Insets in (**b**): dark-field image of α particle and corresponding selected-area diffraction pattern. Figure 4c,d are high-magnification images.

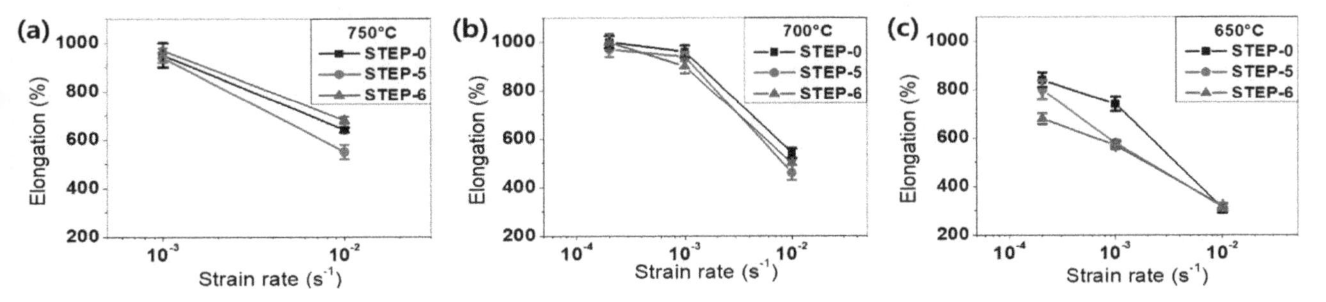

Figure 5. Variation of EL depending on the strain rate for the investigated alloys at (**a**) 750 °C, (**b**) 700 °C, and (**c**) 650 °C.

The stress–strain curves of STEP-0, STEP-5, and STEP-6 were affected by the deformation temperature and strain rate (Figure 6). At 750 °C and 10^{-3} s^{-1}, three specimens showed weak flow hardening behavior. This suggests that continuous dynamic recrystallization (CDRX) could not suppress grain growth, thereby diminishing the superplasticity [20]. The plateau region was confirmed during the deformation at 700 °C, indicating that CDRX strongly suppressed grain growth [21]. To obtain the highest and most efficient superplasticity, this plateau behavior should be achieved. Usually, EL is higher at 750 °C than at 700 °C, but in our results the ELs were similar. In this case, severe grain growth and oxidation degraded the superplasticity at 750 °C. At a deformation temperature of 650 °C, the specimens presented a flow softening followed by the plateau region. Despite the decreasing deformation temperature, they still exhibited significant EL values, suggestive of a low-temperature superplasticity.

The highest superplastic EL of STEP-0 among the investigated step-rolled alloys can be understood in terms of slope m (i.e., strain-rate sensitivity) in a log-scale plot of stress and strain rate (Figure 7). In general, UFG Ti-6Al-4V alloy demonstrates three regions with different slopes in this plot [6,22]. Among them, region II is characterized by (i) an intermediate 10^{-4} $s^{-1} \leq \dot{\varepsilon} \leq 2 \times 10^{-3}$ s^{-1}, (ii) high m, and (iii) significant superplastic EL. The high m of the STEP-0 sample indicates inhibition of the transition from region II to region III at a high strain rate of 10^{-2} s^{-1} [23], supporting a superplasticity obtained by the step rolling. Accordingly, the STEP-0 sample had a uniformly high m in the entire range of investigation employed in this work: m = 0.34 at 650 °C, 0.40 at 700 °C, and 0.55 at 750 °C. These m are comparable to or higher than those of UFG Ti-6Al-4V alloys fabricated by SPD processes: $0.34 \leq \dot{\varepsilon} \leq 0.43$ [4,6,7].

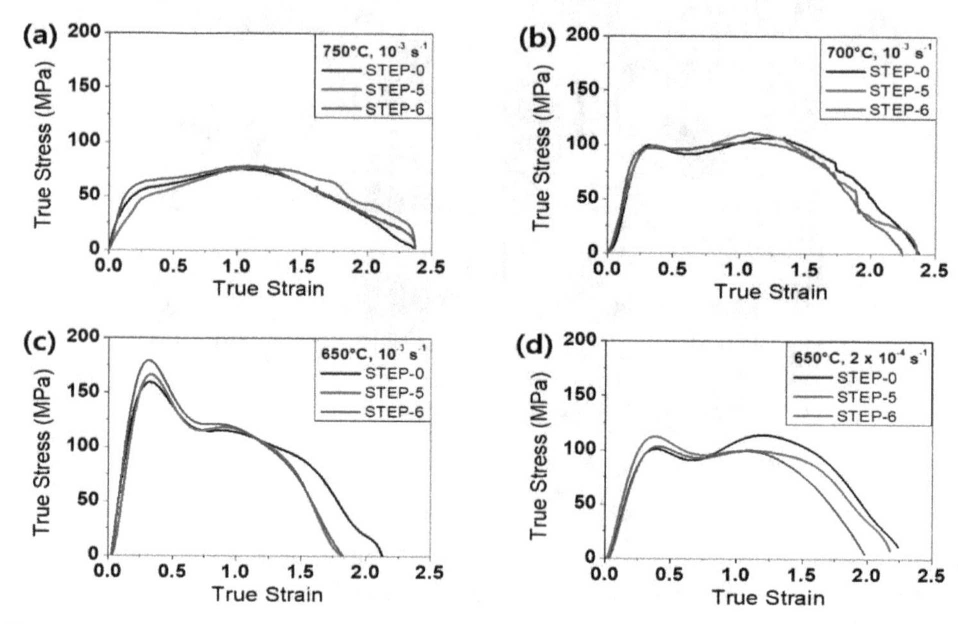

Figure 6. True stress–strain curves of the investigated Ti-6Al-4V alloys at (**a**) 750 °C and strain rate of 10^{-3} s^{-1}, (**b**) 700 °C and 10^{-3} s^{-1}, (**c**) 650 °C and 10^{-3} s^{-1}, and (**d**) 650 °C and 2×10^{-4} s^{-1}.

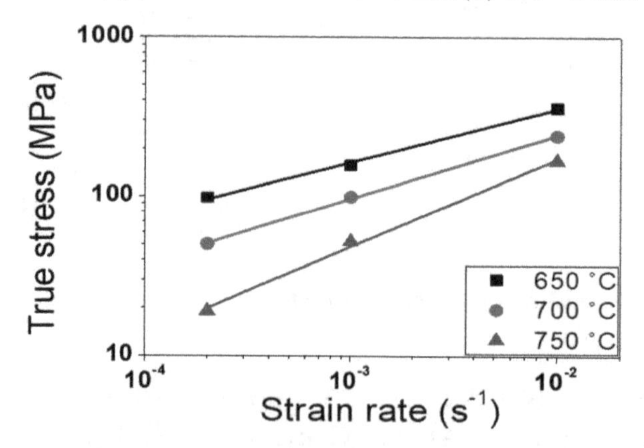

Figure 7. Variation of flow stress as a function of strain rate for STEP-0 specimens.

Step-rolled UFG Ti-6Al-4V yielded excellent superplasticity (i.e., EL up to 960%) that is comparable to the EL obtained using other fabrication methods. ECAP yielded UFG Ti-6Al-4V that had d = 0.3 μm, EL = 356% at 700 °C, and 10^{-3} s^{-1} [6]. HPT produced Ti-6Al-4V alloy that had d ~0.03 μm and EL = 820% under the same conditions [7]. Another HPT study produced d = 0.3 μm and the sample had EL = 676% at 725 °C and 10^{-3} s^{-1} [24]. Another UFG Ti-6Al-4V alloy (d = 0.4 μm) produced by hot rolling had EL = 400% at 700 °C and 10^{-3} s^{-1} [20]. A combination of forging and warm rolling refined the grains to d = 0.3 μm; the sample had EL = 550–900% at 700 °C [25].

Superplasticity is highly dependent on microstructural features. Grain refinement is regarded as the most important factor to activate superplastic behavior. Grain refinement expands the superplastic regime to lower temperatures and higher $\dot{\varepsilon}$, because the increasing fraction of grain boundaries provides increasing sources for GBS [5,9]. Grain morphology also affects superplasticity. Equiaxed grains yield a higher superplastic EL than elongated grains [26,27]. Dislocations distributed at grain boundaries also contribute to superplasticity by providing a fast diffusion path at high temperatures and by increasing the grain-boundary diffusion coefficient [28]. All of these microstructural features that are beneficial to superplasticity appeared in STEP-0, so its significant EL at 650–750 °C is understandable.

Comparison of the microstructural features at strains of ε = 0%, 100%, and 300% provided insight into the microstructural evolution during superplastic deformation at elevated temperatures (Figure 8). STEP-0 and STEP-6 were selected for this comparison, because STEP-6 was expected to show more distinct changes in microstructure than STEP-5 due to the higher annealing temperature of STEP-6. In the EBSD map, the STEP-0 and STEP-6 samples showed similar grain sizes before the deformation, whereas the former exhibited a smaller grain size after applying strains of 100% and 300%.

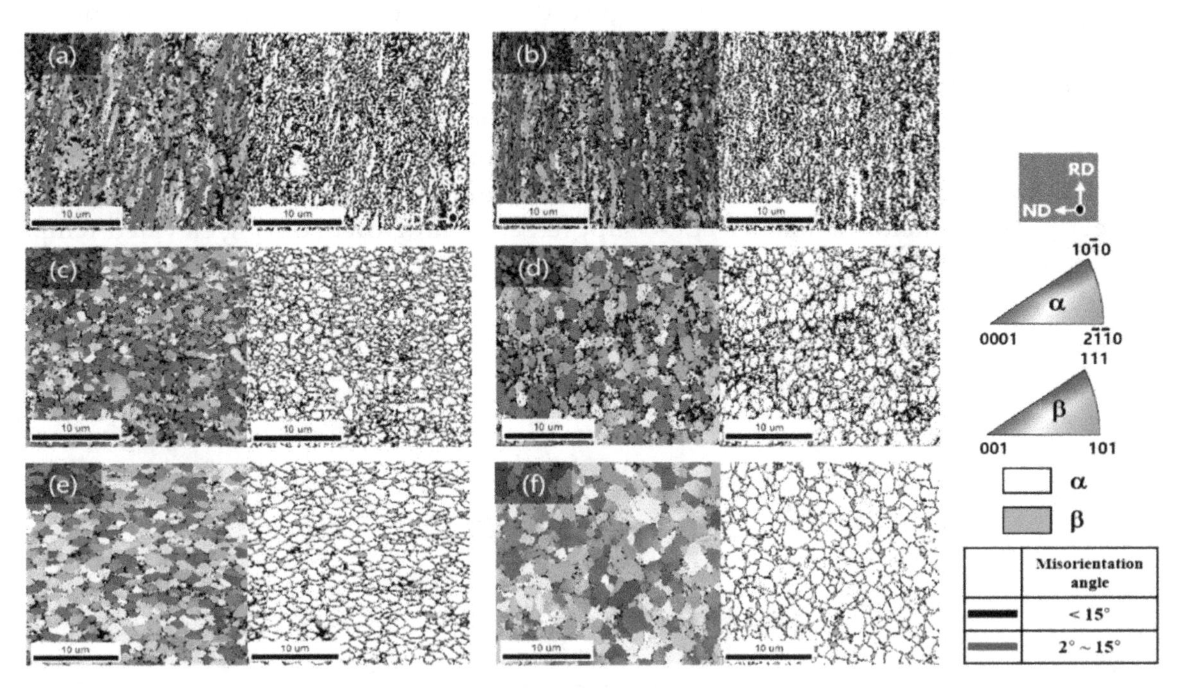

Figure 8. EBSD inverse pole figure maps and phase maps for (**a**,**c**,**e**) STEP-0 and (**b**,**d**,**f**) STEP-6. The micrographs were obtained after a strain of (**a**,**b**) 0%, (**c**,**d**) 100%, and (**e**,**f**) 300%.

Application of ε affected LAB fraction and grain size differently in the two samples (Figure 9). During the early stage of deformation, up to a 100% strain, the LAB fraction was considerably reduced, by 12.2%, in the STEP-0 sample but only by 6.1% in the STEP-6 sample. During this stage, grain size in the STEP-0 sample also increased by only 0.16 μm, whereas in the STEP-6 sample grain size increased by 0.68 μm. Moving from a 100% to a 300% strain, STEP-0 and STEP-6 exhibited a similar reduction in LAB fraction (5.6%) as well as a similar increase in grain size (0.46 μm).

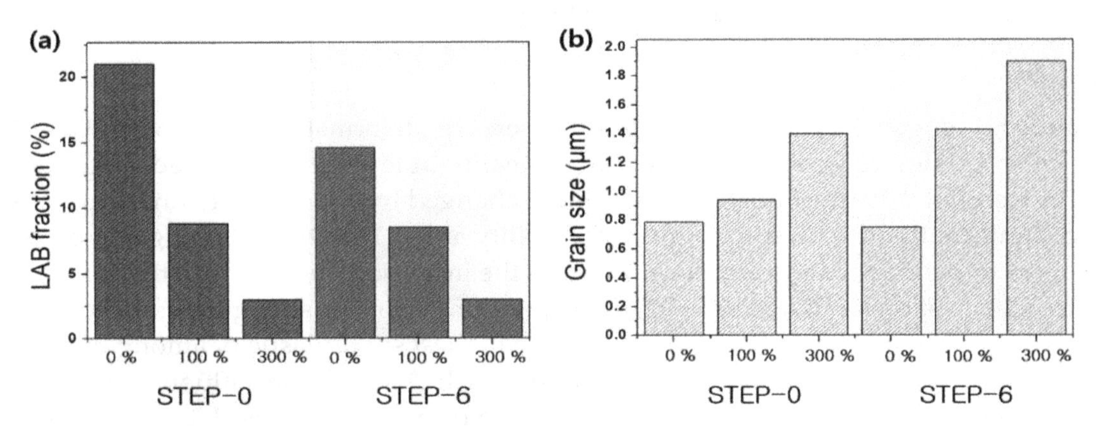

Figure 9. Microstructural features of STEP-0 and STEP-6 samples after strains of 0%, 100%, and 300% at 650 °C and 10^{-3} s^{-1}: (**a**) LAB fraction, (**b**) average grain size.

The distinct difference between the microstructural features of STEP-0 and those of STEP-6 arose from the active occurrence of CDRX in the STEP-0 sample. A continuous accumulation of dislocations during high-temperature deformation drove the formation of subgrains surrounded by LAB. Further straining generated recrystallized grains by increasing the misorientation of the subgrain boundaries [17,29]. The higher amount of pre-existing LAB in STEP-0 than in STEP-6 provided a higher driving force for CDRX in comparison with the annealed samples. This result suggests that CDRX of the STEP-0 sample occurred actively during the early stage of deformation, up to a 100% strain.

The hypothesis of the active occurrence of CDRX in STEP-0 during superplastic deformation is supported by the distribution of grain sizes (Figure 10). The STEP-0 sample showed an increasing fraction of grains with a size of ~1 μm in exchange for a decreasing fraction of coarse grains. In contrast, in the STEP-6 sample, the peak of the distribution curve shifted towards a higher grain size after a 100% strain at 650 °C and 10^{-3} s^{-1}. This is the typical trend that results from dominant grain-growth behavior at an elevated temperature. Such results suggest that the conversion of coarse grains into several fine grains (i.e., CDRX), occurred more actively in STEP-0 than in STEP-6. Consequently, active CDRX during the early stage of deformation in STEP-0 suppressed grain growth during elevated-temperature deformation and thereby contributed to its superior superplasticity.

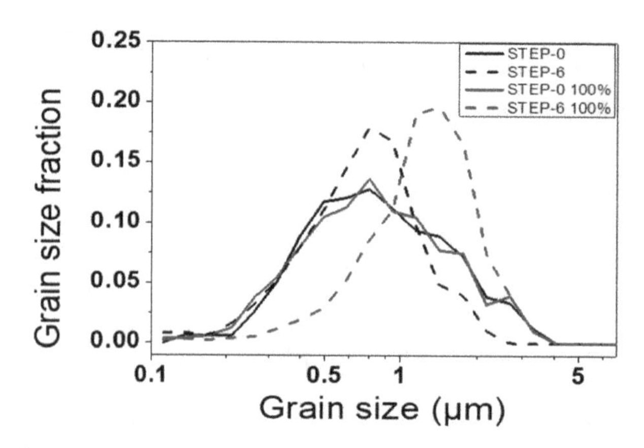

Figure 10. Distribution of grain sizes in STEP-0 and STEP-6 samples before and after applying a strain of 100% at 650 °C and 10^{-3} s^{-1}.

β particles also assisted in the increase in superplasticity in the STEP-0 sample. The sample possessed twice as much β phase as the STEP-5 and STEP-6 samples. Due to the difference in their crystallographic structures, grain-boundary diffusivity is over 100 times higher in the β phase than in the α phase [30]. Such a high diffusivity reduces the resistance to GBS activation. Moreover, resistance to GBS is lower at α/β interfaces than at α/α and α/β interfaces [31]. Therefore, the high β fraction was the secondary factor that increased the superplasticity of the STEP-0 sample.

4. Conclusions

This study investigated the RT and elevated-temperature deformation behaviors of UFG Ti-6Al-4V alloys fabricated by step rolling and subsequent annealing. Step rolling attained a significant grain refinement to a size of 0.70 μm, which was not much changed by subsequent annealing. The STEP-5 sample had significantly increased strength and ductility at RT. The strengthening was a result of the formation of fine α particles and (sub)grains, whereas the increased ductility was the consequence of a decreased frequency of dislocation tangles. The STEP-6 sample lost its strengthening effect because the fine particles disappeared. The STEP-0 sample had the highest superplasticity among the investigated samples; this result suggests that subsequent annealing has a variable influence that depends on the deformation temperature. The superplasticity arose from the equiaxed UFG structure and high dislocation density. Compared to the annealed samples, STEP-0 showed a rapid decrease in LAB fraction during the early stage of deformation. This result suggests that CDRX occurred actively in the STEP-0 sample at this stage and effectively suppressed grain growth during elevated-temperature deformation. The high β fraction of the STEP-0 sample also contributed to the GBS activation due to the considerable grain-boundary diffusivity and the presence of the α/β interface. Overall, active CDRX and a high content of β particles increased the superplasticity of the STEP-0 sample.

Author Contributions: Conceptualization, G.K. and C.S.L.; Writing—original draft, G.K., T.L., and Y.L.; Writing—review and editing, T.L. and C.S.L.; Supervision, C.S.L.; Investigation, G.K., Y.L., and J.N.K.; Resources, S.W.C. and J.K.H. All authors have read and agreed to the published version of the manuscript.

Acknowledgments: The authors gratefully acknowledge the financial supports of the Ministry of Trade, Industry and Energy, Korea for this research.

References

1. Welsch, G. *Materials Properties Handbook: Titanium Alloys*; ASM International: Russell Township, OH, USA, 1993; ISBN 9780871704818.
2. Weisert, E. *Proc. AIME Conference on Superplastic Forming of Structural Alloys*; Paton, N.E., Hamilton, C.H., Eds.; TMS-AIME Publications: Warrendale, PA, USA, 1982.
3. Mishra, R.S.; Stolyarov, V.V.; Echer, C.; Valiev, R.Z.; Mukherjee, A.K. Mechanical behavior and superplasticity of a severe plastic deformation processed nanocrystalline Ti–6Al–4V alloy. *Mater. Sci. Eng. A* **2001**, *298*, 44–50. [CrossRef]
4. Zherebtsov, S.V.; Kudryavtsev, E.A.; Salishchev, G.A.; Straumal, B.B.; Semiatin, S.L. Microstructure evolution and mechanical behavior of ultrafine Ti6Al4V during low-temperature superplastic deformation. *Acta Mater.* **2016**, *121*, 152–163. [CrossRef]
5. Edington, J.W.; Melton, K.N.; Cutler, C.P. Superplasticity. *Prog. Mater. Sci.* **1976**, *21*, 61–170. [CrossRef]
6. Ko, Y.G.; Lee, C.S.; Shin, D.H.; Semiatin, S.L. Low-temperature superplasticity of ultra-fine-grained Ti-6Al-4V processed by equal-channel angular pressing. *Metall. Mater. Trans. A* **2006**, *37*, 381–391. [CrossRef]
7. Shahmir, H.; Naghdi, F.; Pereira, P.H.R.; Huang, Y.; Langdon, T.G. Factors influencing superplasticity in the Ti-6Al-4V alloy processed by high-pressure torsion. *Mater. Sci. Eng. A* **2018**, *718*, 198–206. [CrossRef]
8. Zherebtsov, S.; Kudryavtsev, E.; Kostjuchenko, S.; Malysheva, S.; Salishchev, G. Strength and ductility-related properties of ultrafine grained two-phase titanium alloy produced by warm multiaxial forging. *Mater. Sci. Eng. A* **2012**, *536*, 190–196. [CrossRef]
9. Lee, T.; Shih, D.S.; Lee, Y.; Lee, C.S. Manufacturing Ultrafine-Grained Ti-6Al-4V Bulk Rod Using Multi-Pass Caliber-Rolling. *Metals* **2015**, *5*, 777–789. [CrossRef]
10. Lee, T.; Park, K.-T.T.; Lee, D.J.; Jeong, J.; Oh, S.H.; Kim, H.S.; Park, C.H.; Lee, C.S. Microstructural evolution and strain-hardening behavior of multi-pass caliber-rolled Ti-13Nb-13Zr. *Mater. Sci. Eng. A* **2015**, *648*, 359–366. [CrossRef]
11. Lee, T.; Lee, S.; Kim, I.-S.; Moon, Y.H.; Kim, H.S.; Park, C.H. Breaking the limit of Young's modulus in low-cost Ti–Nb–Zr alloy for biomedical implant applications. *J. Alloy. Compd.* **2020**, *828*, 154401. [CrossRef]
12. Park, C.H.; Kim, J.H.; Yeom, J.-T.T.; Oh, C.-S.S.; Semiatin, S.L.; Lee, C.S. Formation of a submicrocrystalline structure in a two-phase titanium alloy without severe plastic deformation. *Scr. Mater.* **2013**, *68*, 996–999. [CrossRef]
13. Kim, D.; Won, J.W.; Park, C.H.; Hong, J.K.; Lee, T.; Lee, C.S. Enhancing Superplasticity of Ultrafine-Grained Ti–6Al–4V without Imposing Severe Plastic Deformation. *Adv. Eng. Mater.* **2019**, *21*, 1800115. [CrossRef]
14. Jeong, D.; Kwon, Y.; Goto, M.; Kim, S. High cycle fatigue and fatigue crack propagation behaviors of β-annealed Ti-6Al-4V alloy. *Int. J. Mech. Mater. Eng.* **2017**, *12*, 1. [CrossRef]
15. Morita, T.; Hatsuoka, K.; Iizuka, T.; Kawasaki, K. Strengthening of Ti–6Al–4V Alloy by Short-Time Duplex Heat Treatment. *Mater. Trans.* **2005**, *46*, 1681–1686. [CrossRef]
16. Valiev, R.Z.; Alexandrov, I.V.; Enikeev, N.A.; Murashkin, M.Y.; Semenova, I.P. Towards enhancement of properties of UFG metals and alloys by grain boundary engineering using SPD processing. *Rev. Adv. Mater. Sci.* **2010**, *25*, 1–10.
17. Sakai, T.; Belyakov, A.; Kaibyshev, R.; Miura, H.; Jonas, J.J. Dynamic and post-dynamic recrystallization under hot, cold and severe plastic deformation conditions. *Prog. Mater. Sci.* **2014**, *60*, 130–207. [CrossRef]

18. Muga, C.O.; Zhang, Z.W. Strengthening Mechanisms of Magnesium-Lithium Based Alloys and Composites. *Adv. Mater. Sci. Eng.* **2016**, *2016*, 1078187. [CrossRef]
19. Langdon, T.G. Seventy-five years of superplasticity: Historic developments and new opportunities. *J. Mater. Sci.* **2009**, *44*, 5998. [CrossRef]
20. Matsumoto, H.; Velay, V.; Chiba, A. Flow behavior and microstructure in Ti–6Al–4V alloy with an ultrafine-grained α-single phase microstructure during low-temperature-high-strain-rate superplasticity. *Mater. Des.* **2015**, *66*, 611–617. [CrossRef]
21. Semiatin, S.L.; Fagin, P.N.; Betten, J.F.; Zane, A.P.; Ghosh, A.K.; Sargent, G.A. Plastic Flow and Microstructure Evolution during Low-Temperature Superplasticity of Ultrafine Ti-6Al-4V Sheet Material. *Metall. Mater. Trans. A* **2010**, *41*, 499–512. [CrossRef]
22. Alabort, E.; Kontis, P.; Barba, D.; Dragnevski, K.; Reed, R.C. On the mechanisms of superplasticity in Ti–6Al–4V. *Acta Mater.* **2016**, *105*, 449–463. [CrossRef]
23. Lee, T.; Yamasaki, M.; Kawamura, Y.; Lee, Y.; Lee, C.S. High strain-rate superplasticity of AZ91 alloy achieved by rapidly solidified flaky powder metallurgy. *Mater. Lett.* **2019**, *234*, 245–248. [CrossRef]
24. Sergueeva, A.V.; Stolyarov, V.V.; Valiev, R.Z.; Mukherjee, A.K. Superplastic behaviour of ultrafine-grained Ti–6A1–4V alloys. *Mater. Sci. Eng. A* **2002**, *323*, 318–325. [CrossRef]
25. Salishchev, G.A.; Galeyev, R.M.; Valiakhmetov, O.R.; Safiullin, R.V.; Lutfullin, R.Y.; Senkov, O.N.; Froes, F.H.; Kaibyshev, O.A. Development of Ti–6Al–4V sheet with low temperature superplastic properties. *J. Mater. Process. Technol.* **2001**, *116*, 265–268. [CrossRef]
26. Paton, N.E.; Hamilton, C.H. Microstructural influences on superplasticity in Ti-6AI-4V. *Metall. Trans. A* **1979**, *10*, 241–250. [CrossRef]
27. Park, C.H.; Ko, Y.G.; Park, J.-W.W.; Lee, C.S. Enhanced superplasticity utilizing dynamic globularization of Ti-6Al-4V alloy. *Mater. Sci. Eng. A* **2008**, *496*, 150–158. [CrossRef]
28. Ovid'ko, I.A.; Sheinerman, A.G. Grain-boundary dislocations and enhanced diffusion in nanocrystalline bulk materials and films. *Philos. Mag.* **2003**, *83*, 1551–1563. [CrossRef]
29. Huang, K.; Logé, R.E. A review of dynamic recrystallization phenomena in metallic materials. *Mater. Des.* **2016**, *111*, 548–574. [CrossRef]
30. Wert, J.A.; Paton, N.E. Enhanced superplasticity and strength in modified Ti-6AI-4V alloys. *Metall. Mater. Trans. A* **1983**, *14*, 2535–2544. [CrossRef]
31. Kim, J.S.; Chang, Y.W.; Lee, C.S. Quantitative analysis on boundary sliding and its accommodation mode during superplastic deformation of two-phase Ti-6Al-4V alloy. *Metall. Mater. Trans. A* **1998**, *29*, 217–226. [CrossRef]

5

Fatigue Behavior of Non-Optimized Laser-Cut Medical Grade Ti-6Al-4V-ELI Sheets and the Effects of Mechanical Post-Processing

André Reck [1,*], André Till Zeuner [1] and Martina Zimmermann [1,2]

1 Institute of Materials Science, Technical University of Dresden, 01062 Dresden, Germany
2 Department of Materials Characterization and Testing, Fraunhofer-Institute for Material and Beam Technology IWS, 01277 Dresden, Germany
* Correspondence: andre.reck@tu-dresden.de

Abstract: The study presented investigates the fatigue strength of the (α+β) Ti-6Al-4V-ELI titanium alloy processed by laser cutting with and without mechanical post-processing. The surface quality and possible notch effects as a consequence of non-optimized intermediate cutting parameters are characterized and evaluated. The microstructural changes in the heat-affected zone (HAZ) are documented in detail and compared to samples with a mechanically post-processed (barrel grinding, mechanical polishing) surface condition. The obtained results show a significant increase ($\approx$50%) in fatigue strength due to mechanical post-processing correlating with decreased surface roughness and minimized notch effects when compared to the surface quality of the non-optimized laser cutting. The martensitic α'-phase is detected in the HAZ with the formation of distinctive zones compared to the initial equiaxial α+β microstructure. The HAZ could be removed up to 50% by means of barrel grinding and up to 100% through mechanical polishing. A fracture analysis revealed that the fatigue cracks always initiate on the laser-cut edges in the as-cut surface condition, which could be assigned to an irregular macro and micro-notch relief. However, the typical characteristics of the non-optimized laser cutting process (melting drops and significant higher surface roughness) lead to early fatigue failure. The fatigue cracks solely started from the micro-notches of the surface relief and not from the dross. As a consequence, the fatigue properties are dominated by these notches, which lead to significant scatter, as well as decreased fatigue strength compared to the surface conditions with mechanical finishing and better surface quality. With optimized laser-cutting conditions, HAZ will be minimized, and surface roughness strongly decreased, which will lead to significantly improved fatigue strength.

Keywords: Titanium alloys; Ti-6Al-4V-ELI; fatigue; laser cutting; post-processing; α'-martensite; HAZ; barrel grinding; notch; fracture

---

1. Introduction

Titanium alloys are a frequently used material in industrial applications with ongoing market growth in recent years [1]. The industrial applications involve a wide spectrum from the aerospace and automobile sector, the chemical industry, to the field of medical engineering, such as applications in osteosynthesis. The reasons for this broad range of applications are high specific strength even at higher temperatures, excellent corrosion resistance, and the ability to adjust the material properties to a great extent with the optimization of the microstructure, as well as surface properties [1–3].

Titanium alloys possess high sensitivity to processing conditions and the subsequent fatigue loading during the application, strongly depending on the chemical composition, initial microstructure, and thermomechanical history [1,4–6]. Insufficient understanding or negligence of these factors may, therefore, result in an under- or overestimation of the fatigue strength and, as a consequence, either the failure to explore the full potential of titanium alloys or the leading to fatal failure cases.

The processing method of laser-cutting is a frequently used technique to transfer the sheet pre-product to the end geometry for the application. The high cost-effectiveness, economic efficiency, variability, as well as the possibility to produce complex geometries in a short time, are only a few of the many advantages [7–9]. However, the thermal input due to the laser must be adjusted and optimized regarding the surface roughness and resulting heat-affected zone (HAZ). Otherwise, micro- and macro-notch effects can negatively affect the fatigue behavior leading to catastrophic pre-mature failure in their application [4,10–12]. The local changes of the microstructure in the HAZ may also have a significant influence on the mechanical properties and on the fatigue strength, in particular [4,13–15]. However, the mechanical post-processing or finishing methods can improve the surface quality after laser-cutting or related processing methods in a significant way, which was proven in several studies [9,10,16–18]. The effects of surface roughness and the influence of the HAZ of a laser-cut component are, nevertheless, the subject of current research. Depending on the extent of strength reduction, the laser-cutting process has to be adjusted so as to form a favorable surface quality at the expense of the processing speed, while costly post-processing of the surface to remove the HAZ also has to be discussed with regard to its effectiveness on fatigue strength improvement.

The (α+β) alloy Ti-6Al-4V-ELI, which is analyzed in the present study, is one of the standard alloys in the medical sector due to its exceptional combination of high specific strength, good ductility, and remarkable corrosion resistance. As sheet metal, this alloy is mostly processed by laser-cutting and mechanically post-processed with barrel or vibratory grinding methods. While detailed investigations on the surface quality and the HAZ due to laser-cutting and their consequence with regard to the fatigue properties are limited, it is of a common consensus that from a general point of view, a significant negative influence of increasing surface roughness on fatigue behavior is expected [10,14–16,19]. Studies on the general interaction of laser and material surface concentrate on the temperature field, kerf development, and parameter studies (type of laser, power, speed, etc.) [13,20–24]. Fatigue properties of Ti-6Al-4V-ELI have been investigated intensively in the past [10,14–17,25–29]. The sensitivity against surface roughness and underlying surface near microstructure is assessed and confirmed. Da Silva et al. [10] demonstrated a theoretical and experimental decrease in fatigue strength with increasing surface roughness. Morita et al. [27] focused on the influence of short term aging to improve the fatigue performance of notched Ti-6Al-4V-ELI. The development of α'-martensite phase during quenching led to a retardation of crack propagation and, therefore, better fatigue properties [27]. The mechanical post-processing methods to improve the fatigue properties of Ti-6Al-4V-ELI were the subject of investigations, as was the influence of the environment, which is especially important for application in the human body [10,16,17,30,31]. However, to the best of the author's knowledge, no direct studies are found focusing on the interaction between laser-cutting, the subsequent mechanical post-processing, and the endurable stress amplitudes. Furthermore, it is known from related studies concerning laser welding and comparable methods to which extent the laser can change the local microstructure, and therefore, the mechanical and fatigue properties [32–34].

Therefore, the study aims to identify the principle changes that are obtained regarding surface quality and near-surface microstructure due to laser-cutting with intermediate (non-optimized) cutting parameters and to evaluate these effects with respect to the fatigue behavior of Ti-6AL-4V-ELI. The possible local changes in the HAZ-microstructure shall be clarified, and the resulting surface quality assessed. Furthermore, the fatigue strength of the specimens after mechanical post-processing with the method of barrel grinding will be compared to samples with an as-cut surface and with a surface after manual mechanical polishing. Surface roughness is also investigated in conjunction with the crack initiation sites after fatigue loading.

2. Materials and Methods

The studied medical grade Ti-6Al-4V-ELI (ISO 5832-3) sheet pre-product was purchased from the supplier MetSuisse Distribution AG (Zug, Switzerland) and originally produced by RTI International Metals Inc. (Pittsburgh, PA, USA). The sheet thickness was 0.8 mm and the laser-cutting process was realized with a disk laser (TruDisk 5001-Fa. TRUMPF, Ditzingen, Germany) with 3 kW laser power at 25 m/min cutting speed. To minimize the possible chemical reactions due to oxygen or nitrogen, which can lead to hard and brittle TiO_2 or TiN surface layers, laser-cutting was carried out under argon atmosphere with an argon pressure of 6 bar. All laser cutting parameters (Table 1) were chosen to display an average parameter set for the Ti-6Al-4V-ELI alloy. Since the process optimization was not foreseen in the scope of the study presented, the parameters of the laser-cutting process were chosen on the basis of practical knowledge and represent an intermediate condition; however, allowing a general analysis of the possible changes in the microstructure and the extent of surface roughness effects on the fatigue behavior. The geometry used for the laser-cut fatigue samples was developed in previous studies on medical implant alloys [12] and is depicted in Figure 1.

Table 1. Laser cutting parameters applied for Ti-6l-4V-ELI sheets on the TruDisk 5001 laser (Fa. TRUMPF-Series Tru Laser 7025).

Cutting Parameters	Laser Power	Cutting Speed	Spot Size	Laser Beam Quality	Cutting Gas	Nozzle	Focal Distance
Used parameter set	3 kW (λ = 1035 nm)	25 m/min	150 µm (Focal spot on the surface)	$M^2 = 14$	Argon (6 bar)	Single head-Conical (2.0 mm)	6 inch

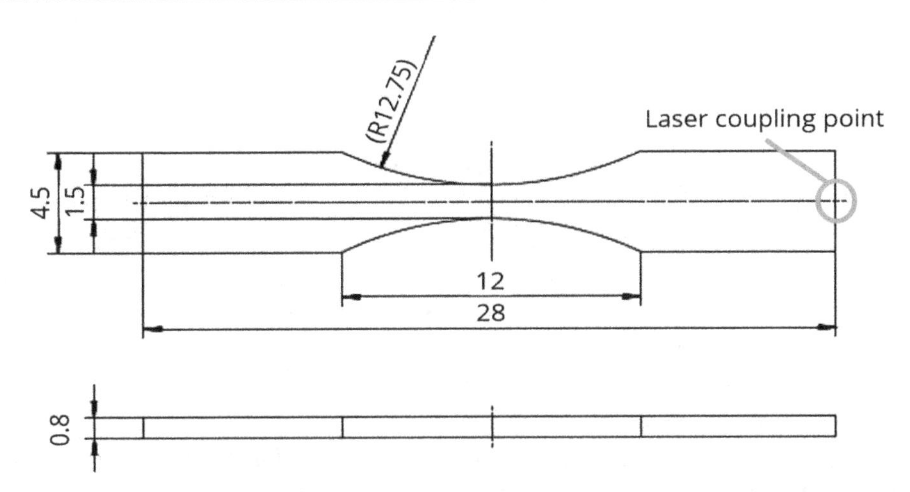

Figure 1. Fatigue sample geometry of Ti-6Al-4V-ELI with a marked laser coupling point. Dimensions in mm.

The samples originally processed by laser-cutting were divided into three series for fatigue testing. Laser-cut samples (Ti-6Al_LC), barrel-grinded samples (Ti-6Al_BG), and mechanically polished samples (Ti-6Al_MP) were investigated in detail before fatigue testing to document and assess the surface quality and near surface microstructure (heat-affected zone—HAZ) after the laser-cutting process, as well as after the mechanical post-processing. Therefore, metallographic cross-sections were prepared and microstructure analysis was executed by means of light microscopy, as well as scanning electron microscopy (SEM–JEOL JSM 7800-JEOL Ltd., Tokyo, Japan). The surface relief and roughness after laser-cutting was evaluated by confocal microscopy (Leica DCM 3D - Leica Microsystems, Wetzlar, Germany) using the software Leica Map Premium 6.2.7487. The element analysis for microstructural mapping in the HAZ was realized by energy dispersive spectroscopy (EDS–Oxford Instruments, Abingdon, Great Britain), which was attached on the SEM column. The AZtecHKL software (version 3.0) was used for evaluation.

All fatigue tests were carried out by a servo-electric load-frame (SEL 010 with software Loadframe SX 2.4–Fa. Thelkin, Winterthur, Switzerland) using an attached cooling device with pressurized air to hold the sample temperature during the fatigue tests at a constant room temperature level. The load-controlled tests were executed under tension-compression mode ($R = -1$) with a frequency of 20 Hz. The fatigue life limit N_G was set to 2×10^6 cycles. To ensure homogeneous sine oscillations several pre-tests were carried out to find the best parameter sets of the test system for the specific vibration behavior of the Ti-6Al-4V-ELI alloy. Subsequently, load-increase tests were executed to find the suitable load horizons. Finally, series testing was realized by means of the staircase method (15–18 samples per series).

The Ti-6Al_LC samples were tested with the original surface condition after laser-cutting. The melting drops were removed in the clamping range to ensure safe testing. The Ti-6Al_BG samples were additionally barrel grinded with standard titanium alloy parameter sets (4 h (rough grit) and 4 h (fine grit) at 1000 RPM, pyramidal polishing stones) in a vibratory grinding machine for small samples (Fa. Rösler, Untermerzbach, Germany). The Ti-6Al_MP samples were manually mechanically polished along the cyclic load direction stepwise from P400, P800; P1200 to a final grid size of P2500 which corresponds to an average roughness value of Sa < 0.500 μm. The mechanical-polishing and barrel-grinding process were, thereby, carried out under constant fluid (water for mechanical polishing grinding fluid for barrel (vibratory) grinding). Metallographic cross-sectioning documented the change in microstructure and HAZ as a consequence of the mechanical post-processing.

The fracture surface analysis was executed after the fatigue tests with SEM to locate crack initiation sites of all samples and to assess the possible connection between surface roughness-related notch effects due to laser-cutting and actual crack initiation after fatigue failure.

3. Results and Discussion

3.1. Surface Quality

The surface quality achieved due to the laser-cutting process with the particular cutting parameters applied (see Material and Methods section) is depicted in Figures 2 and 3. The Ti-6Al-4V-ELI alloy shows a distinctive surface relief with recognizable additional melting drops on the lower cutting edge On the upper cutting edge, an irregular distributed terrace-like surface structure can also be identified (Figure 3a). Both observed features can be traced back to the influence of the laser-cutting parameters applied. The melting drops are caused by an insufficient pressure of the inert argon gas atmosphere Whereas on the upper cutting edge, the occurring melting drops could be sufficiently removed, the lower cutting edge is less accessible, demanding a higher pressurized air stream to successfully remove all residual melting drops. As a consequence, the necessary argon pressure has to be well over 6 bar to homogeneously remove melting drops on the whole cutting edge.

The irregular terrace-like structure can be assigned to an influence throughout the interaction of the laser and Ti-6Al-4V-ELI but not directly explained till now. Further evaluation is, therefore, necessary. However, optimized laser-cutting parameters will improve the surface quality to a great extent, and therefore, also has a positive effect on the resulting fatigue strength of the Ti-6Al-4V-ELI alloy. The average height parameters of the laser-cut surface relief are measured according to ISO 25178 by means of confocal microscopy. Several μm in distance between the relief hills and valleys (Figure 2a) are identified, which concludes a significant surface roughness. Since fatigue failure in the LCF ($< 1 \times 10^3$–10^4 cycles) and HCF (up to 1×10^7 cycles) regime is mostly triggered by the stress concentration at distinct surface flaws leading to early crack initiation and propagation, pronounced surface roughness is detrimental for the expected fatigue strength and endurable cycles [10–12]. The Ti-6Al_LC samples with a surface quality as depicted in Figure 2 could, therefore, be expected to cause a dominant concentration of cyclic stress in the relief valleys during fatigue. In addition, Ti-6Al-4V is known for its high notch sensitivity, which was the subject of detailed studies in the past [3,10,14,19], demonstrating a strong decrease of fatigue strength correlating with higher surface roughness and resulting notch factors. Comparing the surface quality after laser-cutting (Ti-6Al_LC) with the sample series of Ti-6Al-BG and Ti-6Al_MP, the latter ones show a significantly improved surface quality, which correlates also with lower surface roughness. The mechanically polished surfaces of the Ti-6Al_MP samples are completely free of visible micro-notches which is a result of the homogeneous polishing to a grit size of P2500 and corresponds to a surface quality of $S_a < 0.500$ µm. In comparison to the as-cut surface, as well as the barrel grinded surface, this post-processing represents the best surface quality for the fatigue samples in the study presented.

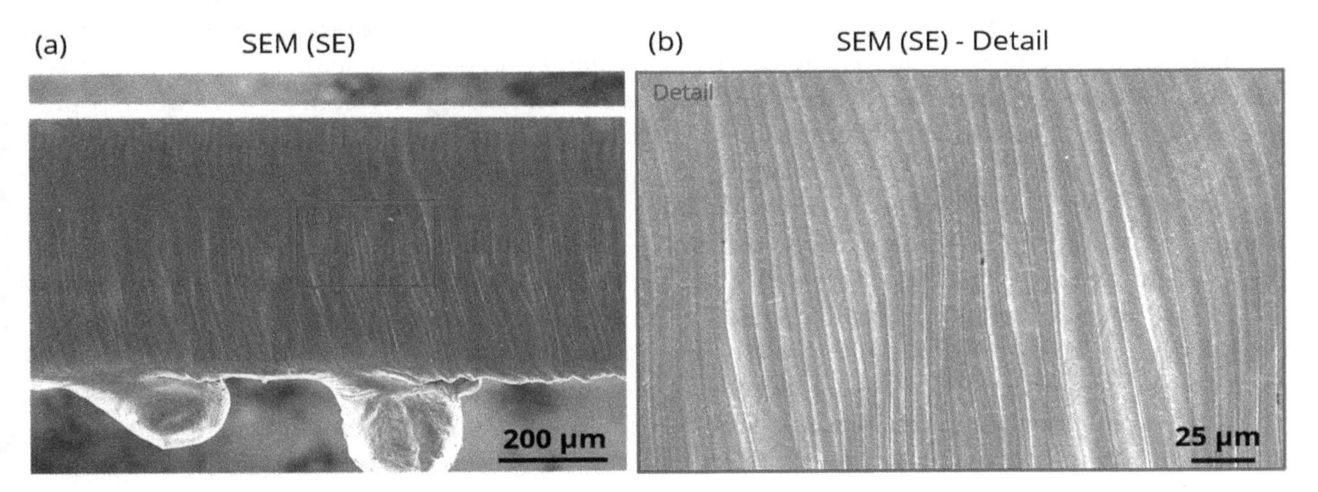

Figure 2. Obtained quality of the laser-cut surfaces of Ti-6Al-4V-ELI with non-optimized laser cutting conditions: (**a**) Overview SEM-image of the cutting edge with residual melting drops at the lower part; (**b**) Detailed SEM-image showing a wavy, irregular surface structure after the laser-cutting process.

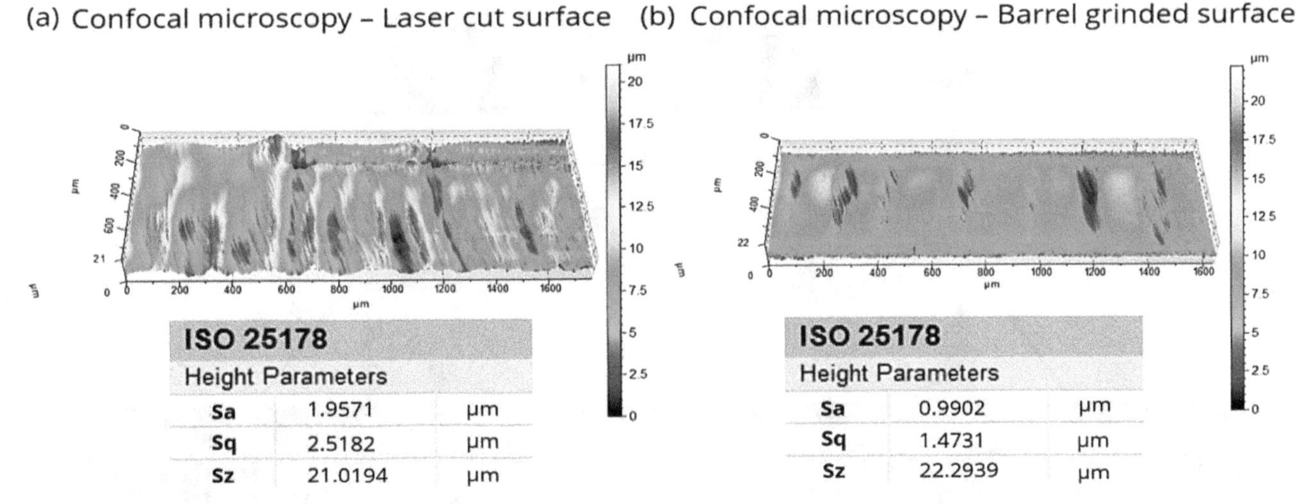

Figure 3. Typical representative surface reliefs by means of confocal microscopy (normalized) of Ti-6Al-4V-ELI with average height parameters: (**a**) Surface after laser-cutting; (**b**) Surface after barrel-grinding—Improvement of surface quality.

Barrel-grinded surfaces, on the other hand, show a strong improvement in surface quality and less roughness, but retained micro-notches are visible in the center area of the cutting edges (Figure 3b). These micro-notches could act as possible crack initiation sites in the application, which was already shown in detail in one of the authors previous studies of laser-cut β-titanium, as well as α-titanium for fatigue samples and osteosynthesis plates [12]. This leads to the conclusion that the barrel-grinding process is an effective tool to reduce detrimental laser-cutting effects, but only under the prerequisite of already optimized laser-cutting parameters for a specific material. Without the improvement of the original laser-cut surface quality, micro-notches seem to remain in the center area of the laser-cutting edges playing a decisive role for achievable fatigue strength in the application.

3.2. Microstructural Development of the Heat-Affected Zone (HAZ)

The microstructural response to the thermal input from processing in the heat-affected zone is alongside the surface quality, the most important factor to define the influence of the laser-cutting process regarding the fatigue behavior. Possible microstructural changes compared to the initial $\alpha+\beta$-microstructure could be expected for Ti-6Al-4V-ELI since the laser-cutting temperature lies well above the β-transus temperature of this alloy ($\approx$980 °C [35]). Hence, the laser-cutting process could change the surface-near microstructure caused by the local heating and subsequent self-quenching.

The HAZ in a cross-sectional view at the upper cutting edge depicted in Figure 4 clearly demonstrates the change in microstructure from equiaxial $\alpha+\beta$ into acicular martensitic α' with retained β, the latter being proven by the EDS-analysis showing a higher V-content as a strong β-phase stabilizer. The transformation into martensitic α' is, thereby, associated with the self-quenching process, which is fast enough to transform the initial equiaxial α-phase (5–10 μm in grain size) to fine acicular α' martensite. The retained β-phase can be explained by the very fast processing time and holding over β-transus not allowing the β-phase also to transform into α' martensite. The lower cutting edge with residual melting drops (Figure 2a) shows, by contrast, two distinctive zones (Figure 5—Cross-sectional view). A small surface layer (average thickness 10–15 μm) consists solely of acicular α' without β-phase. The second zone, which develops with growing distance to the free surface, consists of α' and β-phase, comparable to the upper cutting edge, followed by the initial ($\alpha+\beta$)-microstructure. The development of the different microstructural zones could be explained with the self-quenching gradient from the laser-cutting temperature, which causes different quenching speeds towards the sample interior and, therefore, possible zone formation. However, in order to explain the pronounced microstructural difference of upper and lower cutting edges, more detailed information of the temperature distribution related to the process parameters would be necessary.

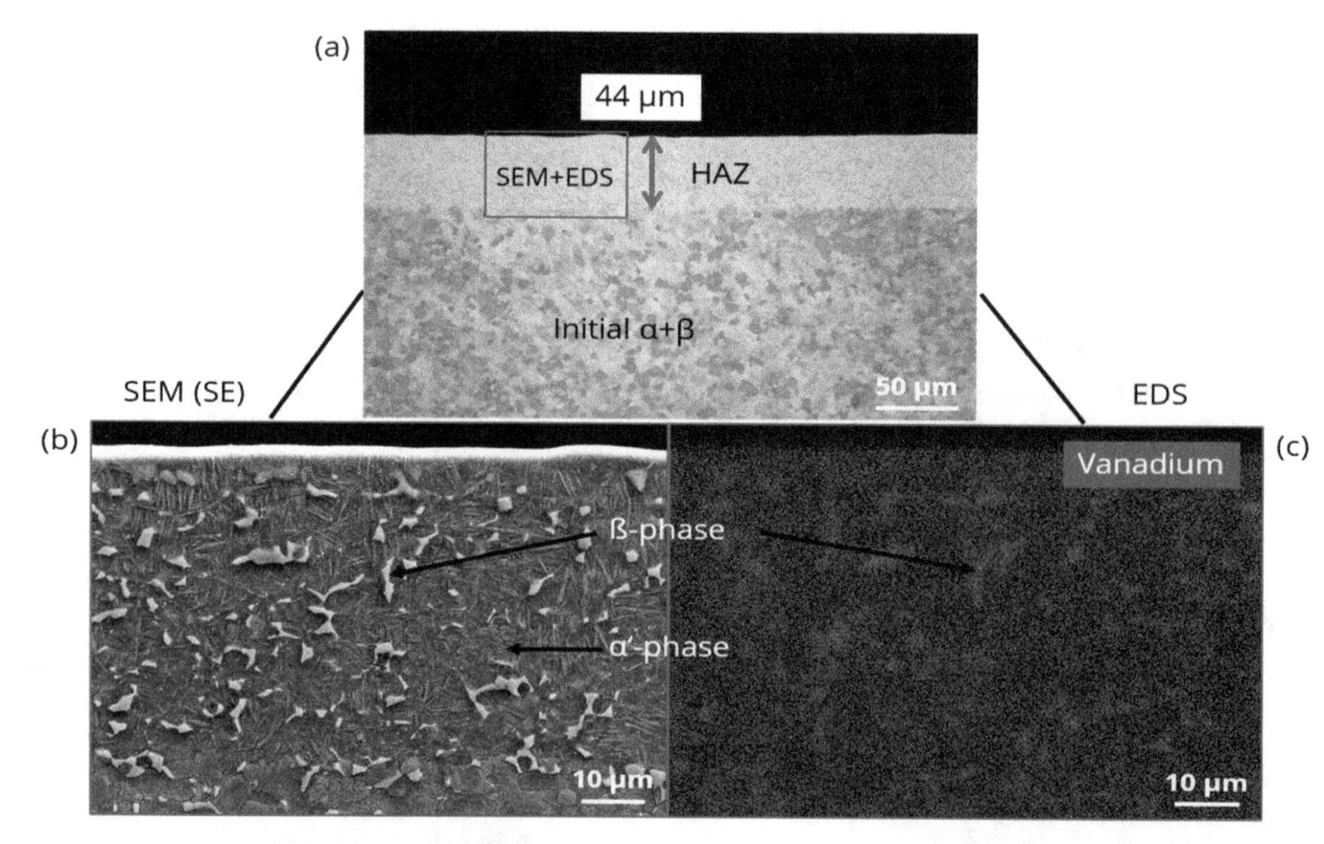

Figure 4. Typical HAZ at the upper cutting edge developed during the laser cutting process (Ti-6Al_LC sample): (**a**) Light microscopic image representing an overview; (**b**) SEM-image (SE cross-section) of the HAZ microstructure with α'-martensite; (**c**) Element analysis (EDS) showing the higher vanadium content (β-stabilizer) in the β-phase of the HAZ.

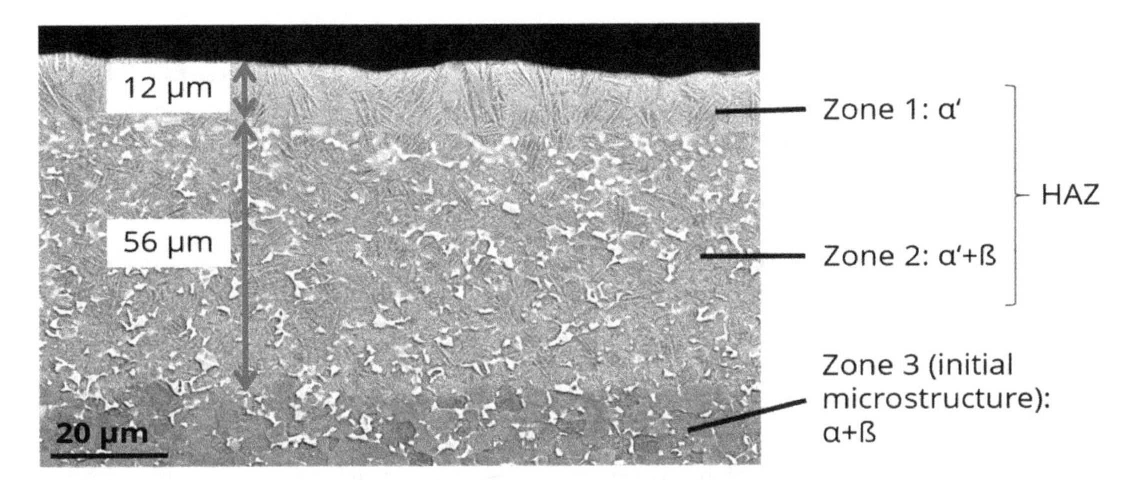

Figure 5. SEM image (BSE cross-section) of the HAZ at the lower cutting edge of a Ti-6Al_LC sample showing three distinctive microstructural zones after laser-cutting consisting of martensitic α′-phase and retained β-phase.

The measured thickness of the HAZ (≈40–70 µm) depends on the occurrence of the different zones and is more pronounced at the lower cutting edge with the additional α′-zone. The optimized laser-cutting parameters, such as speed and power would contribute to a significant less pronounced HAZ. Furthermore, the application of a pulsed laser is expected to have a positive effect on the local microstructural changes due to the very short processing window and heating over the β-transus temperature [36].

With the mechanical post-processing of the laser-cut surface, not only the surface quality is improved, as was discussed in Section 3.1, but also the HAZ is significantly affected. The result is depicted in Figure 6 showing a significant reduction of the HAZ thickness. The barrel-grinding process creates, thereby, a decrease of about 50% HAZ compared to the initially observed HAZ without any mechanical post-processing. The reduction is independent of upper or lower cutting edge but shows irregularities at the crossover region of the radius and clamping range. This phenomenon can be explained by the mechanism of the barrel grinding process itself, which is based on a rotary or vibratory grinding process between the sample and a specific grinding stone, as well as an abrasive medium [37,38]. The careful selection and adaption of this process is a prerequisite for an optimal and homogeneous grinding result. The complex geometries and very poor initial surface quality are prerequisites that may have an adverse effect on the grinding quality. Insufficient grinding results negatively influence the fatigue properties with residual micro-notches on geometrically inaccessible locations on samples or components [12].

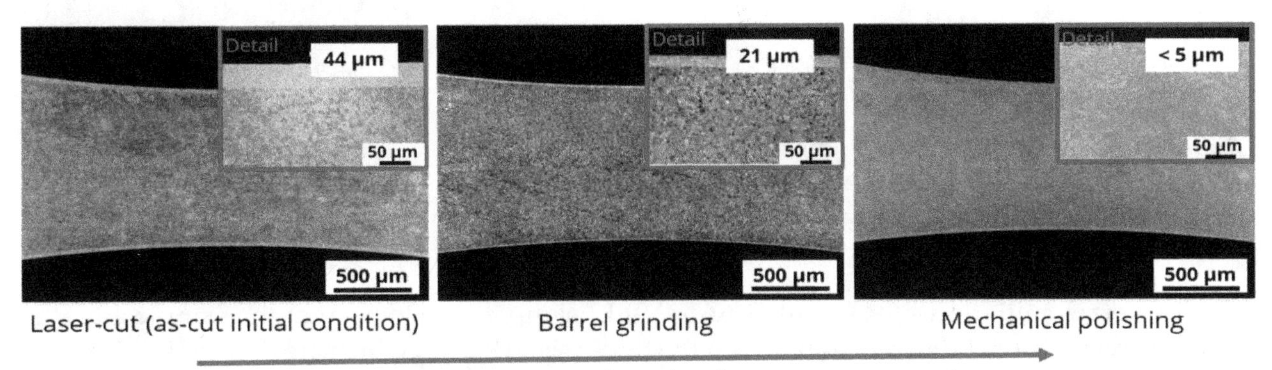

Figure 6. Thickness of the HAZ dependent on the mechanical post-processing showing a significant decrease in measured HAZ-thickness at the upper cutting edge.

The results of mechanical polishing are also depicted in Figure 6. An almost complete removal of the HAZ is observed. However, it has to be pointed out that in the study presented, the mechanical polishing was executed hand-made, leading to slight variations regarding the extent of removal of the HAZ for the overall number of samples.

3.3. *Fatigue Results and Crack Initiation*

The fatigue test results for all the three sample series are shown in Figure 7a. Remarkable differences are recognizable between the samples with the as-cut surface (Ti-6Al_LC) and the sample series with additional post-processing of the surface (Ti-6Al_BG and Ti-6Al_MP). The results for the as-cut samples show a significant scatter and the lowest fatigue strength ($\sigma_{aD,50\%} = 235$ MPa) for a fatigue life limit of $N_G = 2 \times 10^6$. A superior fatigue strength was reached for the series with barrel-grinded surface (351 MPa) and mechanically polished surface (363 MPa). These values represent an increase of around 50% compared to the original laser-cut surface condition. However, while comparing the fatigue strengths obtained for the different sample conditions, it has to be considered that the as-cut condition represents a more or less "worst-case" condition since the process parameters were chosen on the basis of practical knowledge and were not yet optimized. The optimized laser-cutting parameters result in strongly decreased surface roughness and improved fatigue strength. The scattering of the results is less distinctive with mechanical post-processing of the surface and can be attributed to a higher surface quality with less surface roughness, and therefore, less possible notches acting as stress concentrators during fatigue loading. For the barrel grinded surface condition, however, micro-notches remain in the center area of the laser-cut edges, which was exemplarily shown in Figure 3. In comparison to the process of mechanical polishing, which displays the best overall fatigue behavior, barrel grinding of the Ti-6Al-4V-ELI alloy must be optimized, especially in case of homogeneity and removal of all surface notches.

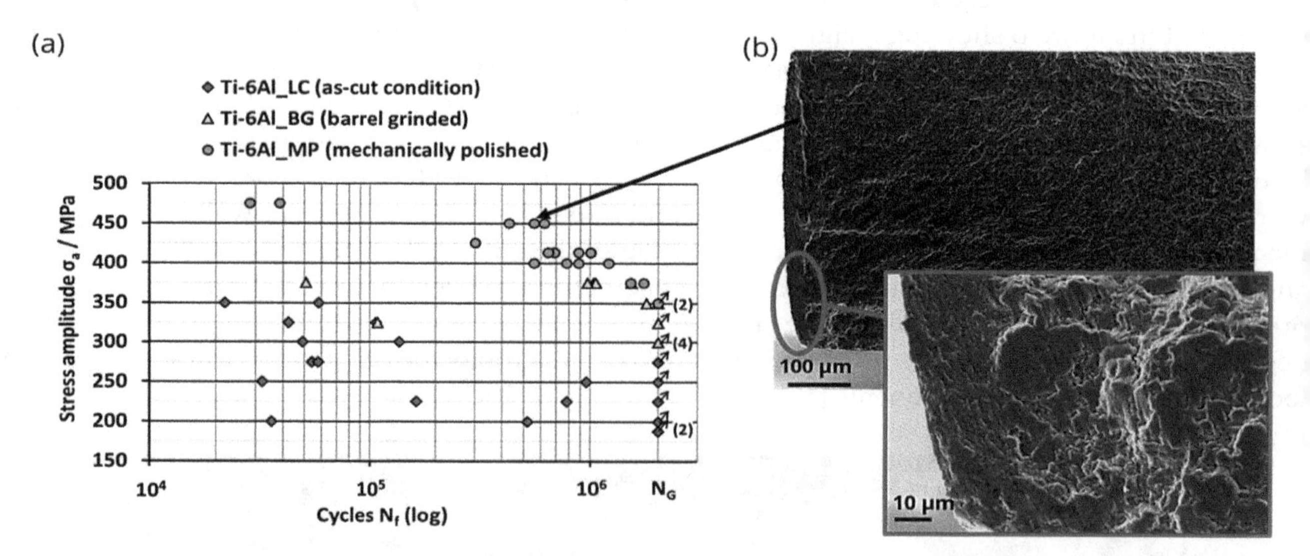

Figure 7. Fatigue results of the Ti-6Al-4V-ELI alloy: (**a**) S-N diagram for all three tested sample series; (**b**) Fracture surface analysis for a Ti-6Al_MP sample with mechanically polished surface conditions including the crack initiation site on the laser-cut edge directly near the sample corner after fatigue failure at 450 MPa and 6.2×10^5 cycles.

Fatigue crack initiation sites for all Ti-6Al-4V-ELI samples are located at the sample edges and corners. There are distinct differences between the three tested series, which are depicted in Figures 7b, 8 and 9. For the mechanically polished surface condition (Figure 7b), fatigue cracks initiate at the laser-cut edges or the flat sample edges, but primarily at or directly near the sample corners. The reason for this behavior is the overall good surface quality and comparable surface roughness of all sample edges with no distinctive micro-notches. In this case, the sample corners act normally as the highest stress concentrator under fatigue loading for flat sample geometries [39]. The consequence is a preferred fatigue crack initiation at these sites for the Ti-6Al_MP samples.

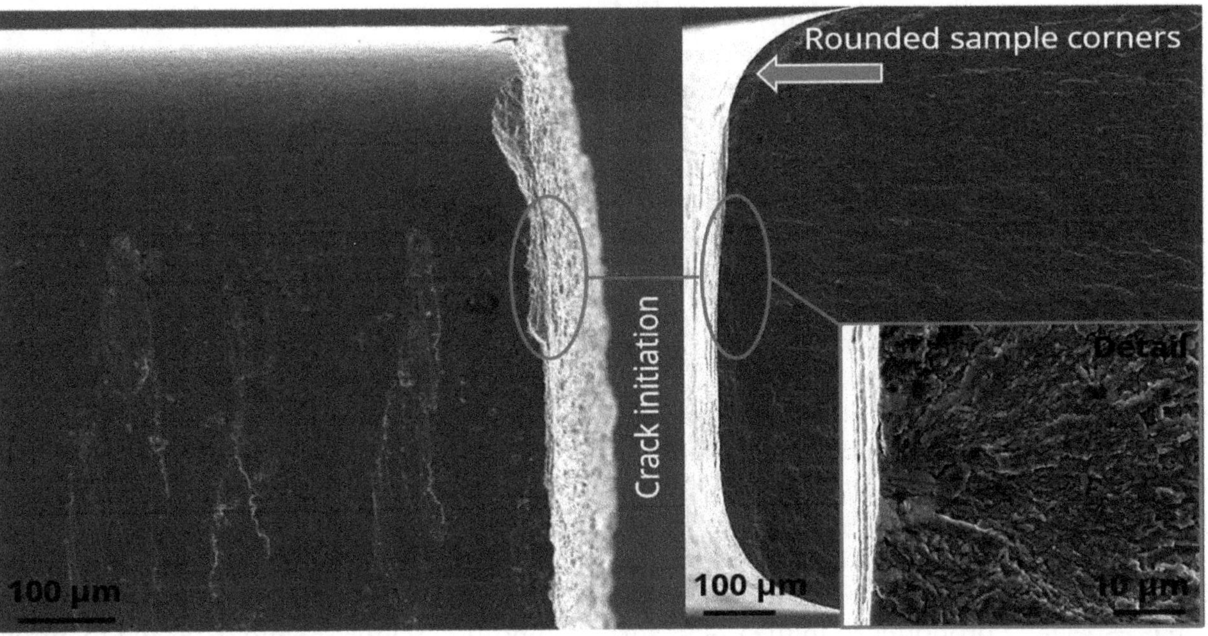

Figure 8. Fractography of a Ti-6Al_BG sample with barrel-grinded surface condition failed at 375 MPa and 5.1×10^4 cycles: (**a**) SEM (SE) analysis of the laser-cut edge with detectable notches; (**b**) The corresponding fracture surface with detailed view on fatigue crack initiation.

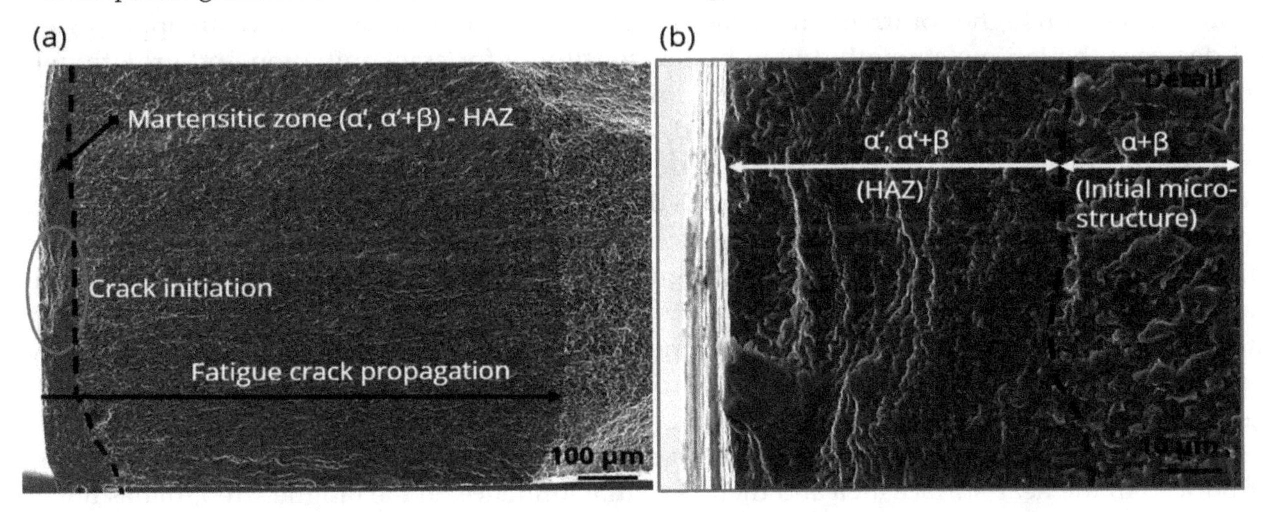

Figure 9. Fractography of a Ti-6Al_LC sample with as-cut surface condition failed at 200 MPa and 5.2×10^5 cycles: (**a**) SEM (SE) analysis of the fractured surface with identification of the martensitic zone; (**b**) The corresponding detailed view at the fatigue crack initiation point.

The sample series with the barrel-grinded surface condition (Ti-6Al_BG) showed, on the contrary, fatigue crack initiation always on the laser-cut edges, which is exemplarily depicted in Figure 8. Furthermore, the crack initiation sites are not always located at the smallest loaded cross-section but rather in the radius section. The cause for this behavior can be assigned to the remaining and detected surface notches after the barrel-grinding process in the middle of the laser-cut edges (Figures 3 and 8a). The irregular size and distribution of these remaining notches leads consequently to stress concentration, crack initiation, and failure at the biggest notches, and these can be located outside the smallest loaded cross-section. Associated with the effective lower overall surface roughness of the laser-cut edges and the observed, particularly rounded, sample corners (Figure 8b—lower stress concentration sites), all Ti-6Al_BG samples are expected to fail at one of the remaining notches.

The sample series with the initial as-cut surface condition (Ti-6Al_LC) reveal almost similar fatigue crack initiation points as the Ti-6Al_BG samples. Due to the significantly stronger local surface roughness (compared to the flat sample edges) caused by the laser-cutting for the process parameters applied, fatigue failure could always be directly assigned to the crack initiation at the laser-cut edges, which is depicted in Figure 9 in the fracture surface analysis. Pronounced irregularity of the surface profile leads subsequently to the high scattering of the fatigue results and the decreased fatigue strength for the Ti-6Al_LC sample series. At this point, it should be pointed out that, irrespective of the severe geometrical irregularity of the dross formation (see Figure 2a) due to a lack of process optimization, fatigue crack initiation always started from an extrusion/intrusion from the surface relief and not from a dross. The optimized cutting conditions improve this behavior and lead to significantly improved fatigue strength, although the formation of a slight surface relief cannot be completely avoided and may have an effect on the fatigue behavior. By comparing the fatigue behavior of the barrel-grinded and the as-cut condition, it becomes obvious that even marginal remains of the surface relief can result in early failure with values for the barrel grinded condition close to the upper scattering range of the as-cut results. The fatigue results obtained for the stress amplitudes between 300 MPa and 350 MPa demonstrate the particularly high sensitivity of the fatigue behavior of Ti-6Al-4V-ELI on surface flaws with early failures ($N < 2 \times 10^5$), as well as run-out samples ($N = 2 \times 10^6$) for the same stress level.

An aspect to be considered for all three sample series, but especially for the as-cut surface condition, is the presence of the HAZ caused by the laser-cutting. The martensitic zone (α' and α'+ß—Figures 4 and 5) is especially pronounced in the as-cut surface condition (Figure 6) and expected to influence the fatigue behavior, although an optimized laser-cutting process will significantly decrease HAZ development. Due to a higher strength and more interphase boundaries, this phase can have a positive effect on the fatigue strength or the fatigue crack growth behavior of Ti-6Al-4V [3,4,29]. However, in connection with higher or irregular surface roughness, α'-martensite can have an opposing effect and decrease the fatigue strength due to higher sensitivity for crack initiation and growth, which can be explained with higher hardness and lower ductility of α' compared with the initial $\alpha+\beta$ microstructure [16,19]. In the study presented, a pronounced HAZ can clearly be observed for all samples with an as-cut surface condition (Figure 9). The specific implication on the resulting fatigue strength seems; however, not significant due to the superior effect of the surface roughness, which is supported by the fact of highly scattered results for Ti-6Al_LC samples. In order to clearly identify an influence of the HAZ on the early stages of fatigue crack initiation and growth, further investigations with pre-notched samples would be necessary in order to quantitatively measure early crack growth rates in the HAZ. For the sample series with mechanical post-processing, HAZ was also observed, but in a less distinctive manner. The superior role of surface roughness is however confirmed with crack initiation on specific notches for the barrel grinded surface condition or near sample corners for the mechanically polished surface condition. It should be mentioned that possible residual stresses introduced in the near-surface area and their potential influence on the fatigue strength has not been part of the investigation of this study. However, this effect must be taken into consideration when further elucidating the superior fatigue behavior of the mechanically polished condition.

A comparison of the findings in the study presented with fatigue properties in the literature show a good correlation with the lower range of possible fatigue strengths for the ($\alpha+\beta$) Ti-6Al-4V alloy [14,15,25,26,29]. The overall fatigue properties of Ti-6Al-4V are, thereby, dependent on various influencial factors regarding the microstructure, the fatigue test conditions, as well as the fatigue geometry, composition and additional surface treatments. Furthermore, the role of thermal and residual stresses and their influence on the fatigue behavior must be taken into account [2–4,16,29]. Although the residual stress measurements were not executed in the study presented, it can be assumed that compressive residual stresses are introduced during the process of mechanical finishing. For the as-cut condition, the higher hardness of the α'-zone indicates an increase in mechanical strength; however, regarding the low surface quality of the as-cut condition, this general increase in strength can be assumed to be more susceptible for early fatigue failure and crack initiation due to an increased notch sensitivity. The compressive residual stresses introduced by means of mechanical polishing, on the other hand, contribute to an enhanced fatigue strength, and combined with an improved surface quality will result in an improved fatigue behavior [3,4,29]. The discrepancies in fatigue strength of Ti-6Al-4V_ELI in the study presented clearly correlates with the surface quality of each processing condition and, in particular, the resulting significant notch effects caused by the non-optimized laser-cutting process. In consequence, an optimization of the used laser-cutting parameters could be expected to significantly improve fatigue behavior.

The results of the as-cut surface condition impressively demonstrate the detrimental effect of the surface roughness with exceptional low fatigue strength ($\sigma_{aD,50\%}$= 235 MPa) and pronounced scatter of the fatigue results. However, optimum laser-cutting parameters decrease this negative effect and strongly enhance fatigue behavior, because of the direct correlation of laser-cutting parameters and resulting surface quality. The barrel-grinding process (Ti-6Al_BG samples) results in a strong increase in fatigue strength, but is, nevertheless, determined by the dominating influence of the remaining notch profile in the center area of the laser-cut edges. The difference in fatigue strength compared to the mechanically polished surface condition (Ti-6Al_MP samples—Figure 7) can mainly be explained by the effect of the flat sample geometry and the consequential influence of the sample corners. These corners act as preferential, and possible early crack initiation sites in case of the Ti-6Al_MP samples with mechanically polished surface, whereas for Ti-6Al_BG samples with barrel grinded surface condition observed significant rounding of the corners (Figure 8) has the opposite effect. Hence, both sample series with mechanical post-processing of the surface exhibit almost similar fatigue behavior. The influence of the underlying microstructure and investigated HAZ seems to play only a circumstantial role compared to the dominant influence of the surface roughness. In consequence, the optimization of the surface roughness has to be the essential goal for laser cutting of Ti-6Al-4V-ELI in order to avoid early fatigue failure in application. Mechanical post-processing by means of barrel (vibratory) grinding or polishing seems to be an excellent choice for achieving this goal due to clear enhancement of fatigue strength caused by the minimization of arising surface roughness during laser cutting. Furthermore, residual compressive stresses could be introduced, positively influencing stress concentration and subsequent crack initiation during fatigue loading. All these factors are controlled by the grinding and polishing parameters, as well as the exact alloy composition, heat treatment condition, and original surface quality due to laser cutting, which, if mutually adjusted, hold the potential for significant fatigue life improvement. The mechanical post-processing, such as barrel grinding, can easily be implemented in industrial process chains and quality control. The downside is, on the other hand, the control of dimensions, especially in the case of maintaining the original edge proportion, which are prone to rounding (Figure 8b). Only recognition of all influencing effects can, therefore, lead to economic and positive post-processing with subsequent utilization of the full material potential under fatigue loading.

4. Conclusions

The study presented investigated the fatigue behavior of the medical implant alloy Ti-6Al-4V-ELI processed by laser-cutting and the consequences of additional surface post-processing on the fatigue properties with regard to surface roughness and HAZ. The fatigue behavior of the as-cut surface condition was compared with mechanically polished surfaces, as well as with barrel-grinded surfaces. The results can be summarized and concluded as follows:

- The surface relief introduced by the non-optimized laser-cutting influences the fatigue behavior of Ti-6Al-4V-ELI significantly. For the process parameters featured in this study, fatigue strength of the as-cut condition results in a drastic decrease of the fatigue strength compared to the additionally surface-treated condition. However, the difference in fatigue strength observed in this study will be controllable by optimizing the laser-cutting parameters.

- The main reason for the superior fatigue strength of the post-processed conditions is a minimized surface roughness, which in turn is responsible for higher resistance against fatigue crack initiation on macro- and micro-notches originally caused by the laser-cutting.
- The process of barrel-grinding after the laser-cutting was effective but revealed retained surface roughness in the center area of the cutting edges, which acts as preferred crack initiation sites compared to the particularly rounded sample corners.
- Mechanically polished samples always failed at or near the sample corners, which is caused by a stress concentration on these sites.
- The HAZ consisting of martensitic α' and β along distinctive surface and subsurface zones was analyzed and does not play a significant role in early fatigue failure, which instead was dominated by the surface roughness. Nevertheless, the applied mechanical post-processing led to an almost complete removal of the HAZ.
- To avoid early fatigue failure in the application, an optimization of the laser-cutting parameters is crucial in order to obtain better surface quality. This allows the required post-processing to improve the surface roughness further and, therefore, the fatigue strength. However, both processes, laser-cutting and mechanical post-processing, have to be optimized in the dependence of the specific alloy composition and fatigue behavior of Ti-6Al-4V-ELI.

Author Contributions: Conceptualization—A.R., M.Z.; Methodology—A.R.; Investigation—A.R., A.T.Z.; Data analysis/curation—A.R., A.T.Z.; Writing—original draft preparation—A.R.; Writing—review and editing—M.Z., A.R.; Project administration—A.R., M.Z.; Funding acquisition—M.Z.

Acknowledgments: The authors thank Nikolai Schröder and the working group laser cutting at IWS for carrying out the laser cutting as well as Robert Kühne and Sebastian Schettler from the department Materials Characterization and Testing for helpful thematic discussions regarding the fatigue testing. Further thanks to Clemens Grahl for confocal microscopy as well as the metallographic team of the IWS, Lars Ewenz and Sebastian Schöne for sample preparation and assistance.

References

1. Banerjee, D.; Williams, J.C. Perspectives on titanium science and technology. *Acta Mater.* **2013**, *61*, 844–879. [CrossRef]
2. Leyens, C.; Peters, M. *Titanium and Titanium Alloys. Fundamentals and Applications*, 1st ed.; Wiley-VCH: Weinheim, Germany, 2003.
3. Lütjering, G.; Williams, J.C. *Titanium*, 2nd ed.; Springer: Berlin, Germany, 2007.
4. Bache, M. Processing titanium alloys for optimum fatigue performance. *Int. J. Fatigue* **1999**, *21*, 105–111. [CrossRef]
5. Chandravanshi, V.; Prasad, K.; Singh, V.; Bhattacharjee, A.; Kumar, V. Effects of $\alpha+\beta$ phase deformation on microstructure, fatigue and dwell fatigue behavior of a near alpha titanium alloy. *Int. J. Fatigue* **2016**, *91*, 100–109. [CrossRef]
6. Ezugwu, E.O.; Wang, Z.M. Titanium alloys and their machinability—A review. *J. Mater. Process. Technol.* **1997**, *68*, 262–274. [CrossRef]
7. Steen, W.M. *Laser Material Processing*, 3rd ed.; Springer: London, UK, 2003.
8. Davim, J.P. *Lasers in Manufacturing*; Wiley-VCH: Chichester, UK, 2013.
9. Yilbas, B.S. *The Laser Cutting Process. Analysis and Applications*; Elsevier Science: San Diego, CA, USA, 2017.
10. da Silva, P.S.C.P.; Campanelli, L.C.; Escobar Claros, C.A.; Ferreira, T.; Oliveira, D.P.; Bolfarini, C. Prediction of the surface finishing roughness effect on the fatigue resistance of Ti-6Al-4V ELI for implants applications. *Int. J. Fatigue* **2017**, *103*, 258–263. [CrossRef]
11. Pessoa, D.F.; Herwig, P.; Wetzig, A.; Zimmermann, M. Influence of surface condition due to laser beam cutting on the fatigue behavior of metastable austenitic stainless steel AISI 304. *Eng. Fract. Mech.* **2017**, *185*, 227–240. [CrossRef]

12. Reck, A.; Pilz, S.; Calin, M.; Gebert, A.; Zimmermann, M. Fatigue properties of a new generation ß-type Ti-Nb alloy for osteosynthesis with an industrial standard surface condition. *Int. J. Fatigue* **2017**, *103*, 147–156. [CrossRef]
13. Yang, J.; Sun, S.; Brandt, M.; Yan, W. Experimental investigation and 3D finite element prediction of the heat affected zone during laser assisted machining of Ti6Al4V alloy. *J. Mater. Process. Technol.* **2010**, *210*, 2215–2222. [CrossRef]
14. Eylon, D.; Pierce, C.M. Effect of microstructure on notch fatigue properties of Ti-6Al-4V. *Metall. Trans. A* **1976**, *7*, 111–121. [CrossRef]
15. Stráský, J.; Janeček, M.; Harcuba, P.; Bukovina, M.; Wagner, L. The effect of microstructure on fatigue performance of Ti-6Al-4V alloy after EDM surface treatment for application in orthopaedics. *J. Mech. Behav. Biomed. Mater.* **2011**, *4*, 1955–1962. [CrossRef]
16. Sonntag, R.; Reinders, J.; Gibmeier, J.; Kretzer, J.P. Fatigue performance of medical Ti6Al4V alloy after mechanical surface treatments. *PLoS ONE* **2015**, *10*, e0121963. [CrossRef] [PubMed]
17. Nalla, R.K.; Altenberger, I.; Noster, U.; Liu, G.Y.; Scholtes, B.; Ritchie, R.O. On the influence of mechanical surface treatments—deep rolling and laser shock peening—on the fatigue behavior of Ti–6Al–4V at ambient and elevated temperatures. *Mater. Sci. Eng. A* **2003**, *355*, 216–230. [CrossRef]
18. Nie, X.; He, W.; Zhou, L.; Li, Q.; Wang, X. Experiment investigation of laser shock peening on TC6 titanium alloy to improve high cycle fatigue performance. *Mater. Sci. Eng. A* **2014**, *594*, 161–167. [CrossRef]
19. Guilherme, A.S.; Henriques, G.E.P.; Zavanelli, R.A.; Mesquita, M.F. Surface roughness and fatigue performance of commercially pure titanium and Ti-6Al-4V alloy after different polishing protocols. *J. Prosthet. Dent.* **2005**, *93*, 378–385. [CrossRef]
20. Arif, A.F.M.; Yilbas, B.S. Thermal stress developed during the laser cutting process: Consideration of different materials. *Int. J. Adv. Manuf. Technol.* **2008**, *37*, 698–704. [CrossRef]
21. Sharma, A.; Yadava, V. Experimental analysis of Nd-YAG laser cutting of sheet materials—A review. *Opt. Laser Technol.* **2018**, *98*, 264–280. [CrossRef]
22. Sheng, P.S.; Joshi, V.S. Analysis of heat-affected zone formation for laser cutting of stainless steel. *J. Mater. Process. Technol.* **1995**, *53*, 879–892. [CrossRef]
23. Tamilarasan, A.; Rajamani, D. Multi-response optimization of Nd:YAG laser cutting parameters of Ti-6Al-4V superalloy sheet. *J. Mech. Sci. Technol.* **2017**, *31*, 813–821. [CrossRef]
24. Pandey, A.K.; Dubey, A.K. Modeling and optimization of kerf taper and surface roughness in laser cutting of titanium alloy sheet. *J. Mech. Sci. Technol.* **2013**, *27*, 2115–2124. [CrossRef]
25. Mower, T.M. Degradation of titanium 6Al–4V fatigue strength due to electrical discharge machining. *Int. J. Fatigue* **2014**, *64*, 84–96. [CrossRef]
26. Carrion, P.E.; Shamsaei, N.; Daniewicz, S.R.; Moser, R.D. Fatigue behavior of Ti-6Al-4V ELI including mean stress effects. *Int. J. Fatigue* **2017**, *99*, 87–100. [CrossRef]
27. Morita, T.; Tanaka, S.; Ninomiya, S. Improvement in fatigue strength of notched Ti-6Al-4V alloy by short-time heat treatment. *Mater. Sci. Eng. A* **2016**, *669*, 127–133. [CrossRef]
28. Akahori, T.; Niinomi, M. Fracture characteristics of fatigued Ti–6Al–4V ELI as an implant material. *Mater. Sci. Eng. A* **1998**, *243*, 237–243. [CrossRef]
29. Wu, G.Q.; Shi, C.L.; Sha, W.; Sha, A.X.; Jiang, H.R. Effect of microstructure on the fatigue properties of Ti–6Al–4V titanium alloys. *Mater. Des.* **2013**, *46*, 668–674. [CrossRef]
30. Papakyriacou, M. Effects of surface treatments on high cycle corrosion fatigue of metallic implant materials. *Inter. J. Fatigue* **2000**, *22*, 873–886. [CrossRef]
31. Roach, M.D.; Williamson, R.S.; Zardiackas, L.D. Comparison of the corrosion fatigue characteristics of CP Ti-Grade 4, Ti-6Al-4V ELI, Ti-6Al-7Nb, and Ti-15Mo. *J. ASTM Int.* **2005**, *2*, 12786. [CrossRef]
32. Gao, X.-L.; Zhang, L.-J.; Liu, J.; Zhang, J.-X. Porosity and microstructure in pulsed Nd:YAG laser welded Ti6Al4V sheet. *J. Mater. Process. Technol.* **2014**, *214*, 1316–1325. [CrossRef]
33. Hong, K.-M.; Shin, Y.C. Analysis of microstructure and mechanical properties change in laser welding of Ti6Al4V with a multiphysics prediction model. *J. Mater. Process. Technol.* **2016**, *237*, 420–429. [CrossRef]
34. Xu, P.-q.; Li, L.; Zhang, C. Microstructure characterization of laser welded Ti-6Al-4V fusion zones. *Mater. Charact.* **2014**, *87*, 179–185. [CrossRef]
35. Factory certification of RTI International Metals Inc., Ingot-No.: 9711830.

36. Shanjin, L.; Yang, W. An investigation of pulsed laser cutting of titanium alloy sheet. *Opt. Lasers Eng.* **2006**, *44*, 1067–1077. [CrossRef]
37. Rowe, W.B. *Principles of Modern Grinding Technology*; William Andrew: Oxford, UK, 2009.
38. Yang, S.; Li, W. *Surface Finishing Theory and New Technology*; Springer: Berlin, Germany, 2018.
39. Schijve, J. *Fatigue of Structures and Materials*; Springer: Dordrecht, The Netherlands, 2009.

6

Effect of Strain Rate on Microstructure Evolution and Mechanical Behavior of Titanium-Based Materials

Pavlo E. Markovsky [1,*], Jacek Janiszewski [2], Vadim I. Bondarchuk [1], Oleksandr O. Stasyuk [1], Dmytro G. Savvakin [1], Mykola A. Skoryk [1], Kamil Cieplak [2], Piotr Dziewit [2] and Sergey V. Prikhodko [3]

1. G.V. Kurdyumov Institute for Metal Physics of N.A.S. of Ukraine, 36 Academician Vernadsky Boulevard, UA-03142 Kyiv, Ukraine; vbondar77@gmail.com (V.I.B.); stasiuk@imp.kiev.ua (O.O.S.); savva@imp.kiev.ua (D.G.S.); mykolaskor@gmail.com (M.A.S.)
2. Jarosław Dąbrowski, Military University of Technology, 2 gen. Sylwester Kaliski str., 00-908 Warsaw, Poland; jacek.janiszewski@wat.edu.pl (J.J.); kamil.cieplak@wat.edu.pl (K.C.); piotr.dziewit@wat.edu.pl (P.D.)
3. Department of Materials Science and Engineering, University of California Los Angeles, Los Angeles, CA 90095, USA; sergey@seas.ucla.edu

* Correspondence: pmark@imp.kiev.ua

Abstract: The goal of the present work is a systematic study on an influence of a strain rate on the mechanical response and microstructure evolution of the selected titanium-based materials, i.e., commercial pure titanium, Ti-6Al-4V alloy with lamellar and globular microstructures produced via a conventional cast and wrought technology, as well as Ti-6Al-4V fabricated using blended elemental powder metallurgy (BEPM). The quasi-static and high-strain-rate compression tests using the split Hopkinson pressure bar (SHPB) technique were performed and microstructures of the specimens were characterized before and after compression testing. The strain rate effect was analyzed from the viewpoint of its influence on the stress–strain response, including the strain energy, and a microstructure of the samples after compressive loading. It was found out that the Ti-6Al-4V with a globular microstructure is characterized by high strength and high plasticity (ensuring the highest strain energy) in comparison to alloy with a lamellar microstructure, whereas Ti6-Al-4V obtained with BEPM reveals the highest plastic flow stress with good plasticity at the same time. The microstructure observations reveal that a principal difference in high-strain-rate behavior of the tested materials could be explained by the nature of the boundaries between the structural components through which plastic deformation is transmitted: α/α boundaries prevail in the globular microstructure, while α/β boundaries prevail in the lamellar microstructure. The Ti-6Al-4V alloy obtained with BEPM due to a finer microstructure has a significantly better balance of strength and plasticity as compared with conventional Ti-6Al-4V alloy with a similar type of the lamellar microstructure.

Keywords: titanium alloy; high strain rate testing; split hopkinson pressure bar technique; microstructure influence; phase transformation; deformation mechanism; strain energy

---

1. Introduction

Titanium alloys are an important structural material used in modern aerospace, automotive, shipbuilding, and military fields, because of the high level of specific strength, fracture toughness, fatigue strength, corrosion resistance, non-magnetization, and other specific physical, mechanical, and service properties [1–4]. Since these alloys are relatively expensive, their advantages over other structural materials become more apparent when they are processed using different methods to as higher as possible values of specific strength. A separate important direction in the application of titanium alloys, which is increasingly being used, is the manufacture of armor elements [4–6], as well as individual elements of military equipment experiencing impact or explosive loading [4,7]. Traditionally the designers of new products are based on the mechanical properties of materials, which have been determined under certain standard, mainly quasi-static test conditions. However, in most cases of machines and devices, it does not correspond to the actual operating conditions of real parts for example, emergency conditions of heavy-loaded structural elements of airplanes subjected to strong turbulence, or the landing gear during extremely hard landings. Therefore, to prevent the destruction of such structures, designers usually follow so-called strength reserve approach, which results in a significant increase in the mass of individual parts and, as a consequence, in an undesirable increase in the total weight of the products, their price and operating costs. Hence, a significant reduction in costs is expected, inter alia, through structure optimization with regard to strength and weight.

The mechanical behavior of materials strongly depends on the loading conditions. The influence of a quasi-static strain rate ranging up to 10 s^{-1} on the mechanical behavior of different titanium alloys has been extensively studied [8–13]. A lot of attention has also been paid to the similar studies on a high-strain-rate (dynamic) testing of Ti-based materials [14–20]. However, some important issues such as an influence of the chemical and phase composition, microstructure, as well as the variation of the strain rate within a wide range, and especially, the combined effect of these have not been studied sufficiently so far. Thus, the goal of present work is a systematic study of a strain rate influence on the microstructure evolution and the mechanical behavior of the commonly used two-phase $\alpha + \beta$ titanium alloy Ti-6Al-4V (wt.%) under a loading condition of compression split Hopkinson pressure bar (SHPB). The results are also compared with the data obtained during the quasi-static compression Moreover, the behavior of commercial pure titanium, considered as single-phase (h.c.p. lattice) material, is studied in similar conditions compared to Ti-6-4 alloy.

2. Materials and Methods

2.1. *Materials and Materials Processing*

The Ti-6Al-4V-alloy was studied in two different states, the first state was the alloy obtained via a conventional cast and wrought technology [1]. The second state alloy was produced using a cost-efficient blended elemental powder metallurgy (BEPM) approach [18,19] and was designated in this study as Ti64BEPM. The cast and wrought alloy was purchased in a shape of 10-mm diameter rods from Perryman Company (Houston, PA, USA). Two different metallurgical states of the Ti-based alloy were obtained using conventional heat treatments. Annealing of the alloy at 850 °C for 2 h resulted in obtaining a globular microstructure (GL), whereas annealing at 1100 °C for 0.5 h resulted in obtaining a lamellar structure (LM). The alloys with two different microstructural conditions were marked as Ti64GL and Ti64LM, respectively. Ti64BEPM alloy was obtained by blending the titanium hydride TiH_2 powder (particles size <100 μm) and 60Al-40V (wt.%) master alloy in a powder form (particles size <63 μm) and then cold pressing of the blend in a die at 640 MPa followed by sintering at 1250 °C for 4 h under the vacuum of 10^{-3} Pa. The used sintering conditions provide removal of hydrogen from the material to an admissible level (0.002–0.003%) and transformation of the powder compacts into bulk homogeneous alloy. More details on materials fabrication using the adopted BEPM protocol can be found in [21,22]. The sintered Ti64BEPM samples were bars with dimensions of 9 × 9 × 60 mm. In order to assist the results evaluation and understand an influence of phase and chemical composition on mechanical behavior of the main object, the Ti64 alloy, a comparative analysis was performed on the commercial pure titanium (c.p.Ti) Grade 1. The alloy was purchased in the shape of rods with a diameter of 10 mm from Titan Ltd. (Kyiv, Ukraine).

2.2. *Quasi-Static and High-Strain-Rate Tests*

The quasi-static tensile properties were determined using INSTRON 3376 strength machine following ASTM E8 standard with specimens having gage diameter 4 mm and gage length 25 mm, whereas the quasi-static compressive tests were carried out with the use of MTS C45 strength machine. For both types of compression tests, quasi-static and dynamic, cylindrical specimens with a diameter and height of 5 mm were used. The Young, shear moduli, and Poisson's ratio of materials were measured with resonance-frequency-damping analysis (RFDA) apparatus (IMCE, Belgium) using impulse excitation technique in accordance with ASTM E1876-15 Standard.

The high-strain-rate compression tests were performed with the use of a split Hopkinson pressure bar (SHPB), also called a Kolsky bar technique [23–25]. The basic parameters of the SHPB system are shown in Figure 1. The length of the input and output bars was 1200 mm, the length of the striker bar was 250 mm, the diameter of all bars was 12 mm. The bars were made of maraging steel (heat-treated MS350 grade: yield strength—2300 MPa; elastic wave speed—4960 m/s). The striker bar was driven by a compressed air system with the barrel length of 1200 mm and the inner diameter of 12.1 mm. The impact striker bar velocities applied during the experiments were in the range from 10 to 25 m/s, which ensures strain rates in the range of 1100 – 3320 s^{-1} for dimensions of the used specimens.

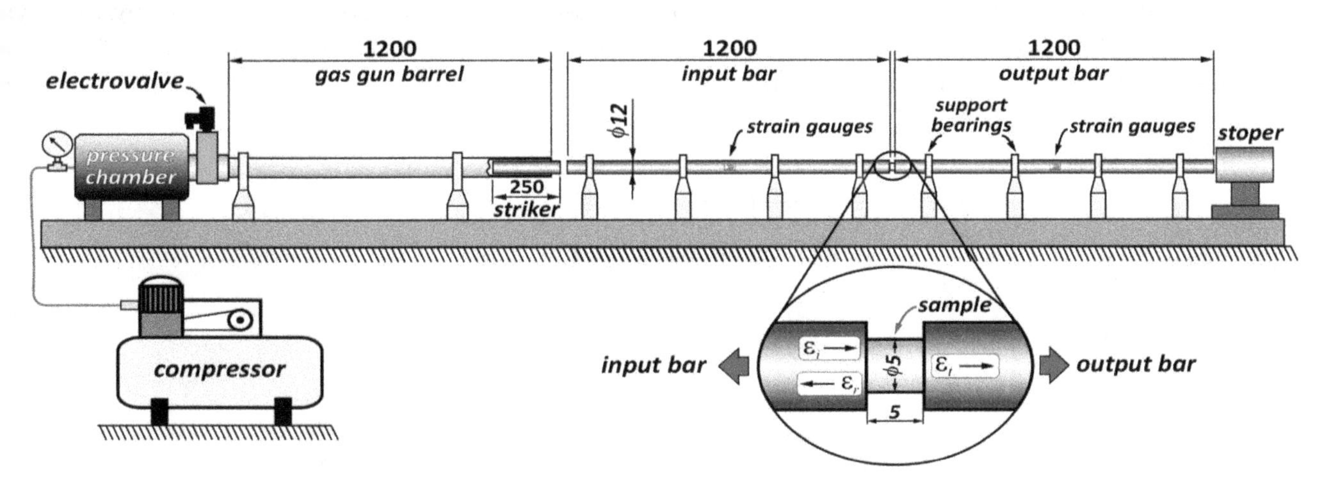

Figure 1. The schematics of the split Hopkinson pressure bar (SHPB) system used in this study.

The plastic flow stress, strain, and a strain rate of the sample were determined according to the classical Kolsky theory [24,25] based on the one-wave analysis method, which assumes stress equilibrium for a specimen under given testing conditions. Wave signals, incident (ε_i), transmitted (ε_t), and reflected (ε_r), were measured by pairs of strain gages (gauge length—1.5 mm; resistance—350 Ohm) glued at the half-length of the input and output bars (Figure 1). The signals from the strain gauges were conditioned with data-acquisition system, composed of Wheatstone bridge, amplifier, and a digital oscilloscope, allowing a high cut-off frequency of 1 MHz.

Strain and a strain rate in the specimens were calculated based on the profile of reflected wave (ε_r) using Equations (1) and (3), respectively, whereas the stress was determined based on transmitted wave (ε_t) (Equation (2))

$$\varepsilon(\mathrm{t}) = -\frac{2c_b}{L}\int_0^t \varepsilon_r(t)dt \tag{1}$$

$$\sigma(\mathrm{t}) = \frac{A_b E}{A_s}\varepsilon_t\,(t) \tag{2}$$

$$\dot{\varepsilon}(\mathrm{t}) = -\frac{2c_b}{L}\varepsilon_r(t) \tag{3}$$

where: L, A_s are the length and the cross-section area of the specimen; E, A_b, c_b are the Young's modulus, the cross-sectional area, and the elastic wave velocity of the pressure bar, respectively.

To minimize dispersion of the wave by damping the Pochhammer-Chree high frequency oscillations and to facilitate the stress equilibrium, a pulse shaping technique was used [26]. The technique consists in placing a small disc made of soft material on the impact end of the input bar. The disc is often called a pulse shaper or a wave shaper. Plastic deformation of the pulse shaper physically filters out the high frequency components in the incident pulse and modifies its profile. The pulse shaper size needs to be chosen for the given striker impact velocity and mechanical response of tested material. The tests were performed to find the proper size of the pulse shaper in order to obtain stress equilibrium during high-strain-rate deformation of titanium alloy specimens. It was found that for a given SHPB test condition, the copper pulse shaper with a diameter of 3 mm and thicknesses in the range from 0.1 to 0.4 mm (depending on impact striker velocity) guarantees damping of the high frequency oscillations and achieving the dynamic stress equilibrium in the specimen (Figure 2). The typical raw signals recorded during the SHPB experiment are shown in Figure 2a, whereas Figure 2b presents the stresses on the front and back face of the Ti64LM specimen cracked under applied compressive loading. As it can be observed in Figure 2a, the presence of pulse shaper limits significantly a stress fluctuation on incident wave profile. The reflected and transmitted signals are also almost smooth without large oscillations. In turn, the stress-state equilibrium condition presented in Figure 2b is satisfied, i.e., front and back stresses are close throughout the loading history, except a peak at the top of the initial rise and a gap occurring at the end of the loading time. This gap is the result of a crack in the Ti64LM specimen.

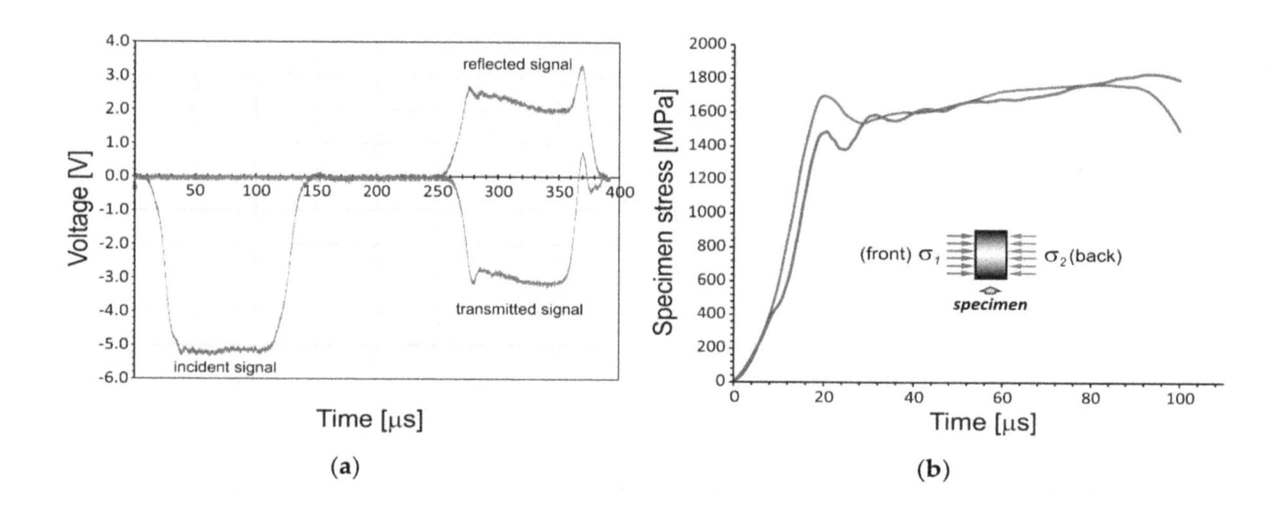

Figure 2. (**a**) Typical raw wave signals from SHPB experiment for Ti64LM; (**b**) stresses on the front and back end of Ti64LM specimen (dynamic stress equilibrium).

Moreover, the grease composed of mineral oil, lithium soap and molybdenum disulfide (MoS_2) was applied to the interfaces between the specimen and the bars to minimize the interfacial friction.

2.3. *Microstructural Characterization*

The structure of all materials was studied with light optical microscopy (LOM) using XL70 (Olympus, Shinjuku, Tokyo, Japan) and scanning electron microscopy (SEM) using Vega 3 and Mira 3 machines (both from Tescan, Czech Republic). SEM Mira 3 was also used to study the fracture surfaces and electron backscatter diffraction (EBSD) was used to evaluate local crystallographic orientation of different microstructural elements. Metallographic specimens were prepared according to standard grinding and polishing methods [1]. For SEM and EBSD, the final polishing was carried out using Saphir Vibro polisher (ATM, Germany). Some samples were additionally ion polished/etched using PECS model 682 (Gatan, CA, USA) or chemically etched using standard Kroll's solution [1]. The gas content in the sintered specimens was measured using a gas analyzer OH900 (Eltra, Germany).

3. Results and Discussion

3.1. Initial Microstructure Characterization

Typical microstructures of the studied materials are presented in Figure 3, and their chemical compositions are listed in Table 1. Pure titanium c.p.Ti is characterized by relatively coarse α-grain structure with an average size of about 600 μm (Figure 3a). The intragrain substructure is highly developed and is characterized both by the presence of twins and the dislocation network, including formation of cells. Such an extensive structure of defects may result from relatively fast cooling rates used after deformation or annealing (Figure 3a).

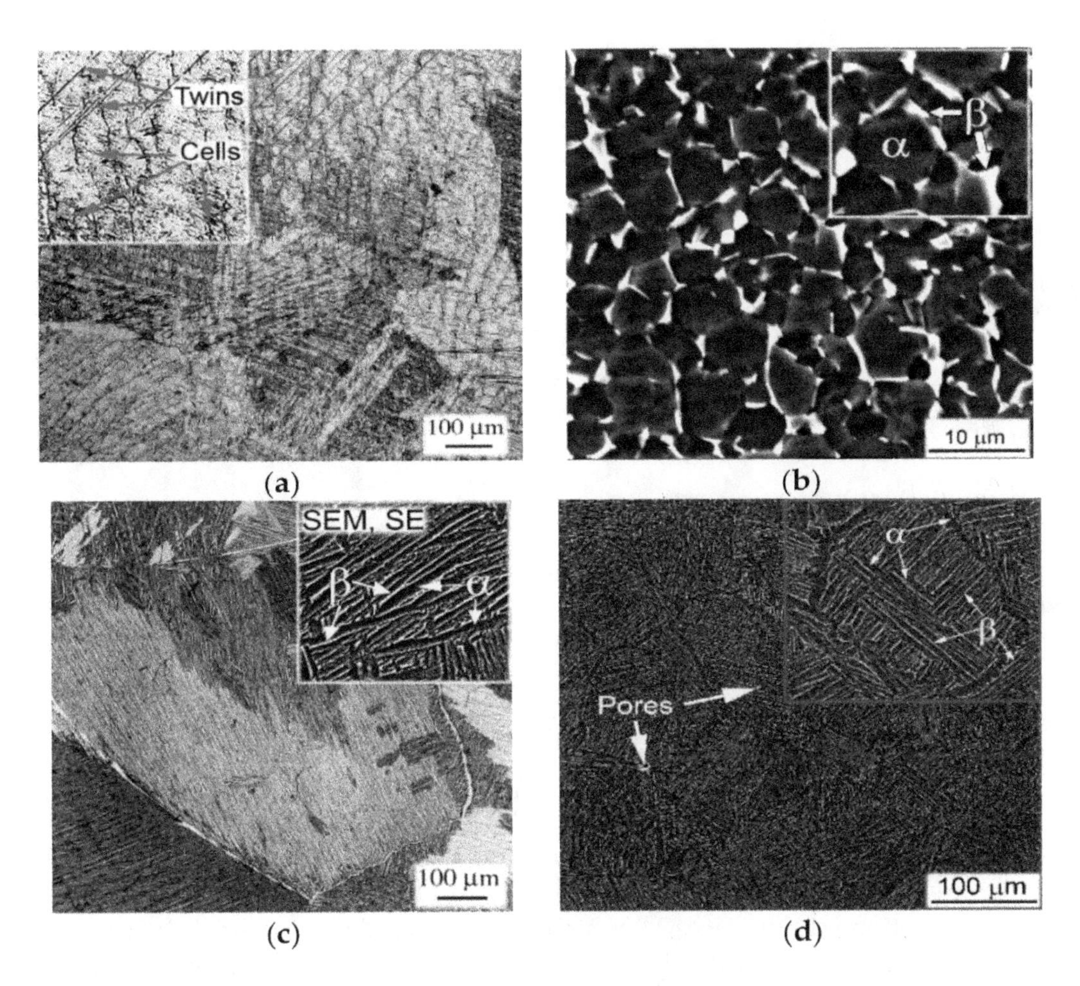

Figure 3. Microstructure of Ti-based materials in the initial state: (**a**) c.p.Ti; (**b**) Ti64GL; (**c**) Ti64LM; (**d**) Ti64BEPM; (**a**,**c**)—LOM; (**b**,**c** (in the corner),**d**)—SEM, SE (secondary electron image).

Table 1. Chemical composition of studied materials.

	Alloying Elements, wt.%					
	Al	**V**	**Fe**	**O**	**N**	**Ti**
c.p.Ti	<0.2	-	<0.08	0.01	0.007	Base
Ti64LM, GL	5.8	3.96	0.21	0.016	0.008	Base
Ti64 BEPM	5.94	4.06	0.16	0.2	0.03	Base

The Ti64GL specimens were of uniform and fine globular microstructure with an average size of α-globules of about 7 μm (Figure 3b), which explains a good balance of the quasi-static tensile strength and ductility (Table 2, #2). The microstructure of the Ti64LM consisted of rather coarse β-grains (average size 800 μm) with coarse (up to 500 μm in some grains) colonies of α-lamellae inside (Figure 3c), which caused a noticeable decrease in both tensile strength and ductility (Table 2, #3). The Ti64BEPM alloy was also of coarse-grained lamellar microstructure (Figure 3d), however, both β-grains (average size of about 100 μm) and colonies of shorter α-plates were much finer (compare Figure 3c,d), because of a pinning role of residual pores in the grain boundary movement which prevents from grain coarsening during sintering [18,19]. The Ti64BEPM demonstrates a higher strength compared to the Ti64LM (Table 2, # 4 vs. #3), because of a higher content of impurities (Table 1, #3), and shows lower ductility, which may be also related to the increased impurities content and presence of about 1.5–2 vol.% of residual pores.

Table 2. Mechanical properties of tested materials (tension rate 8×10^{-4} s^{-1}).

##	Tensile Yield Stress [MPa]	Ultimate Tensile Stress [MPa]	El. [1] [-]	RA [2] [-]	Young Module [GPa]	Shear Module [GPa]	Poisson's Ratio	Vickers Hardness [HV]
#1 c.p.-T	345	408	0.38	0.59	111.5	46	0.253	117
#2 Ti64GL	988	993	0.19	0.42	121.7	47	0.275	312
#3 Ti64LM	824	865	0.15	0.31	121.7	47	0.275	309
#4 Ti64BEPM	932	1033	0.08	0.21	123.0	N/A	N/A	339

[1] El.—elongation, [2] RA—reduction in area.

3.2. *Mechanical Response*

3.2.1. Stress–Strain Behavior

To characterize the base mechanical properties of the tested Ti-based materials, the quasi-static tensile and elastic characteristics are listed in Table 2. The Ti64GL presents the highest yield stress equal to 988 MPa, whereas the Ti64LM presents the lowest one—824 MPa, which is almost 2.4 times higher than the one of c.p.Ti (345 MPa). The Ti64GL material exhibits also the highest ductile properties (elongation 0.19), whereas the Ti64BEPM is characterized by the lowest elongation equal to 0.08. Ductility of all Ti-6-4 alloys tested is relatively low compared to the ductility of c.p.Ti (elongation 0.38).

Stress–strain behavior of the materials tested under uniaxial compression was, as predicted, slightly different. Generally, a compression yield point at a quasi-static strain rate for all materials was at the same or slightly lower level (349, 918, 938, and 879 MPa, for ## 1, 2, 3, and 4 in Table 2, respectively). An exception was the Ti64LM, which demonstrated a higher yield stress in compression than in tension. In turn, fracture of the specimens at the quasi-static compressive regime occurred at significantly higher strain values, i.e., Ti64LM and Ti64BEPM cracked at strain of 0.28 and 0.42, respectively, whereas c.p.Ti and Ti64GL did not fracture before reaching the strain of 0.5, at which the compression was stopped.

Similar dependencies in the specimen damage behavior under compression was observed in dynamic testing conducted at strain rates in the range from 1250 to 3320 s^{-1}. With the exception of c.p.Ti, all other materials cracked under dynamic loading; however, cracking occurred at lower strain values compared to the corresponding quasi-static test results, and it will be discussed in the further part of the paper. In Figure 4a, a few high-speed video frames are presented to illustrate typical successive stages of the specimen deformation process, i.e., the start (Figure 4a-1), the uniform deformation (Figure 4a-2), onset of specimen barreling (Figure 4a-3), intensive local heating, fracturing crack, and the spark flash at the final stage of the deformation (Figure 4a-4). In turn, Figure 4b shows a typical view of the cracked specimens after the quasi-static and dynamic compression.

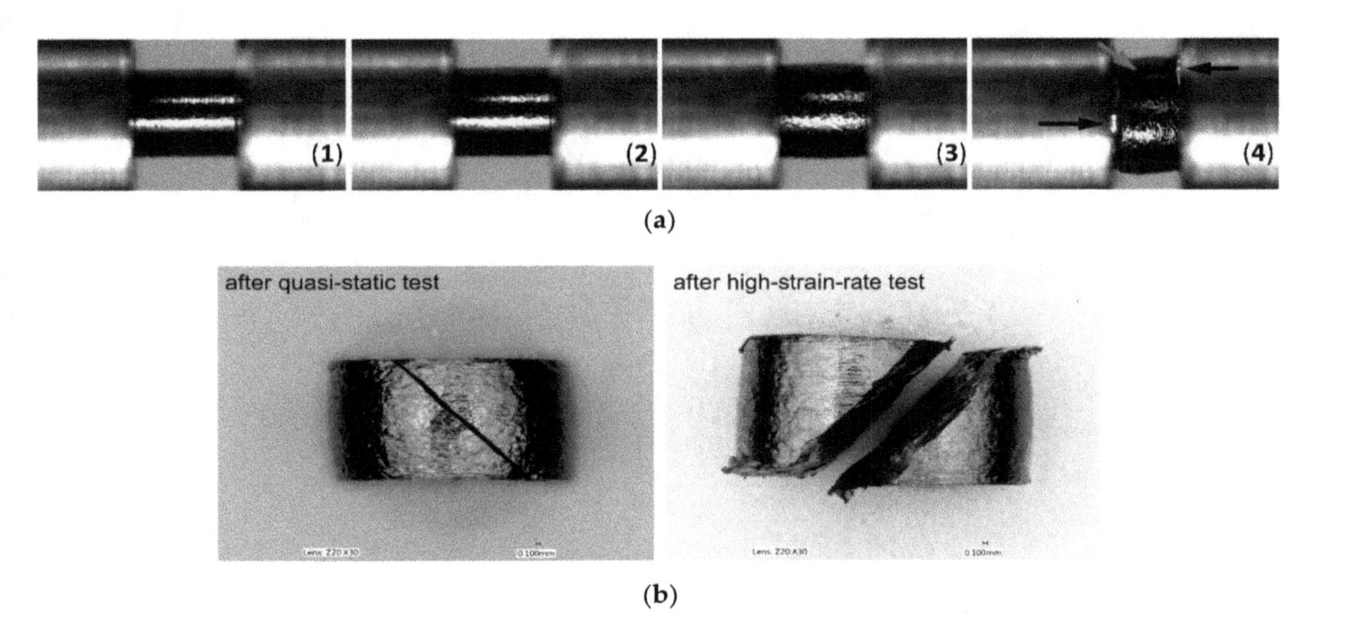

Figure 4. (**a**) High-speed video frames illustrating plastic deformation and fracture of the Ti64BEPM specimen during of the SHPB test at the strain rate of 2100 s^{-1}: the start—(**1**), uniform strain—30 µs (**2**), barreling onset—60 µs (**3**), specimen cracking—120 µs (**4**)—arrows indicate: fracture crack—red arrow, and a areas of intensive local heating on the contact surface with the bar—black arrow; (**b**) view of the cracked specimen of Ti64BEPM after quasi-static and high-strain-rate tests in compression.

The true stress–strain curves of the tested Ti-based materials compressed at quasi-static and high-strain-rates ranges are shown in Figure 5. A number of important observations can be based on these results. First of all, the initial peak stress and oscillations visible on the curve for the highest strain rate are not a real mechanical response of the material, but they result from technical limitations of the SHPB technique. However, the stress–strain curves oscillations were significantly reduced through applying a pulse shaper technique and, in the case of the SHPB experiments with lower strain rates, the obtained stress–strain curves are smooth and almost without oscillation. Second, the quasi-static stress–strain curves reveal differences in the strain hardening behavior of the tested materials (Figure 5e). The strain hardening coefficient n, calculated from a slope of a fitting line of the true stress–strain curve plotted on a logarithmic scale (assumed ranges of plastic strain—0.05–0.2 or 0.05–0.4), is the highest for the Ti64GL (0.148). The values of n for the Ti64LM and Ti64BEPM are relatively lower, and equal to 0.052 and 0.068, respectively. The c.p.Ti demonstrates the highest value of n coefficient, 0.334, as it was expected. Third, the work hardening behavior of the materials tested under high-strain-rate loading is similar to quasi-static loading; however, strain hardening effect is slightly reduced through the heat generation during the dynamic compression, which leads to the flow softening at high strains. This phenomenon seems to be the most pronounced for the Ti64GL and Ti64BEPM. It should be noted here that conversion of the deformation energy to the thermal energy is not uniform throughout the specimen volume, particularly at the end of the sample deformation stage. Careful observation of the high-speed camera films allowed detection of the intense heating and the strain localization near the contact surfaces of the specimen with the front surfaces of the bars (bright areas marked with black arrows in Figure 4a-4). This intense local heating is also manifested by the outflow of a part of the material on the sample side surface (see the image on the right side in Figure 4b).

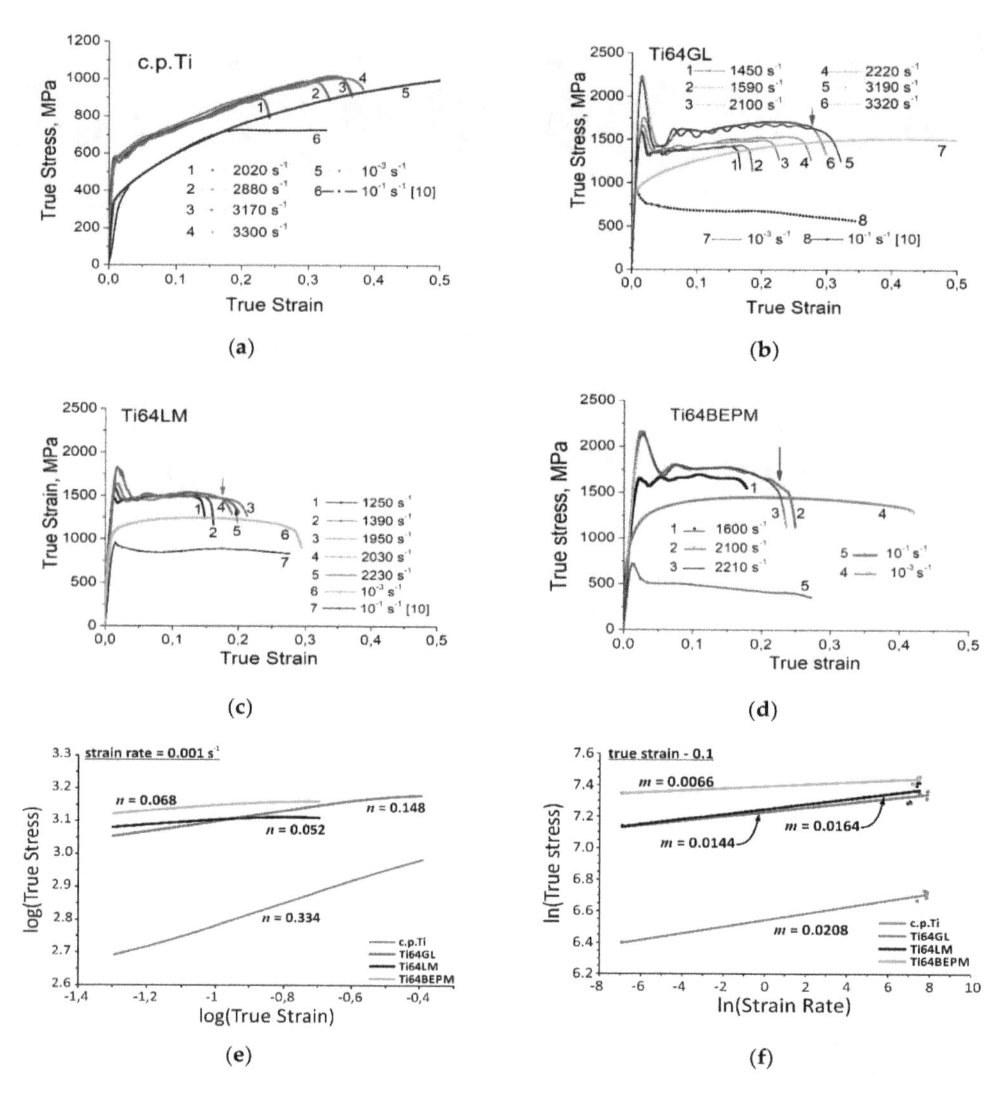

Figure 5. Quasi-static and dynamic stress–strain curves and the calculated mechanical data for: (**a**) c.p.Ti; (**b**) Ti64Gl; (**c**) Ti64LM; (**d**) Ti64BEPM (arrows in (**b**,**d**) indicate possible moment of fracture for relevant curves); (**e**) strain hardening coefficient—*n*; (**f**) strain rate sensitivity exponent—*m*.

As it was expected, the plastic flow stress levels at dynamic regime are significantly higher in comparison to the ones at quasi-static regime. The highest peak flow stress (a maximum stress in the plastic range of deformation) is presented by the Ti64BEPM (1770 MPa at strain of 0.17) and the Ti64GL (1700 MPa at strain of 0.23), whereas the Ti64LM reveals the lowest peak flow stress equal to 1540 MPa at 0.12. In the case of c.p.Ti, the peak flow stress, corresponding to unloading of incident wave, is at the level of 1000 MPa at strain of 0.35. In turn, analysis of values of the strain rate sensitivity exponent *m* (Figure 5f), calculated from the slopes of linear regressions (Equation (4)), shows the highest flow stress increase with an increasing strain rate for the Ti64LM ($m = 0.0164$). The strain rate sensitivity of Ti64GL is slightly lower ($m = 0.0144$), whereas the Ti64BEPM demonstrates the lowest one ($m = 0.0066$). The value of *m* factor for c.p.Ti is relatively high (0.0208) compared to the other alloys examined.

$$m = \mathrm{d}(\ln(\sigma_t))/\mathrm{d}\big(\ln(\dot{\varepsilon})\big) \tag{4}$$

where σ_t is the true flow stress at strain of 0.1, and $\dot{\varepsilon}$ is average strain rates.

As it was noted earlier, cracking the Ti alloys tested under the dynamic loading occurred at lower strain values compared to the corresponding quasi-static test results. However, it was observed (see Table 3) that specimens deformed with the critical strain rates (at which cracking occurred) damage at lower strains (ε_{cr}) than specimens tested with slightly lower strain rates without cracking (strain designation—ε_{max}). For example, the Ti64GL specimen tested at a strain rate of 3190 s^{-1} achieved a strain value equal to 0.30 (curve #5 in Figure 5b), while an analogous specimen deformed at 3320 s^{-1} cracked at a slightly lower strain equal to 0.28 (curve #6 in Figure 5b). The same dependency was found for Ti64LM and Ti64BEPM.

Table 3. Comparison of ε_{cr} and ε_{max} for the tested Ti-alloys.

	Ti64GL	Ti64LM	Ti64BEPM
strain ε_{cr} at $\dot{\varepsilon}$ (s^{-1})	0.28 (3320)	0.17 (2030)	0.23 (2210)
strain ε_{max} at $\dot{\varepsilon}$ (s^{-1})	0.30 (3190)	0.19 (1950)	0.24 (2100)

Based on the data listed in Table 3, it can be concluded that the Ti64LM alloy exhibits the lowest value of ε_{cr} at the strain rates slightly above 2000 s^{-1} compared to Ti64BEPM and Ti64GL materials, which crack at strains of 0.23 and 0.28 and at strain rates of 2210 and 3320 s^{-1}, respectively. It should be emphasized that resistance to cracking of the Ti64GL alloy under the dynamic deformation is significantly higher compared to other Ti-alloys tested.

3.2.2. Material Strain Energy

In order to carry out more in-depth assessment of the mechanical behavior of the materials tested under dynamic loading, an additional parameter, i.e., strain energy (*SE*), was used. It is a convenient parameter that allows comparing the mechanical response of materials tested with various methods and strain rates [11–13]. The *SE* is defined as the internal work performed to deform a material specimen through an action of the externally applied forces. The *SE* was determined by integrating the area under the stress–strain curve (for integration limits from zero to ε_{upper}). In the case of the cracked specimens, a value of ε_{upper} corresponded to a value of strain at fracture, whereas for the non-cracked specimens, ε_{upper} was assumed to be equal to strain at the moment of the specimen unloading (sharp drop in stress–strain curve). The upper integration limit ε_{upper} for the non-cracked specimens under quasi-static loading was assumed to be 0.5.

As it can be seen in Figure 6a, the cast and the wrought alloy Ti64 in both the globular and the lamellar microstructural states have almost the same values of *SE* at the strain rates up to 2000 s^{-1} (curves 2 and 3 in Figure 6a). However, above this strain rate level, the Ti64LM cracked in contrast to the Ti64GL, which demonstrated the highest level of *SE* among all materials studied.

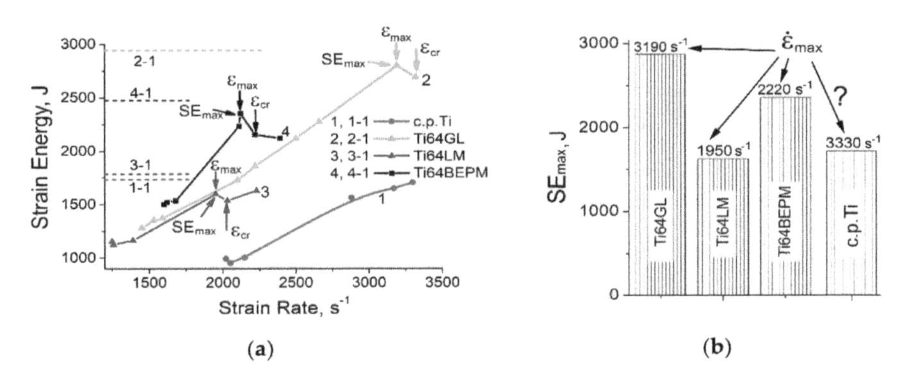

Figure 6. The strain energy for the studied Ti-based materials: (**a**) *SE*-strain rate dependence; (**b**) comparison of the maximum strain energy values (SE_{max}) corresponding to the maximal strain rate $\dot{\varepsilon}_{max}$ (indicated by arrows), for which specimens do not crack during the SHPB tests.

A change from cast and wrought to BEPM in the method for manufacturing the Ti64 alloy significantly affected the measured *SE* values. The Ti64BEPM material revealed a higher ability to store mechanical energy at strain rates up to 2300 s^{-1}. The *SE* values were noticeably higher compared to Ti64GL, and at the strain rate range of 2100–2300 s^{-1}, at which Ti64BEPM specimens broke (compare curves 4, 2 and 3; Figure 6a).

It is also interesting to compare the *SE* values calculated from the quasi-static and the high-strain-rate tests data. The maximal *SE* value determined at quasi-static tests (horizontal (1-1) line in Figure 6a) is approximately equal to the SE_{max} level obtained for the maximum strain rate ($\dot{\varepsilon}_{max}$) for c.p.Ti only (curve #1 in Figure 6a). In turn, *SE* values for Ti64 alloys in all microstructural states from the quasi-static tests (horizontal lines (2-1), (3-1), and (4-1) in Figure 6a) are slightly higher than the SE_{max} values obtained for the whole range of strain rates, at which specimen cracking occurred (marked by arrows on curves 2–4 in Figure 6a).

Since the main difference between the studied Ti64 structures was the β- grain size, the SE_{max} values (Figure 6b) were related to the grain size of the tested Ti-alloy. From the data presented in Table 4 it can be seen that a larger β-grain size in Ti64 alloy structure causes a decrease in SE_{max}.

Table 4. Dependency between SE_{max} parameter and β-grain size of the tested Ti64 alloys.

	Ti64GL	Ti64LM	Ti64BEPM
grain size [μm]	7	800	160
strain energy SE_{max} [J]	2795	1594	2354

It is also worth noting that c.p.Ti has—as predicted—the lowest *SE* value among all studied materials and at all strain rates. The *SE* value for c.p.Ti is almost twice lower compared to the cast and wrought alloys, despite the pure titanium shows very high plasticity, because of which specimen fracture does not occur at the applied strain and strain rates (curve 1 in Figure 6a).

In view of the fact that titanium alloys are often used instead of other structural materials in various critical applications, the results of the present study were additionally compared to similar data of other commonly used structural materials [27] (Figure 7). The lowest level of the *SE* parameter (Figure 7a) demonstrates the aluminum alloy B95 (curve 1), as it was expected. The high-strength steels ARMOX 600T and Docol 1500M are characterized by significantly higher *SE* values (curves 2, and 3), which exceed the corresponding values for Ti64GL and Ti64BEPM alloys at the same strain rates (curves 5, and 6). However, when these curves were converted considering the density of tested materials (Al alloy B95—2850 kg/m^3 [28], for all Ti-based materials—4500 kg/m^3 [1], and for steels of 7850 kg/m^3 [29]), results revealed another dependency between considered materials (Figure 7b).

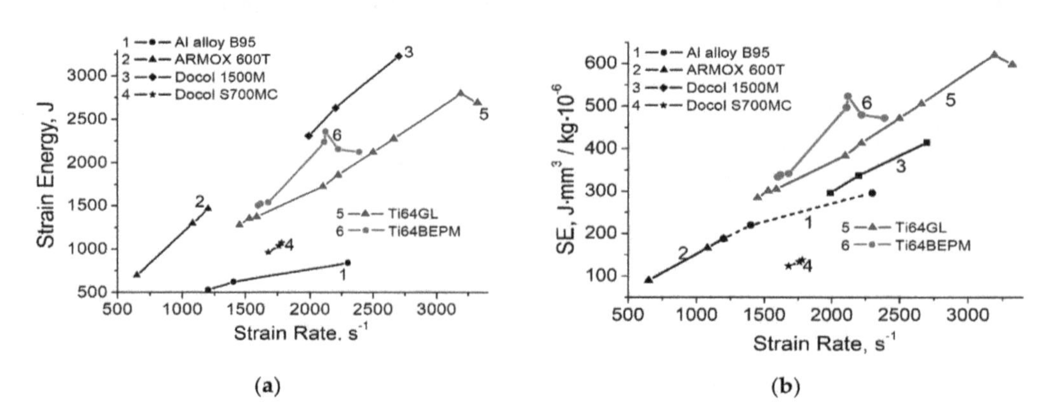

Figure 7. Comparison of the SE-strain rate dependencies for different materials: (**a**) absolute values of the SE; (**b**) relative (specific) values of the SE. The data used for the curves 1–4 were taken from [27].

It can be clearly seen that the aluminum alloy at the strain rates of about 1500 s^{-1} is not worse than ARMOX 600T steel (curves 1, and 2), whereas titanium-based materials are undeniably better than the high-strength steels (curves 5–7). The only, rather serious, drawback of the BEPM made titanium materials is that they could be cracked at relatively low strain rates. However, such a shortcoming could be overcome, at least partially, by incorporating these materials into multilayer structures combining them with high ductility Ti64 alloy layers in the optimized configuration, as it was shown in a few studies published earlier [30,31].

3.3. *Deformed Microstructures Investigation*

Analysis of the microstructure of materials formed during plastic deformation is a primary step to evaluate a difference in their mechanical behavior. The initial examination of the structural features of all the specimens after quasi-static tests and high-strain-rate SHPB deformation allows identification of four main zones distinguished by the stress state and, as a result, having specific differences in the microstructure (Figure 8). Zone I is quite narrow and is characterized by tangential shear stresses caused by the interaction on the contact surfaces of the specimen and SHPB bars. Zone II tracks across the entire specimen at an angle of approximately 45° to the vertical axis of the cylinder and it corresponds to the plane of the maximum shear stresses in accordance with Schmid law [32]. This zone is characterized not only by the maximum shear stresses, but also by the greatest strain localization, due to stress collapse, adiabatic shear band (ASB) initiation, temperature rise and crack formation [11,16,19]. Zone III is adjacent to zone II and it corresponds to the secondary strain localization, where ASBs and secondary (smaller) cracks were also observed. Zone IV is located away from the fields of intense stresses and strain localization; however, the stress state in this zone is more complicated, involving the compressive stress along the vertical axis of the cylinder sample and tensile (or shear) perpendicular to it.

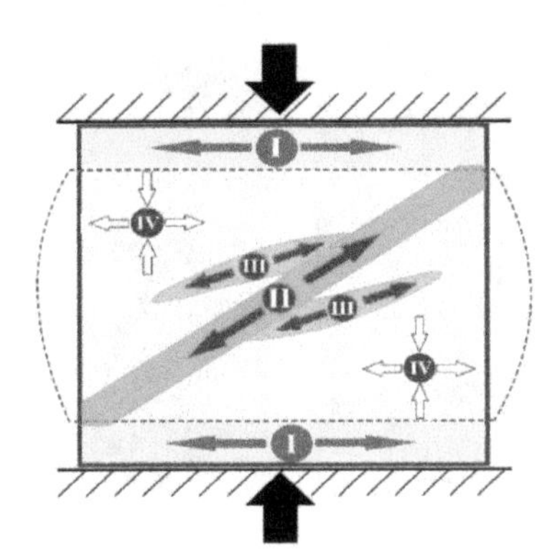

Figure 8. Schematic representation of specific zones distinguished in the longitudinal section of tested cylindrical specimen (Roman number denote a given zone; the arrows indicate the direction of compressive force).

3.3.1. Microstructure Analysis of c.p.Ti

The microstructure of c.p.Ti after SHPB tests is shown in Figure 9a,b. The numerous plastic deformation traces, in the form of slip bands, twins, and well-developed substructure inside α-phase plates, are observed in all zones near the specimen-to-bar contact surfaces (Figure 9a) as well as across the entire bulk of the specimen (Figure 9b). A comparison with the initial not-deformed state (Figure 3a) shows a significant increase in defects density, while a general character of the substructure remains approximately the same. Assuming that titanium with single-phase h.c.p. lattice could easily deform by twinning at sufficiently low temperatures [1,33,34], it was expected that defected areas should contain mainly twins. However, given the probability of a significant local increase in temperature during the deformation [18,35], a few other scenarios should not be excluded, namely, the formation of a well-developed dislocation substructure and the possibility of the phase transformations, including the martensite formation, as it was reported in [36,37], where c.p.Ti was subjected to fast heating and cooling. Eventually, because of the high plasticity of c.p.Ti, introduction of some deformation defects could not be completely excluded for this alloy as a result of sample preparation (after delicate ion polishing and/or etching), although their presence causes a "background" effect in all the samples made of this alloy. Nevertheless, the areas of localized deformation and ASB were not found in c.p.Ti specimens, even in those tested at the highest strain rates (Figure 9b).

The microstructure after the quasi-static tests seems to be uniform in all zones. It is modified by plastic deformation, showing a higher density of deformation defects and a smaller size of cells and twins (Figure 9c,d). This results from different strain rates used at quasi-static tests (0.001 s^{-1}) and SHPB tests (3200 s^{-1}); difference is more than 6 orders. Therefore, under the quasi-static tests, there are more possibilities and longer time for material relaxation compared to SHPB experiments.

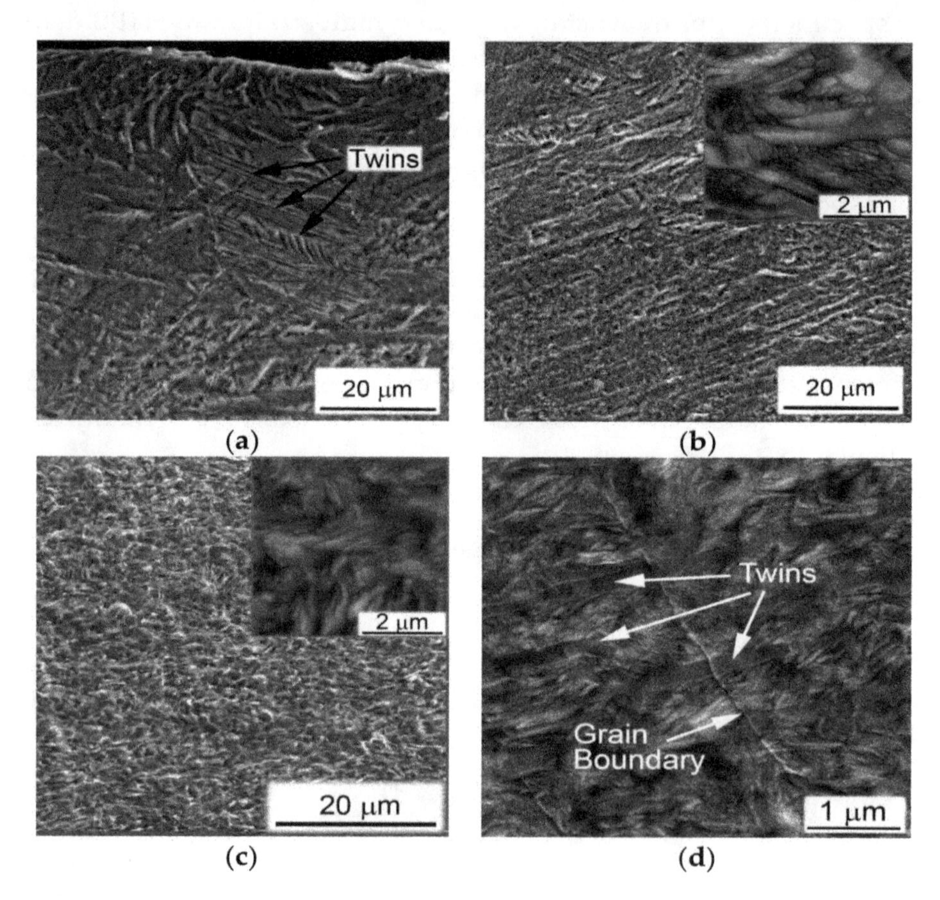

Figure 9. SEM images of c.p.Ti specimens sectioned along the longitudinal direction of the cylinders after: (**a**,**b**) SHPB test at the strain rate of 3200 s^{-1}—(**a**) near the zone I, (**b**) in the center of zone II; (**c**,**d**); quasi-static test with the strain rate 0.001 s^{-1}—zone 4. SEM, SE.

3.3.2. Microstructure Analysis of Ti64GL

Typical images of Ti64GL microstructure after SHPB tests, which did not cause fracture of specimens, can be seen in Figure 10. In general, the microstructure of this material did not change significantly in bulk after compression at 2100 s^{-1}, without failure, compared to the initial state. In some local micro-volumes, a couple of twins in the separate globules were observed closer to the specimen core (zones II and IV, Figure 10b). More significant traces of plastic deformation were observed in thin (5–7 μm) surface layers in zones I (Figure 10a), which could be a consequence of localized deformation, due to interaction between the surfaces of the specimens and the bars. An increase in a compression rate up to 2500 s^{-1} (Figure 10c) or even to 2660 s^{-1} (Figure 10d–f) also did not cause failure of the specimens, but led to a noticeable change in their microstructure. In this case, more than a half of all α-globules (about 65%) contain deformational twins, and they are observed in all zones throughout the bulk of the specimen. According to [1,33,34], during compression, the {11–22} twins are first formed inside the grains with the c-axis oriented parallel to the loading direction, and afterwards the formation of the {10–11} twins occurs in the grains oriented differently. Therefore, it appears that with an increase in compression strain rates, new twinning planes are activated inside the α-globules. Delicate etching of the deformed specimens reveals, in addition to twins, fine dislocation cells (their size is not more than a few tens of nanometers) observed in both α- and β- phases (Figure 10e,f).

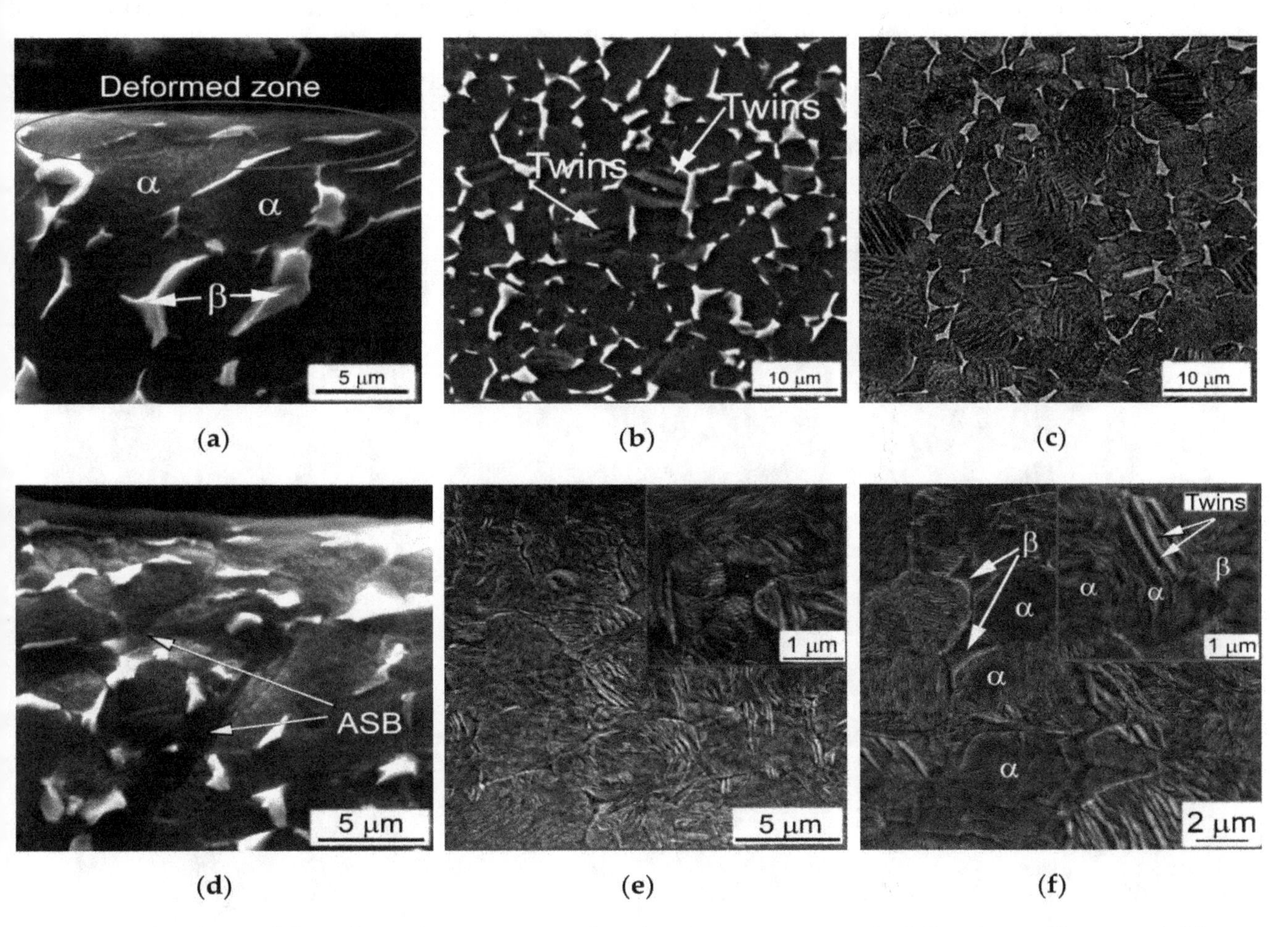

Figure 10. SEM images of Ti64GL specimens after the dynamic tests at the following strain rates: (**a**,**b**) 2100 s^{-1}; (**c**) 2500 s^{-1}; (**d**–**f**) 2660 s^{-1}—(**a**,**d**) show zone I; (**b**,**c**)—zone IV; (**e**,**f**)—zone III; (images (**a**,**b**,**d**–**f**) are taken in SE mode from the etched surface, whereas image (**c**) is in the BSE (back scattered electrons) mode from just polished surface.

A further increase in a strain rate up to 3320 s^{-1} caused cracking of the Ti64GL specimen, and appearance on the fracture surface (Figure 11a) small melted areas (Figure 11b) and dimples (Figure 11c), which are characteristic for ductile fracture. It should be emphasized that the main crack propagated in the unchanged direction (Figure 11a), which is clearly noticeable by the configuration of ductile grooves on the entire fracture surface (Figure 11c). There is also noticeable secondary or lateral crack propagation (indicated by B in Figure 11a) on both sides of the main crack fracture (indicated by A) and forms distinct relief.

The internal microstructure of the samples is distinct from the above-discussed cases of testing with low strain rates. There are multiple primary ASBs observed in zone II, on the edges along the entire main crack (Figure 11d,e). Finer secondary ASBs spreading out of the main crack at a certain direction were also found in zone III (Figure 11f). A much fewer number of twins are observed in zone III (Figure 11d,g) compared to the cases of deformation with low strain rates. Moreover, a lot of structural elements of completely different morphology were revealed. The extremely fine (not larger than 0.1 μm × 2.2 μm) needle-shaped elements are formed within individual globules, as it is shown in the top right corner of Figure 11d. Such crystals could be a result of crystallographically ordered transformation, as it was reported in [34]. Their shape and orientation suggest martensitic needles, considering that the individual needles are located at an angle of approximately 60° relative to each other. It is probable that in proximity of zones of the localized extreme heat, where the temperature can reach the single-phase region (about 1000 °C) [18,19], and primary α- phase was transformed into high-temperature β-phase. In such a case, the transformation takes place without redistribution of alloying elements between neighboring initial crystallites of α- and β- phases, because of extremely high heating and cooling rates. A similar mechanism of the structure formation takes place during laser heating [38,39] and, because of subsequent fast cooling, this metastable β-phase can transform into low-temperature α′-martensite.

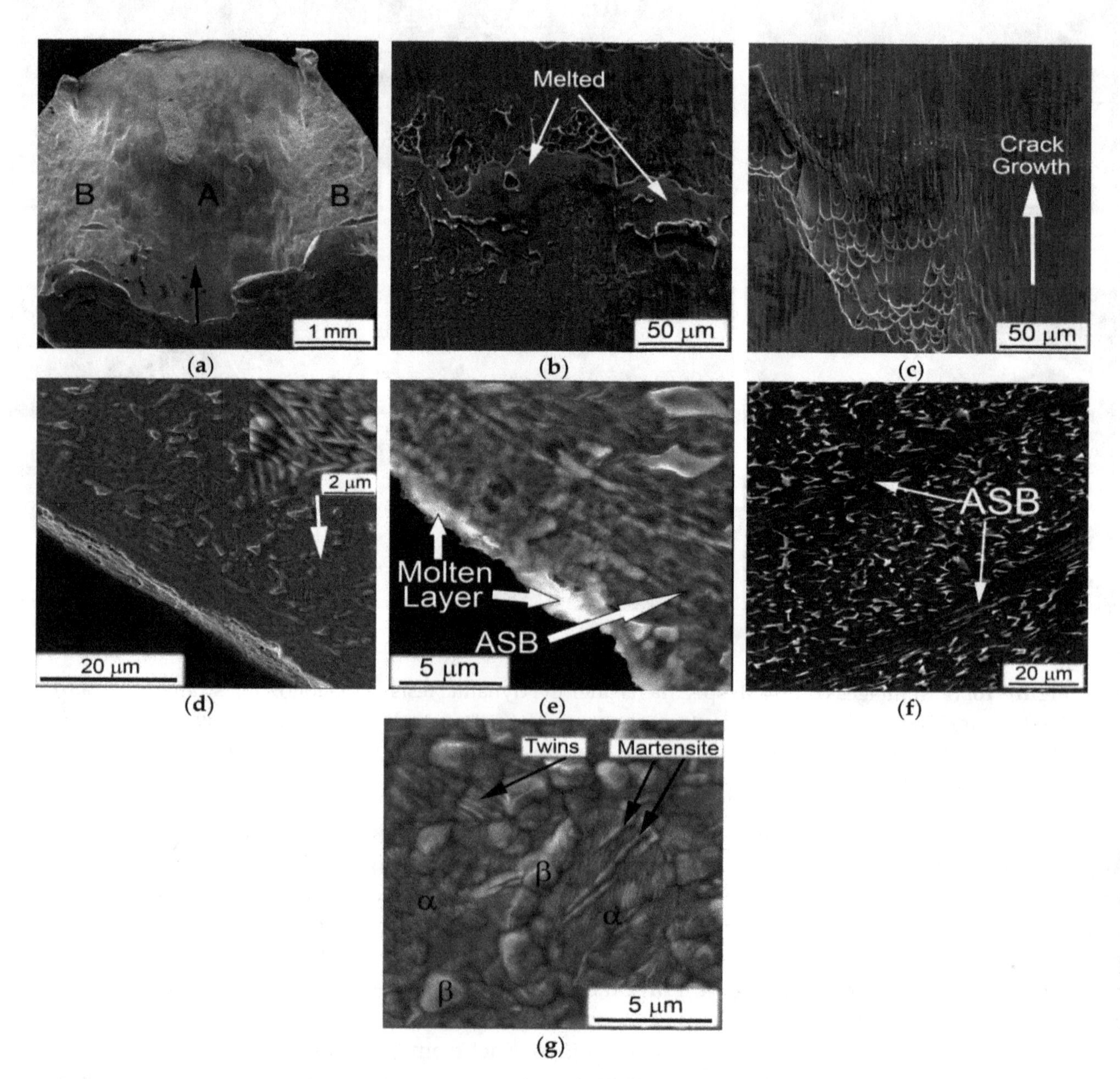

Figure 11. SEM images of the fracture surface (**a**–**c**), and the internal microstructure (**d**–**g**) of the Ti64GL specimen tested under the strain rate of 3330 s^{-1}; A—main crack spread, B—secondary cracking; (**d**) and (**e**)—zone II; (**f**)—zone III with secondary ASBs coming from the main crack; (**g**)—zone IV. The arrow in (**a**) indicates the direction of the crack propagation from its nucleation site. SEM, SE.

Bearing in mind a small size of tested specimens and their contact with massive input and output bars, the cooling process should be relatively rapid. It is probable that a pure (neat shear without any contribution of diffusion) martensitic transformation under these thermo-mechanical conditions does not take place, and a transformation of bainite type could easily take place, when diffusion of alloying elements could have happened during phase transformation, but most importantly Ti atoms move in an ordered way. A similar case was described for the Ti-6Al-4V alloy before experiments when the cooling rate from single β-phase temperatures was changed gradually; however, complete suppression of diffusion was established upon cooling at 400 °C per second [40]. When the cooling rates are lower, the shear transformation occurred with the involvement of diffusion and redistribution of alloying elements. This fact may explain a slightly unusual morphology of the observed martensite crystals (Figure 11d,g).

Microstructure of Ti64GL after the quasi-static tests presents distinct differences compared to the SHPB tests structure. The existed phase constituents, α-phase globules and β-phase interlayers, are flattened perpendicularly to the loading direction through almost the entire specimen volume (Figure 12). All the tested specimens did not crack, and thin bands of localized slip/shear were found in zone II of maximum localized strain (Figure 12a). The microstructure shows the same character of the flattened phase constituents in all other locations (zones III and IV); however, the images of high magnification reveal greater plastic deformation in α-phase compared to β interlayers (Figure 12b).

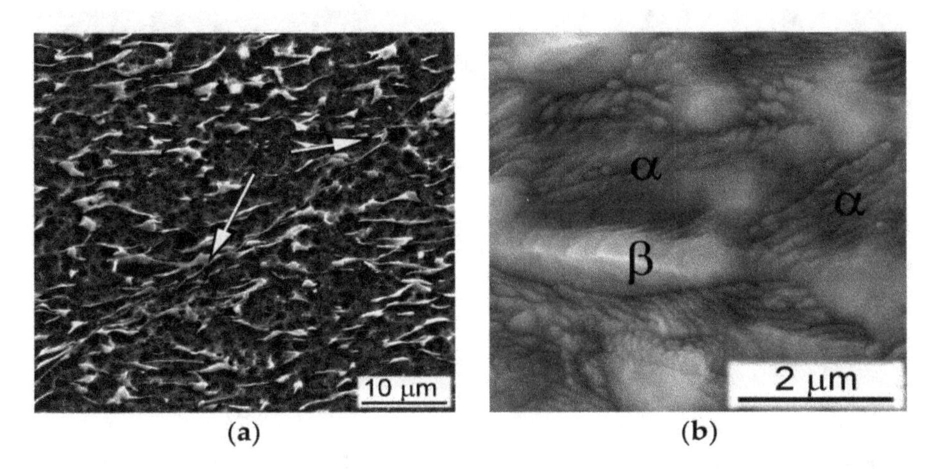

Figure 12. Microstructure of the non-cracked Ti64GL specimen after the quasi-static test at the strain rate of 0.001 s^{-1}: (**a**)—zone II, (**b**) zone IV. SEM, SE. Arrows in (**a**) indicate a slip localization band.

3.3.3. Microstructure Analysis of Ti64LM

The Ti64LM specimens deformed at strain rates for which fracture occurred show distinct plastic deformation in zone I on one side of the specimen (Figure 13a) as well as cracks on the other side (Figure 13b). It is probable that these cracks are initiated at β-grain boundaries reaching the specimen fracture surface and decorated [12] by α-phase layer. It should be noted that outside Zone I there is no significant evidence of plastic deformation in α-plates and their packets; however, small cracks were found within Zones II and IV (Figure 13c,d, respectively). These cracks are not associated with any structural elements such as β-grains boundaries, α-phase colonies, or α/β interlayers. These cracks easily crosscut the α-phase plates, partially deflecting on the α/β interlayers (Figure 13c), and finally move outside the individual plates (Figure 13d). The observed crack propagation insensitivity to structural elements appears to be remarkable since the microstructure of titanium alloys usually plays a pivotal role in the crack nucleation and growth under conditions of the quasi-static tension [1,11–13,41].

The fracture of the Ti64LM specimens under the SHPB test condition was observed at a strain rate of 2030 s^{-1}. A typical image of the crack surface and the microstructure in its vicinity are shown in Figure 14. The surface of the crack shows a typical ductile fracture disclosing multiple tear-off/shear dimples (Figure 14a). However, the zone adjacent to the crack surface on the section cut perpendicular to the crack shows rather unique features (Figure 14b). There are no evidences of plastic deformation at all, even close to the edge of the crack. Except slightly melted edge of the sample in the zone II, the structure in bulk of the sample is unchanged compared to initial condition (compare Figures 14b and 3c). This observation suggests that plastic deformation of the Ti64LM alloy becomes highly localized with an increase in the strain rate. The Ti64LM specimen deformation reaches its critical values slightly above 2000 s^{-1}, when the fracture occurs.

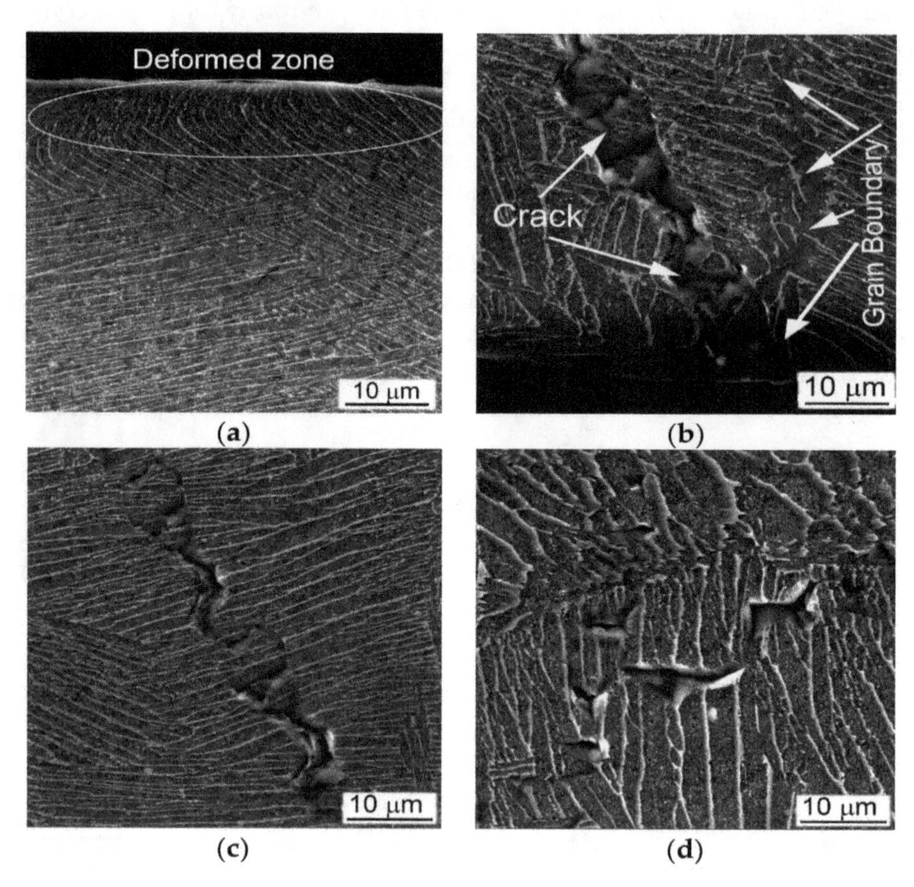

Figure 13. Microstructure of the Ti64LM specimen after SHPB test at the strain rate of 1390 s^{-1}: (**a**,**b**)—zone I, (**c**)—zone II, (**d**)—zone IV. SEM, SE.

The following conclusions can be drawn from comparison analysis of the Ti64GL and the Ti64LM structures subjected to the dynamic loading conditions. The principal difference between these two structural states is defined by configuration and extent of a/β interphase boundaries. The Ti64GL, similarly to c.p.Ti, has predominantly α/α boundaries that appear to facilitate a relatively free propagation of plastic deformation (flow) through the material even at high strain rates despite the fact that majority of neighboring globules present essentially different crystallographic orientation [1]. On the contrary, the Ti64LM boundaries between microstructural elements are almost on 100% presented by interphase α/β boundaries that remarkably prevent a free spread of deformation due to hindering of dislocation movement. It results in an increased strain localization and concentration of the stress, which may enable a cavity nucleation and, conclusively, fracture at an increased strain rate.

The SEM-EBSD analysis of Ti64LM specimen after the SHPB test with a strain rate of 2030 s^{-1} is presented in Figure 15. Kikuchi patterns, required to obtain EBSD data, were distinctive in many points near the contact (fracture) surface. However, Kikuchi lines were not identified by the software, which may result from significant distortions caused by high residual stresses remained in the material after the dynamic plastic deformation. Since the annealing, required to relieve these stresses, would inevitably lead to significant changes in the fine structure of the material resulting in polygonization or recrystallization, it was not carried out. The orientation map shows substantially non-uniform plastic deformation of the structure resulting from the test and clearly reveals few zones where deformation is localized. The β-phase cannot be resolved because of its size and morphology. It is represented by thin layers (from few tens to hundreds of nm) in between relatively thicker (a few m) α-lamellae and considering their likely tilt toward the analyzed surface the β-phase resolution becomes an unsolvable challenge for EBSD even the one operating with field-emission gun SEM, which was the case [42]. A significant number of the noise pixels, mostly aligned along the α-lamellae, result from unsuccessful orientation measurement of the β-phase. Because of the same reason, the individual α-lamellae also cannot be distinguished on the orientation map; however, it is clearly seen in the band contrast image, nonetheless the packers are easier to be traced on the orientation map plot. Some of the packets demonstrate significant misorientation of lamellae within the packet. For instance, a pink and purple packet, 40-m thick, on the left edge of the image shows approximately 10 deg. bent within a few m span. A big yellow and pink packet at the bottom of the image shows even bigger misorientation suggesting big dislocation density accumulated within some small zones. This image also demonstrates short secondary cracks originating outside of the main crack (Figure 15a), which indicate that crack nucleation is not related to such important microstructural elements as grain or interphase boundaries, and crack can nucleate inside a single α-lamella (see the left crack in Figure 15b).

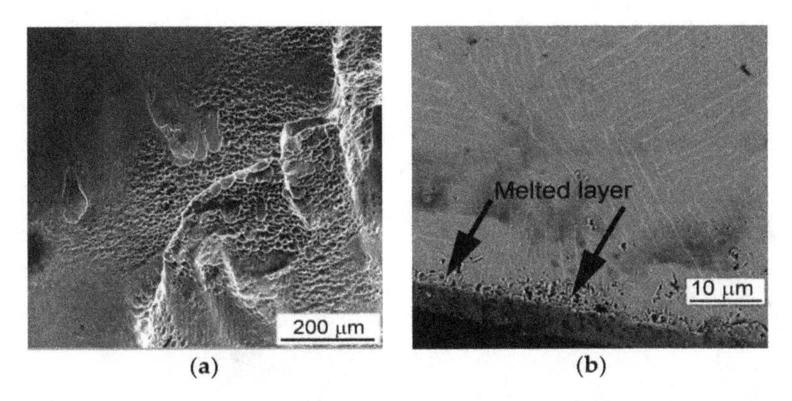

(a) (b)

Figure 14. Microstructure of the cracked Ti64LM specimen after SHPB tested at the strain rate of 2030 s^{-1}: (**a**) fracture surface (SEM, SE); (**b**) microstructure in zone II (polished, not etched) (SEM, BSE).

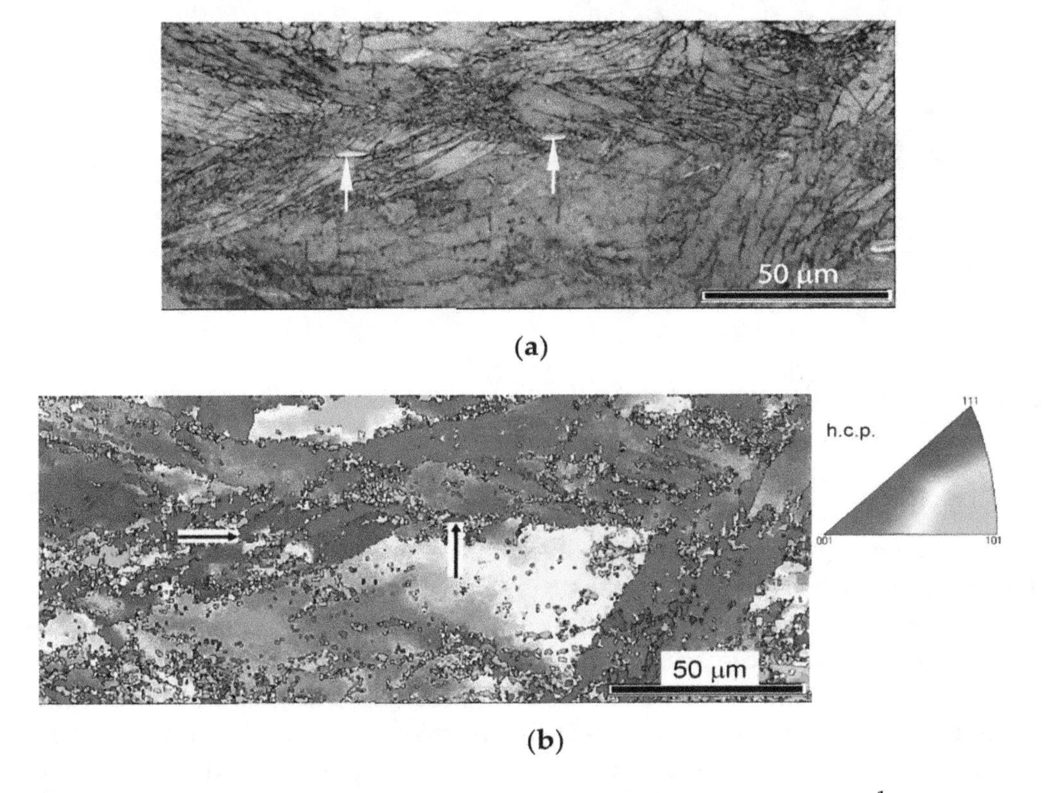

(a)

(b)

Figure 15. The SEM-EBSD images of Ti64LM specimen SHPB tested at 2030 s^{-1}: (**a**) band contrast image, (**b**) EBSD orientation map - arrows indicate small secondary cracks appeared inside α-lamellas.

The decisive impact of the compressive strain rate on plastic deformation localization is underlined by the results of the quasi-static tests of Ti64LM specimens (Figure 16). The plastic strain localization becomes more visible during compression at the strain rate of 0.001 s^{-1}. It resulted in the appearance of a considerably narrow zones crossing the sample at an angle of 45°, where maximum strain was localized, and finally causing the main crack nucleation which, in turn, initiated the secondary cracks (Figure 16a). These secondary cracks often cut and shift the individual grains. The α-layer, labeled with A in Figure 16b, wrapping the initial β-grain, was sheared by the crack in the direction perpendicular to this α-layer. The microstructure after the test (Figure 16c) is not much different compared to the initial structure (Figure 3c), except a few cracks seen in zones IV; however, a detailed examination reveals a complex substructure of multiple twins and dislocation slip traces inside the α-phase plates (Figure 16d).

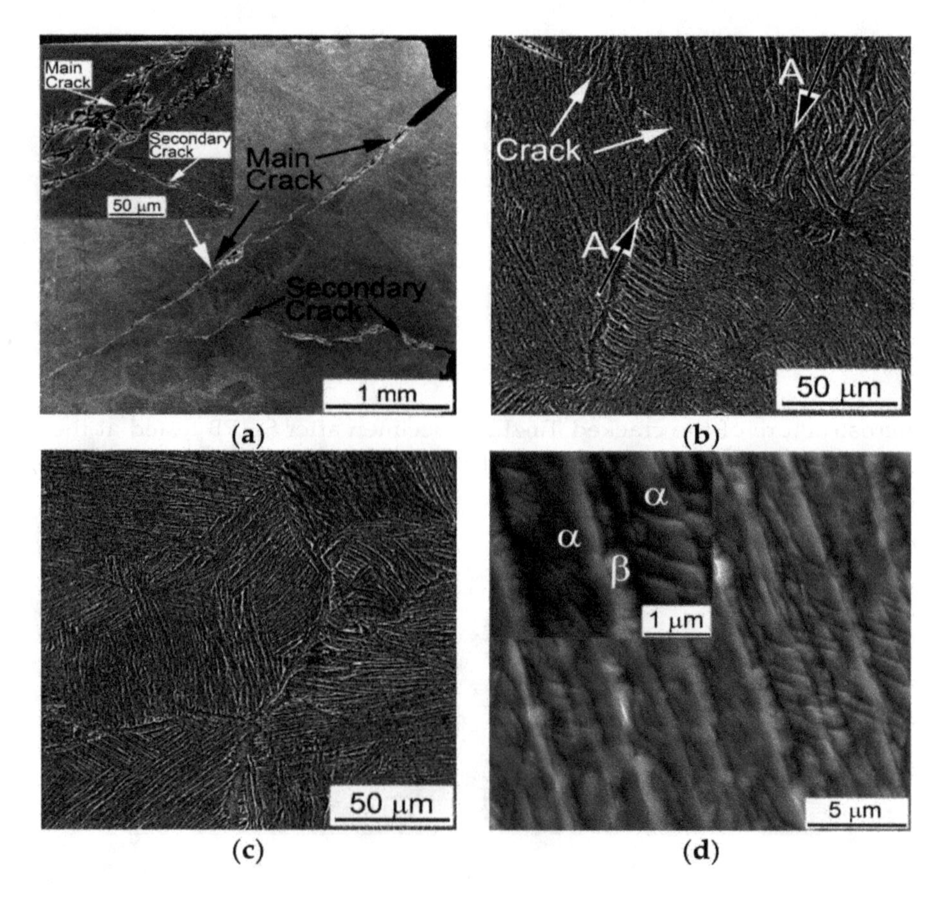

Figure 16. Microstructure of the Ti64LM specimen after the quasi-static test at the strain rate of 0.001 s^{-1}: (**a**) general view with zones I and II, (**b**)—zone III, (**c**,**d**)—zone IV. SEM, SE.

The processes of plastic deformation on micro- (inside separate α- and β-phases crystallites) and macro-level (at least a group of crystallites involved in cooperative response) should be distinguished which means that the dissipation of the total strain energy can be prioritized differently depending on the real structure. The alloy Ti64LM under slow quasi-static compression technically deforms in the same way as in the case of dynamic impact loading. Under such a condition, a smaller strain energy portion is apprehended at the micro-level in a form of plastic deformation localized inside α-lamellas, while its major fraction dissipates at the macro-level and is localized within a small zone. On the contrary, in the Ti64GL alloy, the strain energy is more apprehended at the micro-level, and no distinction between micro- and macro-level of energy dissipation is observed in c.p.Ti.

3.3.4. Microstructure Analysis of Ti64BEPM

Similarly to the cast and wrought Ti64LM, the Ti64BEPM alloy deformed at high strain rates demonstrates plastic deformation mainly in zone I, primarily on the samples that did not crack (Figure 17a). Moreover, the intense plastic deformation also took place locally, for example, in grain boundaries, and it was recognized as shear deformation (Figure 17b). Furthermore, the deformed (collapsed) residual pores were observed in various locations (Figure 17a). The diagonal zone II across the specimens was not found.

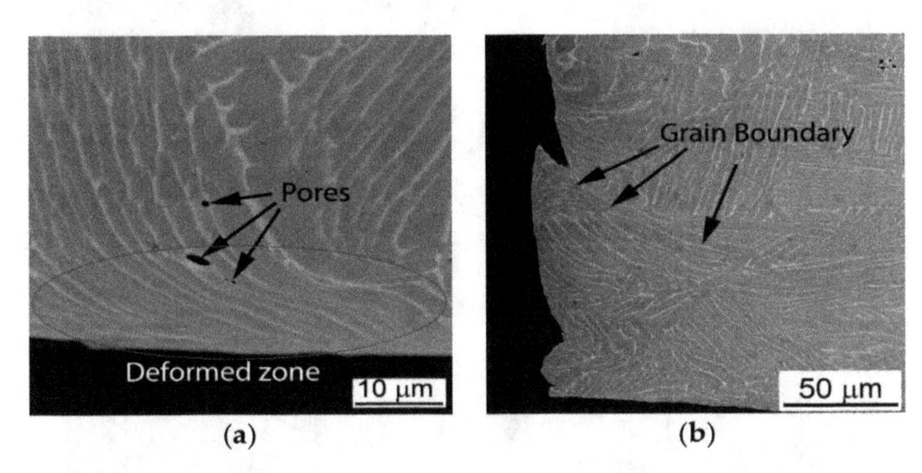

Figure 17. Microstructure of the non-cracked Ti64BEPM specimen tested at the strain rate of 2100 s^{-1}: (**a**) plastic deformation in zone I; (**b**) shear deformation in the vicinity zone I. SEM, BSE.

The microstructure of the Ti64BEPM alloy, after the test, was almost the same as the one of the Ti64LM discussed before. In zone I, similarly to the previous cases of the Ti64LM and Ti64BEPM (not fractured specimen—Figure 17), a plastically deformed layer of approximately the same depth (up to 20–30 m) was observed (Figure 18c). The main crack propagates in zone II through the entire specimen at an angle of 45° toward its vertical axis. Zone II itself is thin, not more than 5–6 m, and the ASB are formed on the sample fracture edges (Figure 18d). A number of pores collapsed and multiple α-lamellae and β-interlayers are curved due to the plastic flow of the material in vicinity of the crack (Figure 18d). Small secondary cracks initiated on the β-grain boundaries and α-colony boundaries are observed in zone III (Figure 18e). There is no evident plastic deformation in zone IV, except some slightly deformed pores partially flattened perpendicular to the direction of the applied load (Figure 18f). However, more detailed images of the slightly etched ion-beam samples reveal a fine needles microstructure inside the α-plates (Figure 18g,h). The formation of such an inner plate substructure is unique, and the observation of the uniform triangle arrangement of the needles with a uniform 60° angle between them (Figure 18h) is typical for martensite in h.c.p. lattice metals [37]. As it was mentioned before, a similar microstructure was also observed in the Ti64GL specimen fractured at the strain rate of 3330 s^{-1} (Figure 11d). Such a structure may result from rapid heating of the material during the impact, when the α-phase plates transformed into the high-temperature β-phase in such a fast way that the redistribution of alloying elements between the initial α- and β-phases did not occur. It is confirmed by the clear interphase boundary between the β-phase and the α-phase lamellar region, which includes the needles (Figure 18g). The area outlined by the circle in Figure 18h is of particular interest since it shows the intersection of the needles lying in the plane of the polished surface and perpendicular to it.

A general view of the fracture surface of the cracked Ti64BEPM specimen compressed at the strain rate of 2220 s^{-1} (Figure 18a) is similar to the fractures of the Ti-6-4 alloy with both globular (Figure 11a) and lamellar microstructure (Figure 14a). The fracture also demonstrates a rectilinear zone of the main crack growth (A in Figure 18a) and two side zones of the secondary crack propagation (B, ibid.). The images of higher magnification show the residual pores (Figure 18b), and their presence in the structure makes Ti64BEPM alloy essentially different from the cast and wrought Ti64LM alloy (Figure 14a).

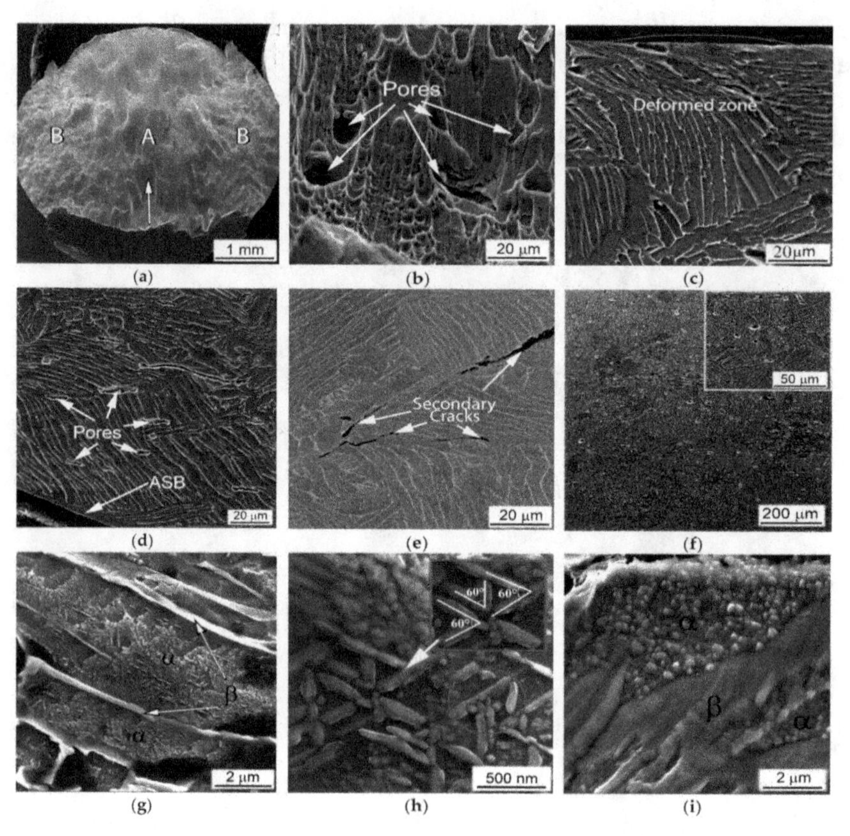

Figure 18. The SEM images of the fracture surface of the Ti64BEPM specimen tested at the strain rate of 2220 s^{-1}: (**a**); general view; (**b**) residual pores at fracture surface; (**c**–**i**) internal microstructure; (**c**) zone I, (**d**) zone II; (**e**,**f**) zone III; (**g**–**i**) zone IV; (the arrow in (**a**): indicates direction of the crack growth from nucleation site; (A) and (B) label the fields of main and secondary cracks propagation, respectively). (**a**–**e**,**g**–**i**) SE; (**f**) BSE.

It should be noted that the needles, due to the ion etching, do not demonstrate perfectly smooth boundaries as they should for the martensite structure. The formation of a martensite structure during plastic deformation at the strain rate of up to 2000 s^{-1} was also reported on titanium Ti-8.5Cr-1.5Sn alloy [35], although in that case the observed structure was deformation-induced high-alloyed α''-martensite that formed simultaneously with the twins. The present study case is also different from the case of a fast laser heating [35], because formation of martensitic needles appeared inside of the α-plates with their apparently unchanged outlines (Figure 18g). The observed result can be explained by the fact that, in addition to rapidly changing temperature, a complex stress state also acted on particular zones within the specimen. Such a complex and combined effect is probably responsible for the formation of the martensite in localized volumes and, as a result, at least three variants of Burgers orientation relationships from 12 allowed are clearly seen. It should be emphasized that martensite-like needle crystals were not frequently well revealed. They do not appear to be distinct and they are similar to those shown in Figure 18i. However, it is implicit that the clear view of the martensite needles depends on the "successful" coincidence of planes of their location with the prepared specimen's section that is not highly probable.

The microstructures of the Ti64BEPM alloy after the quasi-static test were characterized by the main crack appeared in zone II surrounded by the secondary cracks appeared on different microstructural elements (Figure 19a). The other deformation features uniformly distributed throughout the specimen were the collapsed residual pores and strongly bent α-lamellae (Figure 19b). Moreover, a cellular dislocation substructure was observed inside the curved lamellae of the α- phases (Figure 19c), similar to that observed in the case of the Ti64LM specimen after the identical quasi-static test (Figure 16b). This was expected since there is no significant heating taking place in material deformed at a low strain rate, and there is also enough time for more complete stress relaxation in the areas outside of the zone II.

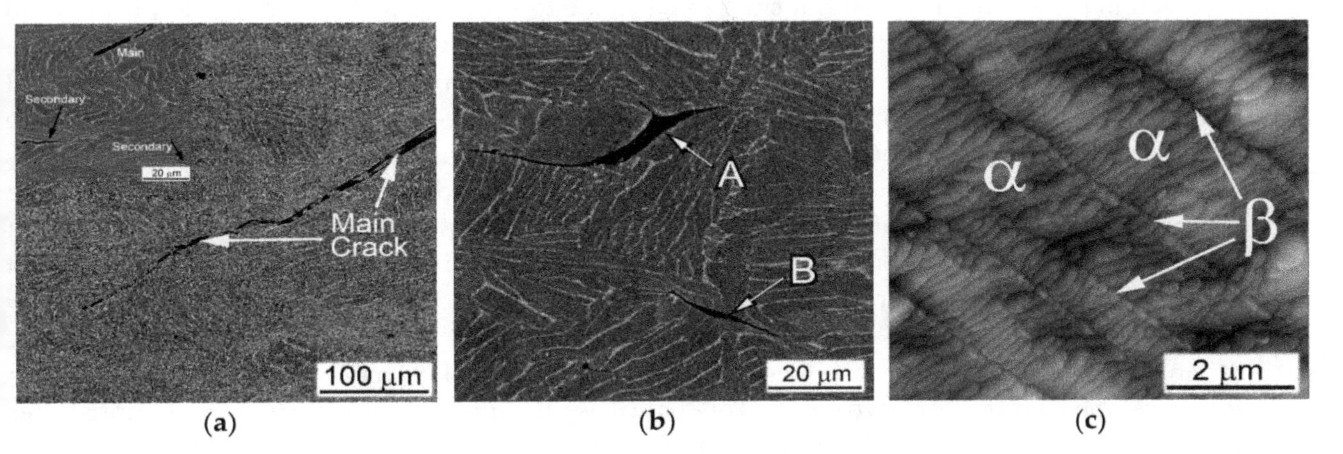

Figure 19. The SEM images of the fractured Ti64BEPM specimen after the quasi-static tests at the strain rate of 0.001 s^{-1}: (**a**) main crack—zone II; (**b**) collapsed residual pores—zone II; (**c**) cellular dislocation substructure inside the curved lamellae of the α-phases—zone IV. (**a**,**b**) BSE, (**c**) SE.

As it was shown in Section 3.1, there is a significant difference in the mechanical behavior of the Ti64LM and Ti64BEPM alloys. Based on the microstructural analysis presented above, different mechanical behavior of the tested Ti-based materials could be explained as follows. Both alloys present a similar type of microstructure, which is represented by relatively coarse β-grains with the colonies of lamellar α-phase inside. However, in detail, they are considerably different with a number of features, such as different sizes of both β- grains (more than 500 m in the Ti64LM vs. 100 m in the Ti64BEPM) and intragrain α-lamellae, the presence of residual pores in the Ti64BEPM, as well as the content of impurities (see Table 1). The high oxygen content as well as high nitrogen content in the Ti64BEPM led to an increase in strength, compared to the Ti64LM, and combined with residual pores of the Ti64BEPM significantly reduces its ductility under the tensile loading (Table 2). However, under quasi-static and dynamic compression the porous Ti64BEPM alloy shows noticeably better characteristics compared to the Ti64LM (Figure 5d,c). It could be explained by the fact that, under the tension loading, the pores work as the stress concentrators and/or crack initiation sites, particularly, when the pore shape is not globular, and the matrix alloy lacks the ductility. However, the pores impact on a deformation mechanism and the material damage process under compression is slightly different. The pores become flattened or even completely collapsed and, thence, play the role of additional "soft" phase without generating stress concentration zones, therefore general compressive plasticity of the porous material is determined by the plasticity of the matrix. Moreover, according to studies [43–45], a positive effect of porosity on the structure sustainability under compression can be amplified by their higher content; the most effective deformation energy absorption was reported at a porosity value of around 60% [46,47].

4. Conclusions

Based on presented experimental data and their analysis the following conclusions are drawn:

(a) Compressive mechanical behavior of titanium alloys is strongly dependent on the phase composition and microstructure of both the studied materials and the applied strain rate level.
(b) The mechanical behavior of a two-phase α + β Ti-6-4 alloy strongly depends on the type and coarseness of the microstructure. The fine-grained Ti-6-4 alloy with a globular (equiaxed) microstructure is more ductile and has the high reserve of plasticity, which allows it to deform

without fracture at the strain rate below 3320 s^{-1}. The critical compression strain rate, at which the fracture occurred, falls to 2030 s^{-1}, when the microstructure changed from globular to coarse-grained lamellar. The observed significantly different mechanical behavior of two structures can be explained by the nature of the interface boundaries between the structural constituents involved in plastic deformation transmission, i.e., the α/α interphase boundaries are prevalent in the globular microstructure, while α/β boundaries are predominant in the lamellar microstructure

(c) The Ti-6-4 alloy fabricated using BEPM demonstrates the reduced the size of β-grains and intragrain α- lamellae compared to the alloy with a coarse-grained lamellar microstructure produced using a conventional cast and wrought approach. The Ti64BEPM alloy demonstrates a considerably better balance of strength and plasticity under the quasi-static and dynamic compression tests, because of its finer microstructure despite of the presence of about 2% (vol.) of residual pores and higher content of impurities (oxygen and nitrogen). The residual pores do not play any negative role under compression loading in contrary to tension, since they do not work as stress concentrators.

(d) Strain energy was used as a parameter to compare mechanical behavior of the studied materials. It was established that the two-phase Ti-6-4 alloy with a globular microstructure demonstrates the highest value of SE_{max}, which implies the largest reserve of deformability of this alloy under the compression impact at the strain rates. The Ti64 alloy produced using BEPM demonstrates a lower value of the SE_{max} parameter. The Ti64 alloy with coarse lamellar microstructure reveals the lowest values SE_{max}.

(e) It was found that the strain rates increase up to 2200 s^{-1} cause a change in the strain localization mechanism in Ti64BEPM alloy from the macro-level (plastic flow in the sample volume, formation of adiabatic shear bands and cracks) to the micro-level (deformation within individual α-phase lamellae).

(f) The structures of all the studied materials demonstrate more uniform plastic deformation and the absence of its micro-level strain localization after quasi-static compression compared to the dynamic loaded structures.

(g) The Ti-6-4 alloys with a globular microstructure, fabricated using ingot metallurgy, and the Ti64BEPM alloy demonstrate higher relative (specific) *SE* values than B95 aluminum alloy, ARMOX 600T armor steel, or AHSS steel Docol 1500M.

Author Contributions: Conceptualization, validation and project administration, P.E.M., J.J.; methodology, P.E.M., J.J., D.G.S., M.A.S.; formal analysis, V.I.B., S.V.P.; experiments and investigation, P.E.M., J.J., O.O.S., V.I.B., K.C., P.D., M.A.S.; writing—original draft preparation, P.E.M.; writing—review and editing, J.J., D.G.S., V.I.B., S.V.P.; visualization, O.O.S.; All authors have read and agreed to the published version of the manuscript.

References

1. Luetjering, G.; Williams, J.C. *Titanium*, 2nd ed.; Springer: Berlin/Heidelberg, Germany, 2007.
2. Williams, J.C.; Boyer, R.R. Opportunities and Issues in the Application of Titanium Alloys for Aerospace Components. *Metals* **2020**, *10*, 705. [CrossRef]
3. Niinomi, M. Recent metallic materials for biomedical applications. *Met. Mater. Trans. A* **2002**, *33*, 477–486. [CrossRef]
4. Fanning, J. Military Application for b Titanium Alloys. *J. Mater. Eng. Perform.* **2005**, *14*, 686–690. [CrossRef]
5. Montgomery, J.S.; Wells, M.G.H.; Roopchand, B.; Ogilvy, J.W. Low-cost titanium armors for combat vehicles. *JOM* **1997**, *49*, 45–47. [CrossRef]
6. Fanning, J. Ballistic Evaluation of Titanium Alloys Against Handgun Ammunition. In *Ti-2007, Science and Technology, Proceedings of the 11th World Conference on Titanium, Kyoto, Japan, 3–7 June 2007*; The Japan Institute of Metals Publish.: Tokyo, Japan, 2007; Volume 1, pp. 487–490.
7. Gooch, W. Potential Applications of Titanium Alloys in Armor Systems. In *Titanium-2011*; International Titanium Association: San Diego, CA, USA, 2011; Available online: https://www.researchgate.net/

publication/292328353_Potential_Applications_of_Titanium_Alloys_in_Armor_Systems_-2011 (accessed on 20 October 2020).
8. Stefansson, N.; Weiss, I.; Hutt, A.J. *Titanium'95: Science and Technology*; Blenkinsop, P.A., Evans, W.J., Flower, H.M., Eds.; The University Press: Cambridge, UK, 1996; Volume 2, pp. 980–987.
9. Bhattacharjee, A.; Ghosal, P.; Gogia, A.; Bhargava, S.; Kamat, S. Room temperature plastic flow behaviour of Ti–6.8Mo–4.5Fe–1.5Al and Ti–10V–4.5Fe–1.5Al: Effect of grain size and strain rate. *Mater. Sci. Eng. A* **2007**, *452*, 219–227. [CrossRef]
10. Markovsky, P.; Matviychuk, Y.; Bondarchuk, V. Influence of grain size and crystallographic texture on mechanical behavior of TIMETAL-LCB in metastable β-condition. *Mater. Sci. Eng. A* **2013**, *559*, 782–789. [CrossRef]
11. Markovsky, P.; Bondarchuk, V.; Herasymchuk, O. Influence of grain size, aging conditions and tension rate on the mechanical behavior of titanium low-cost metastable beta-alloy in thermally hardened condition. *Mater. Sci. Eng. A* **2015**, *645*, 150–162. [CrossRef]
12. Markovsky, P.; Bondarchuk, V.I. Influence of Strain Rate, Microstructure and Chemical and Phase Composition on Mechanical Behavior of Different Titanium Alloys. *J. Mater. Eng. Perform.* **2017**, *26*, 3431–3449. [CrossRef]
13. Markovsky, P.E. Mechanical Behavior of Titanium Alloys under Different Conditions of Loading. *Mater. Sci. Forum* **2018**, *941*, 839–844. [CrossRef]
14. Peirs, J.; Verleysen, P.; Degrieck, J.; Coghe, F. The use of hat-shaped specimens to study the high strain rate shear behaviour of Ti–6Al–4V. *Int. J. Impact Eng.* **2010**, *37*, 703–714. [CrossRef]
15. Zheng, C.; Wang, F.; Cheng, X.; Liu, J.; Fu, K.; Liu, T.; Zhu, Z.; Yang, K.; Peng, M.; Jin, D. Failure mechanisms in ballistic performance of Ti–6Al–4V targets having equiaxed and lamellar microstructures. *Int. J. Impact Eng.* **2015**, *85*, 161–169. [CrossRef]
16. Morrow, B.; Lebensohn, R.; Trujillo, C.; Martinez, D.T.; Addessio, F.; Bronkhorst, C.A.; Lookman, T.; Cerreta, E. Characterization and modeling of mechanical behavior of single crystal titanium deformed by split-Hopkinson pressure bar. *Int. J. Plast.* **2016**, *82*, 225–240. [CrossRef]
17. Yin, W.; Xu, F.; Ertorer, O.; Pan, Z.; Zhang, X.; Kecskes, L.; Lavernia, E.J.; Wei, Q. Mechanical behavior of microstructure engineered multi-length-scale titanium over a wide range of strain rates. *Acta Mater.* **2013**, *61*, 3781–3798. [CrossRef]
18. Zhou, T.; Wu, J.; Che, J.; Wang, Y.; Wang, X. Dynamic shear characteristics of titanium alloy Ti-6Al-4V at large strain rates by the split Hopkinson pressure bar test. *Int. J. Impact Eng.* **2017**, *109*, 167–177. [CrossRef]
19. Guo, Y.; Ruan, Q.; Zhu, S.; Wei, Q.; Lu, J.; Hu, B.; Wu, X.; Li, Y. Dynamic failure of titanium: Temperature rise and adiabatic shear band formation. *J. Mech. Phys. Solids* **2020**, *135*, 103811. [CrossRef]
20. Sreenivasan, P.R.; Ray, S.K. Mechanical Testing at High Strain Rates. In *Encyclopedia of Materials: Science and Technology*, 2nd ed.; Elsevier: New York, NY, USA, 2001; pp. 5269–5271.
21. Ivasishin, O.M.; Anokhin, V.M.; Demidik, A.N.; Savvakin, D.G. Cost-Effective Blended Elemental Powder Metallurgy of Titanium Alloys for Transportation Application. *Key Eng. Mater.* **2000**, *188*, 55–62. [CrossRef]
22. Ivasishin, O.; Moxson, V. Low-cost titanium hydride powder metallurgy. *Titan. Powder Metall.* **2015**, *8*, 117–148. [CrossRef]
23. Chen, W.; Song, B. *Split Hopkinson (Kolsky) Bar: Design, Testing and Applications*; Springer: Berlin/Heidelberg, Germany, 2011.
24. Kolsky, H. Propagation of Stress Waves in Linear Viscoelastic Solids. *J. Acoust. Soc. Am.* **1965**, *37*, 1206. [CrossRef]
25. Kolsky, H. Stress waves in solids. *J. Sound Vib.* **1964**, *1*, 88–110. [CrossRef]
26. Panowicz, R.; Janiszewski, J.; Kochanowski, K. The non-axisymmetric pulse shaper position influence on SHPB experiment data. *J. Theor. App. Mech.* **2018**, *56*, 873–886. [CrossRef]
27. Janiszewski, J. *Unpublished Experimental Data Report*; Jarosław Dąbrowski Military University of Technology: Warsaw, Poland, 2020.
28. Database of Steel and Alloy (Marochnik). Available online: http://www.splav-kharkov.com/en/e_mat_start.php?name_id=1448 (accessed on 21 September 2020).
29. Nilsson, M. *Constitutive Model for Armox 500T and Armox 600T at Low and Medium Strain Rates*; Technical Report FOI-R-1068-SE; Swedish Defence Research Agency: Stockholm, Sweden, 2003; Available online: https://www.foi.se/rest-api/report/FOI-R--1068--SE (accessed on 21 September 2020).

30. Markovsky, P.E.; Savvakin, D.G.; Ivasishin, O.M.; Bondarchuk, V.I.; Prikhodko, S.V. Mechanical Behavior of Titanium-Based Layered Structures Fabricated Using Blended Elemental Powder Metallurgy. *J. Mater. Eng. Perform.* **2019**, *28*, 5772–5792. [CrossRef]
31. Prikhodko, S.V.; Ivasishin, O.M.; Markovsky, P.E.; Savvakin, D.G.; Stasiuk, O.O. Chapter 13: Titanium Armor with Gradient Structure: Advanced Technology for Fabrication. In *Advanced Technologies for Security Applications*; Springer: Dodrecht, The Netherlands, 2020; pp. 127–140.
32. Schmid, E.; Boas, W. *Plasticity of Crystals, Special Reference to Metals*; Springer US: New York, NY, USA, 1968.
33. Partridge, P.G. The crystallography and deformation modes of hexagonal close-packed metals. *Metall. Rev.* **1967**, *12*, 169–194. [CrossRef]
34. Ma, C.; Wang, H.; Hama, T.; Guo, X.; Mao, X.; Wang, J.; Wu, P. Twinning and detwinning behaviors of commercially pure titanium sheets. *Int. J. Plast.* **2019**, *121*, 261–279. [CrossRef]
35. Yang, H.; Wang, D.; Zhu, X.; Fan, Q. Dynamic compression-induced twins and martensite and their combined effects on the adiabatic shear behavior in a Ti-8.5Cr-1.5Sn alloy. *Mater. Sci. Eng. A* **2019**, *759*, 203–209. [CrossRef]
36. Markovsky, P.; Semiatin, S. Microstructure and mechanical properties of commercial-purity titanium after rapid (induction) heat treatment. *J. Mater. Process. Technol.* **2010**, *210*, 518–528. [CrossRef]
37. Banerjee, S.; Mukhopadhyay, P. Phase Transformations: Examples from Titanium and Zirconium Alloys. *Pergamon Mater. Series* **2007**, *12*, 1–813.
38. Markovsky, P.E. Two-stage transformation in ($\alpha+\beta$) titanium alloys on non-equilibrium heating. *Scr. Met. Mat.* **1991**, *25*, 2705–2710. [CrossRef]
39. Semiatin, S.L.; Obstalecki, M.; Payton, E.J.; Pilchak, A.L.; Shade, P.A.; Levkulich, N.C.; Shank, J.M.; Pagan, D.C.; Zhang, F.; Tiley, J.S. Dissolution of the Alpha Phase in Ti-6Al-4V During Isothermal and Continuous Heat Treatment. *Met. Mater. Trans. A* **2019**, *50*, 2356–2370. [CrossRef]
40. Gridnev, V.N.; Ivasishin, O.M.; Markovsky, P.E.; Svechnikov, V.L. The role of the cooling rate in the formation of the structure of titanium alloys thermally hardened with incomplete homogenization of the β phase. *Metallofiz (Phys. Met.)* **1985**, *7*, 37–44. (In Russian)
41. Lütjering, G. Influence of processing on microstructure and mechanical properties of ($\alpha+\beta$) titanium alloys. *Mater. Sci. Eng. A* **1998**, *243*, 32–45. [CrossRef]
42. Isabell, T.C.; Dravid, V.P. Resolution and sensitivity of electron backscattered diffraction in a cold field emission gun SEM. *Ultramicroscopy* **1997**, *67*, 59–68. [CrossRef]
43. Kumar, P.; Chandran, K.S.R.; Cao, F.; Koopman, M.; Fang, Z.Z. The Nature of Tensile Ductility as Controlled by Extreme-Sized Pores in Powder Metallurgy Ti-6Al-4V Alloy. *Met. Mater. Trans. A* **2016**, *47*, 2150–2161. [CrossRef]
44. Biswas, N.; Ding, J. Numerical study of the deformation and fracture behavior of porous Ti6Al4V alloy under static and dynamic loading. *Int. J. Impact Eng.* **2015**, *82*, 89–102. [CrossRef]
45. Banhart, J. Manufacture, characterization and application of cellular metals and metal foams. *Prog. Mater. Sci.* **2001**, *46*, 559–632. [CrossRef]
46. Suzuki, A.; Kosugi, N.; Takata, N.; Kobashi, M. Microstructure and compressive properties of porous hybrid materials consisting of ductile Al/Ti and brittle Al3Ti phases fabricated by reaction sintering with space holder. *Mater. Sci. Eng. A* **2020**, *776*, 139000. [CrossRef]
47. Garcia-Avila, M.; Portanova, M.; Rabiei, A. Ballistic performance of composite metal foams. *Compos. Struct.* **2015**, *125*, 202–211. [CrossRef]

7

Effect of Hybrid Reinforcements on the Microstructure and Mechanical Properties of Ti-5Al-5Mo-5V-Fe-Cr Titanium Alloy

Shuyu Sun [1,*] and Weijie Lu [2]

[1] School of Mechanical Engineering, Taizhou University, Taizhou 318000, China
[2] State Key Laboratory of Metal Matrix Composites, Shanghai Jiao Tong University, Shanghai 200240, China; luweijie@sjtu.edu.cn
* Correspondence: sunshuyu@tzc.edu.cn

Abstract: In order to investigate the different effects of trace TiB and TiC on the microstructure and the mechanical properties of Ti-5Al-5Mo-5V-1Fe-1Cr Ti alloy, two different modified Ti-5Al-5Mo-5V-1Fe-1Cr Ti alloys are fabricated via a consumable vacuum arc-remelting furnace in this work. Though the volume fractions of the reinforcements are the same in the two alloys, the molar ratio of short fibers to particles is different. The materials are subjected to thermomechanical processing and heat treatment. The effects of TiB short fibers and TiC particles on the spheroidization of α phase or the refinement of β phase have no obvious difference during heat treatment. Subsequently, the room temperature tensile test is carried out. The area covered by the σ-ε curve of the tensile test is used to compare toughness. It is revealed that the refinement of the β phase and the load bearing of TiB play key roles in promoting the toughness of the alloys. TiB tends to parallel the external load during tensile tests. The distribution of TiB also changes during isothermal compression test. Owing to the competition of dynamic softening with dynamic hardening, the length direction of TiB tends to parallel to the direction of maximum shear stress during the compression, which makes TiB play the role of load bearing better.

Keywords: Ti-5Al-5Mo-5V-Fe-Cr titanium alloy; trace TiB and TiC; microstructure; toughness

---

1. Introduction

Ti-5Al-5Mo-5V-1Fe-1Cr titanium alloy is an important aviation structural material. The alloy composition results in the tendency of the spheroidization of α phase during heat treatment. Owing to the special microstructural characteristics, the alloy presents a good combination of strength and plasticity after heat treatment [1,2]. The most common processing method for this alloy is isothermal forging. However, it may appear serious microstructural heterogeneity during isothermal compression. The experiment shows that reinforcement can solve the problem by accelerating the spheroidization of α phase [2].

While continuous-SiC fiber-reinforced Ti alloys have significantly superior mechanical properties, widespread application of them has been hindered by economic factors associated with high processing costs, as well as design limitations imposed by the anisotropy of properties (exacerbated by the presence of relatively weak fiber/matrix interfaces) [3,4]. Conventional Ti alloys modified with discontinuous reinforcements, on the other hand, have been widely used in recent years [5–7]. The application of the hybrid reinforcements can achieve better results in improving mechanical properties [8–10]. However, little information is available to date concerning the different effects of short fibers and particles on Ti-5Al-5Mo-5V-Fe-Cr Ti alloy.

Two different compositions of Ti-5Al-5Mo-5V-Fe-Cr Ti alloys are fabricated in this work One alloy is modified with only B_4C, the other is modified with B_4C and C. The volume fractions of the reinforcements in the two alloys are the same. Addition of B_4C to Ti produces TiB and TiC during solidification by in situ chemical reaction. C to Ti also produces TiC. The reinforcements are both thermomechanically stable and essentially insoluble in Ti at all temperatures in the solid state. The in situ synthesized intermetallic reinforcements have good interfacial bonding strength due to the orientation relationships [11–13]. The advantages of the reinforced alloys also include reaction-free interfaces and ease of processing. In order to obtain good toughness, the addition of B_4C or C is very small.

The solid solubility of B and C is low in Ti alloy during the solidification. Solute enrichment results in constitutional supercooling which in turn provides the driving force for nucleation and increases the nucleation rate. Furthermore, excess B and C in the solid-liquid interface also lower the growth rate of the grains [11]. The refinement of β grains and intersection of different orientated acicular α colonies within β grains retard the further growth of α [1,14].

The area covered by σ-ε curve of room temperature tensile test is used to compare toughness The experiments show that the toughness of the modified alloys is superior to that of the alloy The different toughening effects of trace short fibers and particles are investigated by studying the improvement of the plasticity and the strength of the modified alloys.

The β phase with BCC structure is considered with high stacking fault energy, which could accelerate dislocation climbing and crossing-slipping for dynamic softening process during isothermal compression [15]. The dynamic softening dominates the competition with the work hardening. The distribution of the reinforcements changes during isothermal compression. For comparison, the alloy specimen with the two ends not polished is used. The specimen preserves the shallow circular groove caused by turning machining on the two ends, which causes serious microstructural heterogeneity during the compression. Some α grains are elongated obviously along the direction of maximum shear stress [16]. The length direction of TiB in the B_4C-modified alloy is almost the same as that of the elongated α phase in the alloy after the compression. The distribution variation of the reinforcements can promote the effective aspect ratio (length/diameter) of TiB. Therefore, the distribution variation law of the reinforcements is also the key research of this work.

2. Experimental Procedure

The modified alloys are fabricated by in situ synthesis method. The stoichiometric weight fractions of the raw materials including sponge Ti, B_4C powder, Al, Al-Mo, Al-V, Fe, and Cr are blended uniformly, and then are compacted into pellets by forging press. The pellets are melted in a consumable vacuum arc-remelting furnace (Model VCF-10, Shanghai, China). Small addition of B_4C to Ti produces TiB and TiC during solidification by chemical reaction [12]:

$$5Ti + B_4C = 4TiB + TiC \tag{1}$$

The molar ratio of TiB to TiC is 4:1. In addition to B_4C, small addition of C powder is added in the other modified alloy. The molar ratio of TiB to TiC is 1:1. C to Ti produces TiC by chemical reaction [12]:

$$Ti + C = TiC \tag{2}$$

The volume fractions of the reinforcements in the two alloys are the same. The weight fractions of the reinforcements are shown in Table 1.

The as-cast ingots are forged at 1150 °C and are rolled at 840 °C into rods with a diameter of 15 mm. Then the rod is subjected to heat treatment. In order to control the overgrowth of grains and obtain good mechanical properties, the triplex heat treatment is employed with the following process: 830 °C/1.5 h + furnace cooling, 750 °C/1.5 h + air cooling, 600 °C/4 h + air cooling.

Ti-5Al-5Mo-5V-1Fe-1Cr Ti alloy is also prepared with the same method. The ingot of the alloy is subjected to the same thermomechanical processing and heat treatment as the modified alloys.

Table 1. Program materials and weight fractions of reinforcements.

Sample	B_4C/wt %	C/wt %	TiB/wt %	TiC/wt %	TiB/TiC Molar Ratio
Alloy A	0.1	0	0.4	0.1	4:1
Alloy B	0.06	0.04	0.27	0.27	1:1

The specimens for the tensile test and the isothermal compression test are machined from the heat-treated rods.

For tensile testing, 30 mm gage length cylindrical specimens (6 mm wide) is used. Tensile tests are performed in a servohydraulic testing machine (Model YF28A-315/200, Shanghai, China) with a strain rate of 0.001 s^{-1}. An extensometer is mounted on the specimens to measure the tensile strains.

The compression specimens are 8 mm in diameter and 12 mm in height. The two ends of the alloy compression specimen are not polished. The isothermal compression tests are performed on a Gleeble simulator (Model GLEEBLE 3500, Shanghai, China). The compression condition is 840 °C/0.01 s^{-1}/60% (temperature/strain rate/deformation reduction). The specimens are water quenched at room temperature immediately after the compression and then are axially sectioned.

Microstructure observations by optical microscope (OM, Model CM12, Shanghai, China), scanning electron microscope (SEM, Model JSM-6460, Shanghai, China) are conducted after the specimens are polished and etched. Moreover, transmission electron microscope (TEM, Model JZM-100CX, Shanghai, China) is also used to observe the microstructure of the specimens.

3. Results and Discussion

3.1. Microstructure of the Materials after Heat Treatment

Figure 1a shows the microstructure of alloy A after thermomechanical processing. The reinforcements can accelerate the recrystallization of β grains by providing nucleation sites and accelerating diffusion. However, the microstructure is still very heterogeneous due to the segregation of the alloying elements.

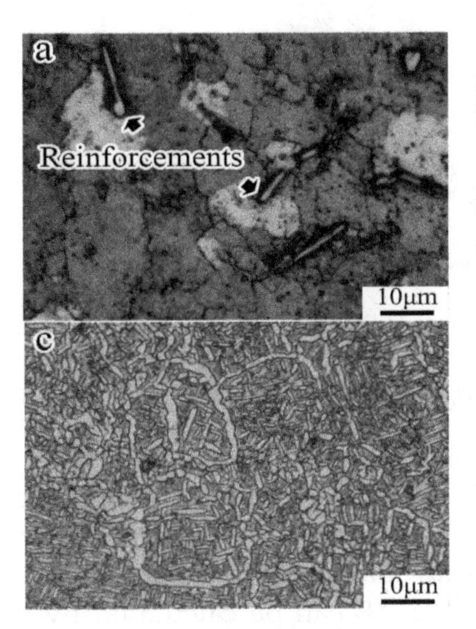

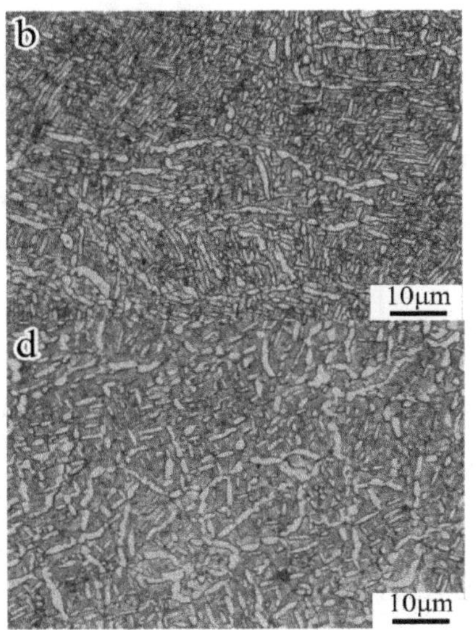

Figure 1. Optical microscope (OM) images of (**a**) alloy A after thermomechanical processing; (**b**) unmodified alloy after heat treatment; (**c**) alloy A after heat treatment; and (**d**) alloy B after heat treatment.

In order to measure the average grain size, the other heat treatment is used with the following process: 910 °C/1.5 h + furnace cooling, 750 °C/1.5 h + air cooling, 600 °C/4 h + air cooling. 910 °C is higher than the phase transformation temperature. Therefore, it shows the characteristics of the Widmannstatten structure in the alloys after the heat treatment [1]. The average grain size of β in alloy A is decreased by about 56%, and in alloy B it is decreased by about 59%. This is mainly ascribed to the Zener dragging force exerted by the reinforcements [2]. The length scales of the reinforcements are nearly the same since the additions of B_4C and C are very small. TiB has an average length of 7.9 μm after heat treatment, while TiC 2.6 μm. The size of TiC ranges from less than 200 nm to more than 4 μm. A broad size distribution of particles gives rise to a larger Zener dragging force than a narrow size distribution of particles [17]. Moreover, the segregation can also exert a pinning effect on the boundary migration of β grain during heat treatment.

Figure 1b–d show the microstructures of the materials after heat treatment. The phase contrast is α phase, white; β phase, black. The dislocation density in α decreases significantly during the heat treatment. It appears the tendency of the spheroidization process to decrease interfacial energy due to the dispersivity of α [1]. However, it is observed that trace reinforcements have no obvious effect on the spheroidization of α phase during heat treatment. The decisive factor affecting the spheroidization of the α phase is the first stage temperature during triplex heat treatment [18]. The increase in temperature can accelerate the boundary migration of α grain, which helps to overcome the dragging force of the segregation on the boundary migration of the α grain. Since the weight fraction of the reinforcements is very small, the segregation caused by the reinforcements counteracts the promoting effect of the reinforcements on the spheroidization of α phase. When the temperature approaches to the phase transformation temperature, the degree of the recrystallization of α phase is enhanced greatly. The reinforcements may overcome their negative effect on the spheroidization of the α phase by accelerating recrystallization [18].

Figure 2 shows the microstructure of the alloy after thermomechanical processing. The phase contrast is α phase, black; β phase, white. Primary α phase almost shows as rod shaped. The α phase cannot be spheroidized due to insufficient recrystallization. It appears the division of α grain during heat treatment [18]. It is suggested that the rod-shaped α formed during thermomechanical processing should be an important source of spheroidization grain after splitting.

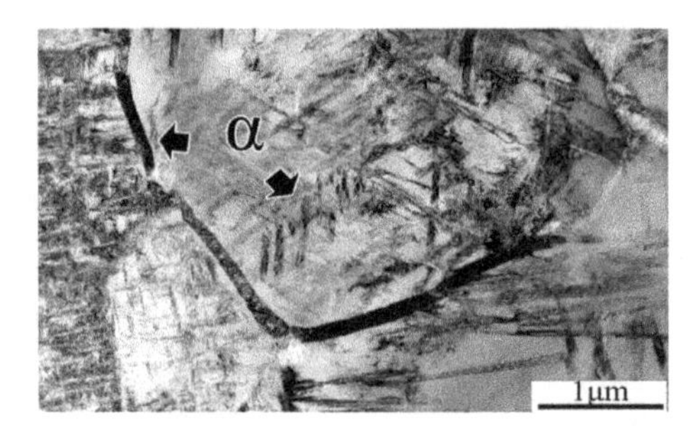

Figure 2. Transmission electron microscope (TEM) image of the alloy after thermomechanical processing.

3.2. Effect of Reinforcements on the Toughness of the Alloy

Figure 3 shows σ-ε curves of the alloy, alloy A, and alloy B. The results in Figure 3 are the average values of three tensile tests for each material. The yield stresses of the materials are 1195, 1232, and 1224 MPa, respectively. If the area covered by the curve is used to compare toughness, the toughness of the reinforced alloys is superior to that of the alloy (in Figure 3). The average aspect ratio of TiB is 7.1. The critical aspect ratio of TiB can be calculated by the Kelly formula, approximately [19]:

$$l_c/d = \sigma_f/(2\sigma_{ym}) \tag{3}$$

where l_c is critical length of TiB, d is radius of TiB, σ_f is tensile strength of TiB (3500 MPa), and σ_{ym} is yield stress of matrix (1195 MP). The critical aspect ratio of TiB is about 1.46, which is lower than the average aspect ratio of TiB, significantly.

The promotion of the strengths of the modified alloys is ascribed to the load bearing of TiB, the dispersion strengthening of TiC and the grain refinement strengthening [18,20–22].

TiB short fibers tend to parallel the external force during the tensile test, which promotes the load bearing effect. The load bearing of TiB extends the strengthening process of the modified alloys. Meanwhile, the load bearing of TiB and the increase of the number of the β grains can improve the homogeneity of the loads applied to each β grain. The deformation homogeneity helps to decrease the crack nucleation.

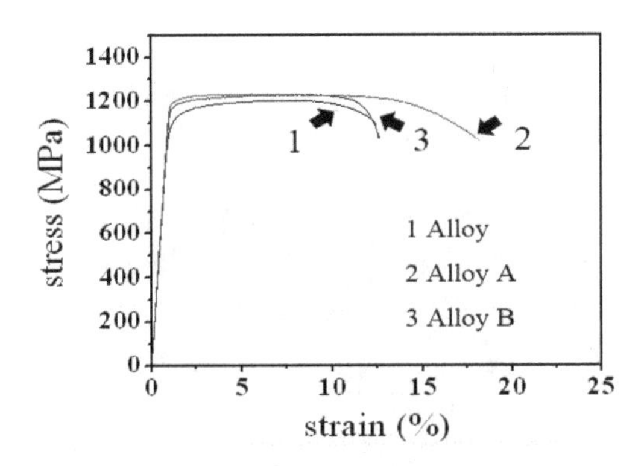

Figure 3. Stress-strain curves of the alloy, alloy A and alloy B.

Since the weight fraction of the reinforcements is very small, the negative effect of the reinforcements on the plasticity is limited. The elongation of the modified alloys is still increased during the strengthening process.

The broken TiB short fibers and TiC particles increase the propagation of crack during the necking process. Therefore, the toughening effect of the load bearing of the short fibers is mainly reflected in the strengthening process. The improvement of the reduction of area is mainly attributed to microstructural refinement during the necking process. The increase of β grain boundary contributes to retard crack propagation.

The tensile properties of alloy A are different from those of alloy B. The weight fraction of TiB in alloy A is higher than that in alloy B. Therefore, the load bearing of short fibers plays a more important role in alloy A. The effect of the microstructural refinement plays more important role in alloy B.

3.3. *Microstructure of the Modified Alloy after Isothermal Compression*

Figure 4 shows the microstructure of the alloy after isothermal compression. The phase contrast is α phase, black; β phase, white. The compression specimen shows as a drum shape due to the friction between the specimen and the push rod. The macrosegregation in Figure 4a is induced by the shallow circular groove on the two ends of the specimen [16]. The metal at the center of the specimen is extruded towards the side of the specimen. Therefore, in addition to the spheroidized α, some α grains are elongated. Moreover, the distribution of the reinforcements in the modified alloys changes for the same reason.

The elongation of α phase in Figure 4b is associated with vacancy flow. It produces a vacancy chemical potential gradient in the direction parallel to the maximum shear stress during the compression. The atomic motion is opposite to the vacancy flow. The dynamic softening dominates the competition with the dynamic hardening in α. Therefore, the obstacles to the vacancy flow that caused by the dislocations is greatly reduced. The length direction of the elongated α is approximated as the direction of maximum shear stress.

As shown in Figures 5 and 6, the distribution of TiB is similar to the elongated α in the alloy specimen. The distribution of TiB continuously changes in order to seek the equilibrium of forces until the length direction tends to parallel to the direction of maximum shear stress during the compression.

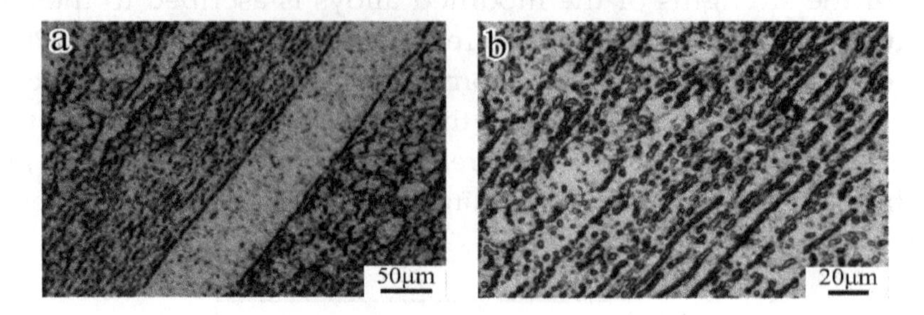

Figure 4. OM images of the longitudinal section of the alloy specimen after isothermal compression (**a**) the macrosegregation; and (**b**) the spheroidized and the elongated α.

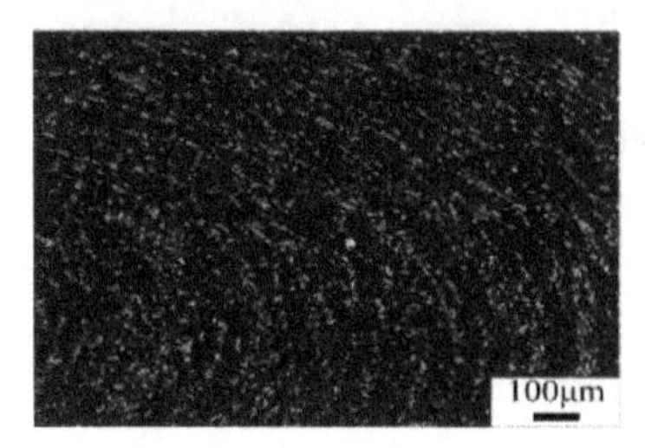

Figure 5. Scanning electron microscope (SEM) image of the distribution of the reinforcements in alloy A after isothermal compression.

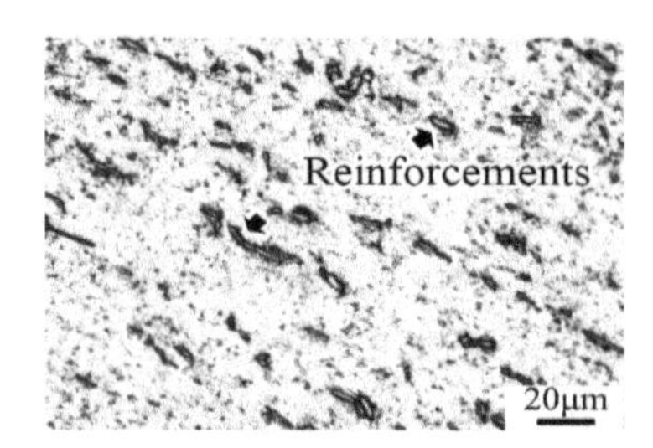

Figure 6. OM image of the distribution of the reinforcements in alloy A after isothermal compression.

Figure 7 shows TEM images of TiB and TiC in alloy A after isothermal compression. There also exists the competition of dynamic softening with dynamic hardening in β phase. As shown in Figure 7a, the end of TiB induces relatively large lattice distortion. The lattice distortion is nearly the largest when the length direction approximately parallels to the direction of maximum shear stress. Thus, the dynamic hardening gets the advantage in the competition at the ends of TiB. This plays a pinning effect on the distribution variation of TiB.

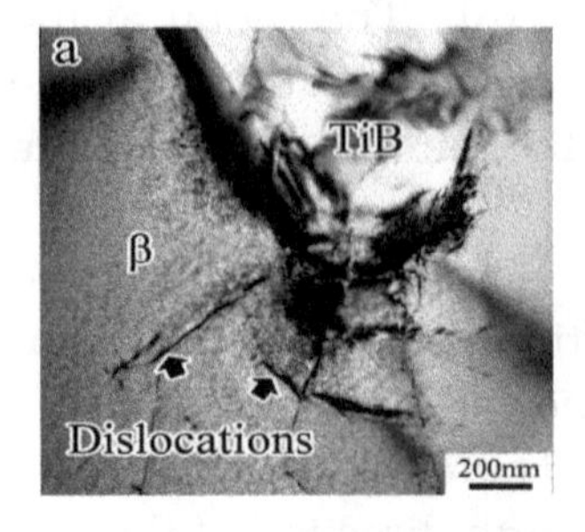

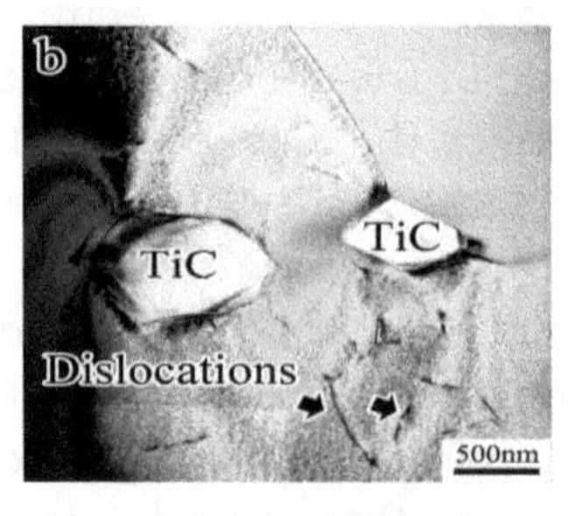

Figure 7. TEM images of the reinforcements in alloy A after isothermal compression (**a**) TiB (**b**) TiC.

The distribution variation of the reinforcements increases the effective aspect ratio of TiB, which allow TiB short fibers to play the role of load bearing better.

4. Conclusions

(1) The load bearing of TiB extends the strengthening process of the modified alloys. Moreover, the load bearing of TiB and the increase of the number of β grains can improve deformation homogeneity of the modified alloys, which decreases the crack nucleation. Therefore, the reinforcements increase the plasticity of the modified alloys. The toughening effect of the load bearing of TiB is mainly reflected in the strengthening process. The improvement of the toughness of the modified alloys is mainly attributed to the microstructural refinement during the necking process.

(2) The distribution of TiB constantly changes in order to seek the equilibrium of forces until the length direction tends to parallel to the direction of maximum shear stress during isothermal compression. Moreover, the lattice distortion is nearly the largest when the length direction of TiB approximately parallels to the direction of maximum shear stress. This plays a pinning effect on the distribution variation of TiB. Therefore, the length direction of TiB tends to parallel the direction of maximum shear stress during the compression, which can promote the effective aspect ratio of TiB.

Acknowledgments: We would like to acknowledge the financial support provided by Foundation of Zhejiang Educational Committee (grant No. Y201533396).

Author Contributions: Shuyu Sun and Weijie Lu designed and conducted the experiments. Shuyu Sun analysed the results.

References

1. Sun, S.-Y.; Lv, W.-J. Microtructure and mechanical properties of TC18 Titanium alloy. *Rare Met. Mater. Eng.* **2016**, *45*, 1138–1141.
2. Sun, S.-Y.; Lv, W.-J. Effects of trace reinforcements on microstructure and tensile properties of in-situ synthesized TC18 Ti matrix composite. *J. Comp. Mater.* **2017**. [CrossRef]
3. Tamirisakandala, S.; Bhat, R.-B.; Miracle, D.-B.; Boddapati, S.; Bordia, R.; Vanover, R.; Vasudevan, V.-K. Effect of boron on the beta transus of Ti-6Al-4V alloy. *Scr. Mater.* **2005**, *53*, 217–222. [CrossRef]
4. Tanaka, Y.; Kagawa, Y.; Liu, Y.-F.; Masuda, C. Interface damage mechanism during high temperature fatigue test in sic fiber-reinforced Ti alloy matrix composite. *Mater. Sci. Eng. A* **2001**, *314*, 110–117. [CrossRef]
5. Ni, D.-R.; Geng, L.; Zhang, J.; Zheng, Z.-Z. Effect of B_4C particle size on microstructure of in situ titanium matrix composites prepared by reactive processing of Ti-B_4C system. *Scr. Mater.* **2006**, *55*, 429–432. [CrossRef]
6. Wang, B.; Huang, L.-J.; Geng, L.; Rong, X.-D. Compressive behaviors and mechanisms of TiB whiskers reinforced high temperature Ti60 alloy matrix composites. *Mater. Sci. Eng. A* **2015**, *648*, 443–451. [CrossRef]
7. Qi, J.-Q.; Chang, Y.; He, Y.-Z.; Sui, Y.-W.; Wei, F.-X.; Meng, Q.-K.; Wei, Z.-J. Effect of Zr, Mo and TiC on microstructure and high-temperature tensile strength of cast titanium matrix composites. *Mater. Des.* **2016**, *99*, 421–426. [CrossRef]
8. Zhang, C.; Li, X.; Zhang, S.; Chai, L.; Chen, Z.; Kong, F.; Chen, Y. Effects of direct rolling deformation on the microstructure and tensile properties of the 2.5 vol% (TiB_w + TiC_p)/Ti composites. *Mater. Sci. Eng. A* **2016**, *684*, 645–651. [CrossRef]
9. Shufeng, L.-I.; Kondoh, K.; Imai, H.; Chen, B.; Jia, L.; Umeda, J. Microstructure and mechanical properties of P/M titanium matrix composites reinforced by in-situ synthesized TiC–TiB. *Mater. Sci. Eng. A* **2015**, *628*, 75–83.
10. Rahoma, H.-K.-S.; Chen, Y.-Y.; Wang, X.-P.; Xiao, S.-L. Influence of (TiC + TiB) on the microstructure and tensile properties of Ti-B20 matrix alloy. *J. Alloys Compd.* **2015**, *627*, 415–422. [CrossRef]

11. Sen, I.; Tamirisakandala, S.; Miracle, D.-B.; Ramamurty, U. Microstructural effects on the mechanical behavior of B-modified Ti-6Al-4V alloys. *Acta Mater.* **2007**, *55*, 4983–4993. [CrossRef]
12. Lu, W.; Zhang, D.; Zhang, X.; Wu, R.; Sakata, T.; Mori, H. HREM study of TiB/Ti interfaces in a TiB-TiC in situ, composite. *Scr. Mater.* **2001**, *44*, 1069–1075. [CrossRef]
13. Ozerov, M.; Klimova, M.; Vyazmin, A.; Stepanov, N.; Zherebtsov, S. Orientation relationship in a Ti/TiB metal-matrix composite. *Mater. Lett.* **2016**, *186*, 168–170. [CrossRef]
14. Lütjering, G. Influence of processing on microstructure and mechanical processing. *Mater. Sci. Eng. A* **1998**, *243*, 32–45. [CrossRef]
15. Sun, Z.-C.; Yang, H.; Han, G.-J.; Fan, X.-G. A numerical model based on internal-state-variable method for the microstructure evolution during hot-working process of TA15 titanium alloy. *Mater. Sci. Eng. A* **2010**, *527*, 3464–3471. [CrossRef]
16. Sun, S.-Y.; Lv, W.-J. Microstructure heterogeneity of TC18 Ti alloy during hot deformation. *Rare Met. Mater. Eng.* **2016**, *45*, 1545–1548.
17. Eivani, A.-R.; Valipour, S.; Ahmed, H.; Zhou, J.; Duszczyk, J. Effect of the size distribution of nanoscale dispersed particles on the zener drag Pressure. *Metall. Mater. Trans. A* **2011**, *42*, 1109–1116. [CrossRef]
18. Sun, S.-Y.; Wang, L.-Q.; Qin, J.-N.; Chen, Y.-F.; Lv, W.-J.; Zhang, D. Microstructural characteristics and mechanical properties of in situ synthesized (TiB + TiC)/TC18 composites. *Mater. Sci. Eng. A* **2011**, *530*, 602–606. [CrossRef]
19. Baxter, W.-J. The strength of metal matrix composites. *Metall. Mater. Trans. A* **1992**, *23*, 3045–3053. [CrossRef]
20. Boehlert, C.-J.; Tamirisakandala, S.; Curtin, W.-A.; Miracle, D.-B. Assessment of in situ TiB whisker tensile strength and optimization of TiB-reinforced titanium alloy design. *Scr. Mater.* **2009**, *61*, 245–248. [CrossRef]
21. Tjong, S.-C.; Ma, Z.-Y. Microstructural and mechanical characteristics of in situ metal matrix composites. *Mater. Sci. Eng. R* **2000**, *29*, 49–113. [CrossRef]
22. Soboyejo, W.-O.; Shen, W.; Srivatsan, T.-S. An investigation of fatigue crack nucleation and growth in a Ti-6Al-4V/TiB in situ composite. *Mech. Mater.* **2004**, *36*, 141–159. [CrossRef]

8

Titanium Powder Sintering in a Graphite Furnace and Mechanical Properties of Sintered Parts

Changzhou Yu, Peng Cao * and Mark Ian Jones *

Department of Chemical and Materials Engineering, The University of Auckland, Private Bag 92019, Auckland 1142, New Zealand; cyu060@aucklanduni.ac.nz
* Correspondences: p.cao@auckland.ac.nz (P.C.); mark.jones@auckland.ac.nz (M.I.J.)

Academic Editor: Mark T. Whittaker

Abstract: Recent accreditation of titanium powder products for commercial aircraft applications marks a milestone in titanium powder metallurgy. Currently, powder metallurgical titanium production primarily relies on vacuum sintering. This work reported on the feasibility of powder sintering in a non-vacuum furnace and the tensile properties of the as-sintered Ti. Specifically, we investigated atmospheric sintering of commercially pure (C.P.) titanium in a graphite furnace backfilled with argon and studied the effects of common contaminants (C, O, N) on sintering densification of titanium. It is found that on the surface of the as-sintered titanium, a severely contaminated porous scale was formed and identified as titanium oxycarbonitride. Despite the porous surface, the sintered density in the sample interiors increased with increasing sintering temperature and holding time. Tensile specimens cut from different positions within a large sintered cylinder reveal different tensile properties, strongly dependent on the impurity level mainly carbon and oxygen. Depending on where the specimen is taken from the sintered compact, ultimate tensile strength varied from 300 to 580 MPa. An average tensile elongation of 5% to 7% was observed. Largely depending on the interstitial contents, the fracture modes from typical brittle intergranular fracture to typical ductile fracture.

Keywords: titanium alloys; sintering; powder metallurgy; fracture

---

1. Introduction

Structural applications of titanium and its alloys are limited because of the high cost of production despite the fact that titanium exhibits admirable combined properties such as high strength-to-weight ratio and excellent corrosion resistance [1,2]. In comparison with the mainstream titanium production processes such as casting and wrought, the major advantage of powder metallurgical (PM) methods including conventional press-and-sinter and novel additive manufacturing [3,4] is the potential of cost reduction resulting from its near-net-shaping ability [5]. In recent major milestone developments in the PM titanium industry, near-net-shaped Ti-6Al-4V products have been approved for commercial airplane components by Boeing (Seattle, WA, USA) [5]. The manufacture of titanium powder metal products has received standardized quality management system AS9100/ISO9001 certification in April 2013 [6]. Titanium powders are generally sintered under vacuum rather than in an atmosphere such as argon or helium so as to obtain high densification and superior mechanical properties [7]. Vacuum sintering with a pressure of the order of 10^{-2} Pa or lower [8–11] can effectively control the active reaction between titanium and interstitial impurities such as oxygen, nitrogen and carbon. The mechanical properties of as-sintered titanium are highly sensitive to the interstitials [5,12]. For example, in samples with a similar sintered density of 98%, the ductility of press-and-sintered commercial purity titanium with an oxygen content of 3000 ppm (elongation: 11%) is less than 1/3 of that for Ti containing 700 ppm oxygen (elongation: 37.1%) [7,13]. Vacuum sintering can also remove detrimental volatiles such as chlorides that pre-exist in titanium sponges and help with densification [13,14]. However, vacuum sintering is a batch process and has high capital equipment requirement. Possible continuous mass production is suggested to be realized by atmospheric sintering so as to reduce the production cost [14].

In 1937, Kroll conducted the first sintering trials of 14 Ti binary alloys in argon atmosphere [13]. Subsequent work in the late 1940s and early 1950s saw some other examples of the sintering of titanium mill products [7,15]. Limberg et al. [16] studied the influence of sintering atmosphere on sintering behavior of a titanium aluminide Ti-45Al-5Nb-0.2B-0.2C (at. %) through altering the argon atmosphere pressure (from 10 to 80 kPa) in comparison with sintering under vacuum (10^{-3} Pa). In their study, the residual porosity of sintered specimens increased proportionally with applied argon pressure to 1.1% at the maximum pressure (80 kPa), but no other pronounced microstructural difference such as grain size was observed due to the various sintering atmosphere pressures [16]. Limberg et al.'s study also shows that the pressure of the sintering atmosphere had no obvious effect on the tensile properties of the as-sintered compacts and all sintered specimens exhibited similar ultimate tensile strength (around 630 MPa) and elongation (0.15%–0.19%) [16]. Since titanium has a high affinity with nitrogen, oxygen, hydrogen and carbon, atmospheric sintering typically has to be conducted in an inert atmosphere such as argon [17]. Even then, purification of the inert atmosphere is still required [15] before a titanium powder compact enters into the hot zone because even small amounts of reactive gases (nitrogen, oxygen and hydrogen) are detrimental and make sintered specimens brittle. To address this issue, Arensburger et al. developed a purification setup by allowing argon flowing through pre-heated titanium sponges (900–1000 °C) [17]. An alternative solution is to reduce the oxygen partial pressure in the argon gas further by utilizing reactions with graphite materials in the furnace such as in a KYK Oxynon furnace (Kanto Yakin Kogyo, Shinomiya, Japan) [18–20]. As a result, extremely low oxygen pressure can be generated by the reaction between oxygen and carbon, which is deemed to be able to reduce oxides such as titanium oxide (TiO_2) into metal. The Oxynon furnace has been used for titanium sintering in argon since 2002. However, a recent comparative study revealed that the ductility of commercially pure (C.P.) Ti sintered in the Oxynon furnace is much lower than that sintered under vacuum, although the oxygen and carbon contents in titanium sintered in Oxynon furnace were lower than those in vacuum [18]. This signifies the necessity of investigating what causes the difference between sintering in argon and vacuum.

The reports on the atmospheric sintering of titanium powder are very limited. Our previous work using a furnace constructed of graphite heating elements and liners showed that the presence of impurities resulted in a thick porous contaminated scale, which was identified as titanium oxycarbonitride (Ti(CNO)) [21]. This is a follow-up study, which aims to understand the effects of possible contaminants on the sintering and properties of C.P. Ti during sintering in a common graphite furnace.

2. Materials and Methods

2.1. Materials, Compaction and Sintering

Hydrogenated-dehydrogenated (HDH) C.P. titanium powder (particle size <75 μm, nominal impurities provided by the supplier: 0.4060 wt % O, 0.0470 wt % C, 0.0070 wt % N) was supplied by Xi'an Lilin Ltd., Xi'an, China. The powder morphology and particle size distribution have been reported previously [22]. The powders were pressed uniaxially under a compaction pressure of 400 ± 50 MPa into small disc samples 12 mm in diameter and 5 to 8 mm thick. These compacts had a green density of ~77% theoretical density. The green compacts were then sintered at three different temperatures (1100, 1250 and 1400 C) and three different holding times (4, 6 and 8 h). In order to track possible interstitial diffusion from the atmosphere and understand its effects on mechanical properties, a large cylindrical compact (32 mm in diameter and 30 to 40 mm high) was uniaxially pressed at a pressure of 200 MPa followed by cold isostatic pressing (CIP) under a pressure of 600 MPa. This large green compact had a density of 87%. The large cylindrical compact was then sintered at a temperature of 1250 °C for 4 h. Tensile specimens were machined from different positions of this large cylinder.

The graphite furnace employed in this work (Mellen, WI, USA) has three sets of graphite heating elements and uses graphite liners for insulation. The hot zone is 20 cm × 20 cm × 20 cm. The compacts were placed on a boron nitride plate at the center of the hot zone. After the compacts were loaded, the furnace was evacuated to a vacuum level of 0.1 to 0.01 Pa and heated to 850 °C with a heating rate of 5 $°C{\cdot}min^{-1}$. During this preheating, argon was purged a few times to remove possible volatiles before argon gas backfilling. In order to avoid possible gas leaking at high temperatures, at this stage, argon gas (purity 99.99%) was backfilled to the furnace and a pressure of 1.1 atm (1.1×10^5 Pa) was maintained until completion of sintering. The positive net furnace pressure is designed to avoid possible inwards airflow from ambient atmosphere.

2.2. *Characterization*

The sintered density was measured by the Archimedes method according to the ASTM B962-14 standard. Metallographic specimens were prepared as per the standard metallographic procedures. The polished cross sections were etched with Kroll's reagent (2 mL HF, 4 mL HNO_3, 100 mL H_2O). Microstructural observations were carried out using an environmental scanning electron microscope (ESEM, Quanta 200F, FEI, Hillsboro, OR, USA), equipped with energy dispersive spectrometry (EDS) and an Olympus BX60M optical microscopy (Olympus, Waltham, MA, USA) coupled with polarized light optics. The average grain size was measured as per the linear intercept method described in the ASTM E112-12 standard. The pore size was estimated using an Image-J software (ImageJ Developers, National Institutes of Health, Bethesda, MD, USA). A Bruker D2 PHASER (Bruker AXS, Karlsruhe, Germany) X-ray diffraction (XRD) machine was employed to identify phase constituents. The patterns were collected over a 2θ range of 20° to 80° with a step size of 0.02°. Oxygen, nitrogen and carbon contents in the as-received titanium powder and the sintered compacts were measured using a LECO oxygen-nitrogen analyzer (TCH-600, Leco, St Joseph, MI, USA) and carbon-sulfur analyzer (CS-444, Leco, St Joseph, MI, USA).

2.3. *Mechanical Testing*

Tensile slices were cut from the as-sintered large cylinder after removal of the contaminated scale, as shown in Figure 1. The contaminated scale layer is designated as S1 and this porous scale contains many cracks and can be easily removed to expose the interior bulk titanium. The S2 specimen cut 1.5 mm beneath the surface scale contains visible pores and is believed to contain too high impurity levels and therefore is not suitable for tensile testing. The slices S3 to S7 do not have visible pores with naked eyes and were then further machined into standard tensile specimens (~2 mm × 2 mm cross-section, 16 mm shoulder-to-shoulder length, and 10 mm gauge length). Tensile testing was performed on an Instron 3367 machine (Instron, Norwood, MA, USA) with a crosshead speed of 0.1 $mm{\cdot}min^{-1}$ (initial strain rate: 1.67×10^{-4} s^{-1}). The precise displacement was determined using an extensometer with a gauge length of 8 mm (model: 2630-120, Instron Co., Norwood, MA, USA).

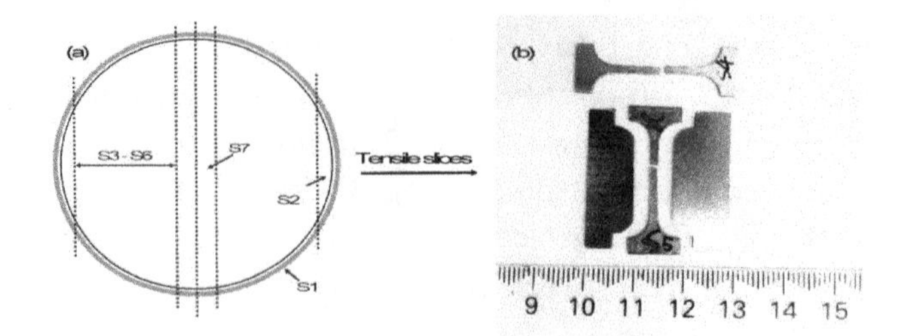

Figure 1. (**a**) Schematic illustration of the cylinder and the seven slices cut from the cylinder, only slices S3 to S7 were further machined for tensile test and (**b**) the tensile specimens after fracture. S1 was the outer scale, S2 to S7 were the slices cut from outer slice to the interior slices. S7 slices were cut from the central of the sintered large cylinder.

3. Results

3.1. Sintering Densification

The sintered density (relative to theoretical density) and densification parameter are illustrated in Figure 2, as a function of sintering temperature and holding time. Densification parameter, φ, is calculated according to Equation (1) [23]

$$\varphi = \left(\rho_s - \rho_g\right) / \left(\rho_{th} - \rho_g\right) \quad (1)$$

where ρ_s, ρ_g and ρ_{th} are sintered density, green density and theoretical density, respectively. The value of φ represents how much porosity in the green compact has been removed by sintering. In general, the sintered densities in all small samples (i.e., the 12 mm-in-diameter samples) were low and in the range of 82.5% to 92.5% of theoretical density (see Figure 2a). A higher sintering temperature and/or longer hold increased the sintered density and densification (Figure 2b). However, it must be noted that the density was measured by weighing the entire sintered compact and therefore included the contaminated porous surface scale.

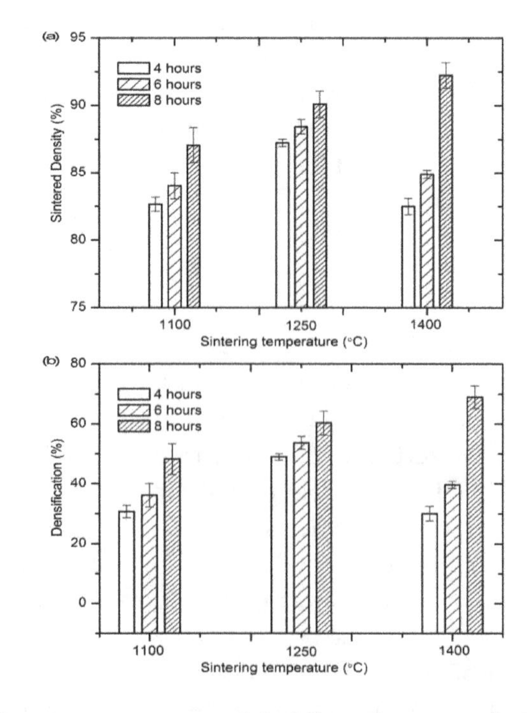

Figure 2. Effects of sintering temperature and holding time on relative sintered density (**a**) and densification parameter (**b**).

3.2. Microstructural Observation

Figure 3 presents an example of the compacts sintered at the three temperatures for 4 h. The metallographic observations of the cross sections of these compacts clearly show a fairly thick contamination scale on the compact surface. This porous layer was 300–500 μm thick, regardless of the sintering temperature. Separation between this layer and the remaining compact is visible. XRD and EDS analyses suggest that this layer is Ti(CNO) [21] and the formation mechanism will be discussed in subsequent Section 4.2.

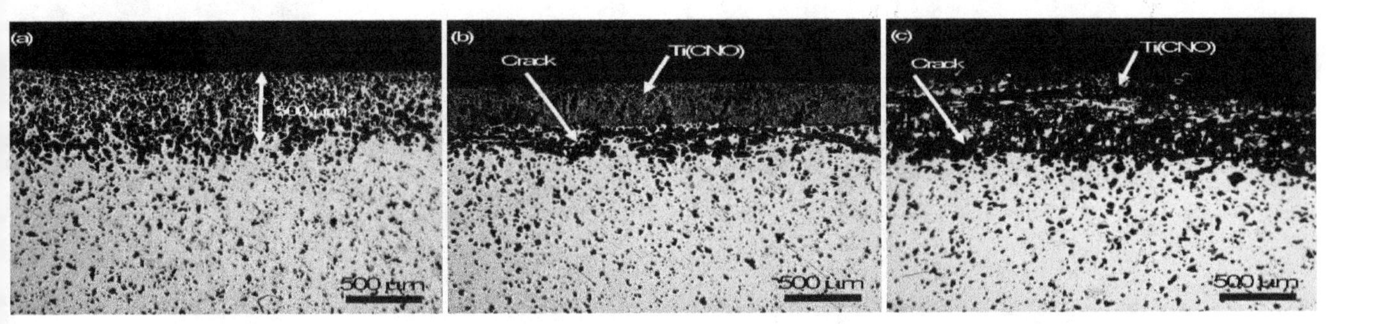

Figure 3. Micrographs of the cross-sections of sintered small compacts at different sintering temperatures: (**a**) 1100 °C; (**b**) 1250 °C; and (**c**) 1400 °C. All compacts were sintered for 4 h at the respective temperature.

The microstructures in the interior of the sintered compacts were observed after the outermost layer was removed. Under the polarized light microscope, macropores were evident in all the cases, as shown in Figure 4. The pore size and morphology are dependent on the sintering temperature. For instance, after being sintered at 1100 °C for 4 h, the pore shape is irregular and interconnected. This implies that, under this sintering condition, the compact might be still in the early stage of densification (Figure 4a). By increasing the holding time to 8 h, the number and size of pores decreased, although the majority of pores was still of irregular shape (Figure 4d). Pore spheroidizing occurred at higher sintering temperatures of 1250 and 1400 °C, in which cases the pores were spherical and isolated. Sintering at 1250 °C led to most pores being located either along grain boundaries or at the triple junctions (Figure 4b). At 1400 °C, some of these pores became trapped within grains as a result of grain growth in the later stages of sintering (Figure 4c). In addition to the pore morphology, the pore size is also related to the sintering conditions. The compact sintered at 1250 °C for 4 h contained pores of typically 20–40 μm in size (Figure 4b). The average pore size observed in the compact sintered at 1400 °C increased to 50 μm (Figure 4c). When the holding time was increased to 8 h, these small pores vanished and the number of pores decreased in both 1250 and 1400 °C sintered compacts (Figure 4e,f).

3.3. Close-Up Observation of Surface Contaminants

The surfaces of the sintered compacts, regardless of sintering temperature, show powdery morphology (Figure 5). To a much lesser content, some sinter-necks could be found. Shiny carbon particles were also visible on the sample surface. This is validated by the EDS analysis. Further evidence presented in the subsequent Section 3.4 suggests that this thick layer of surface contaminant is a compound containing Ti, C, N and O.

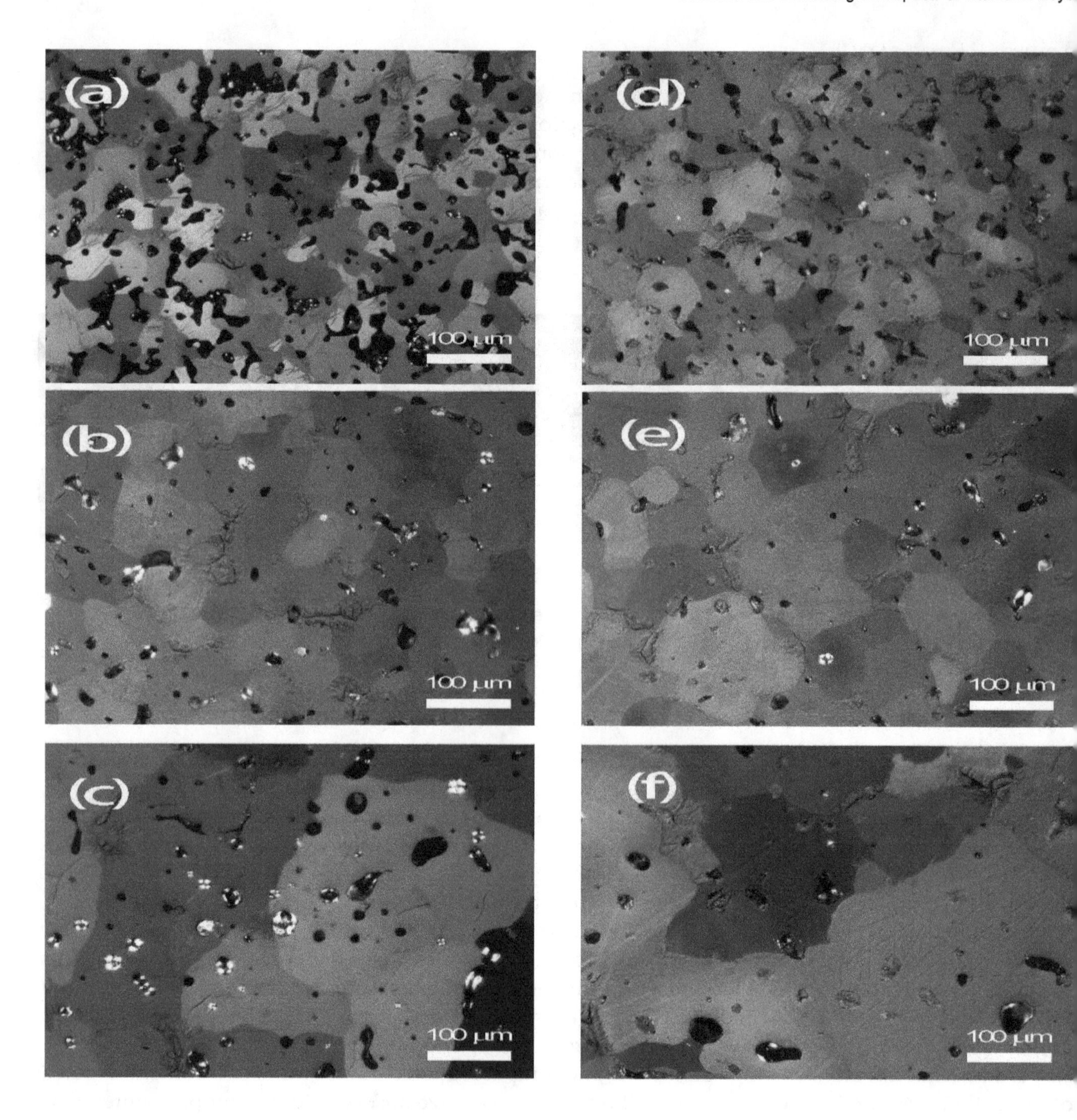

Figure 4. Microstructure of the cross sections of sintered small compacts at various sintering conditions. The micrographs were taken under polarized light microscopy. The compacts were sintered at various temperatures for different holding times: (**a**) 1100 °C/4 h; (**b**) 1250 °C/4 h; (**c**) 1400 °C/4 h; (**d**) 1100 °C/8 h; (**e**) 1250 °C/8 h; and (**f**) 1250 °C/8 h.

Figure 5. Surface morphologies of the small compacts sintered at 1250 °C (**a**) and 1400 °C (**b**), respectively. Sintering time was 4 h.

3.4. Phase Characterization of the Surface and the Interior

Figure 6 compares the XRD patterns taken from the surface and the cross section of the compact interior for samples sintered for 4 h. It is evident that the main phase constituents depend on the sintering temperature. At 1100 °C, three main phases are identified, i.e., Ti_2O (JCPDS # 73-1582), TiC (JCPDS # 73-0472) and $TiC_{0.2}N_{0.8}$ (JCPDS # 76-2484), while at 1250 and 1400 °C, only $Ti(C_{0.53}N_{0.32}O_{0.19})$ is present (JCPDS # 50-0681). In contrast, the only phase observed in the interior of samples sintered at all temperatures is α-Ti (JCPDS # 44-1294) without any other phases present.

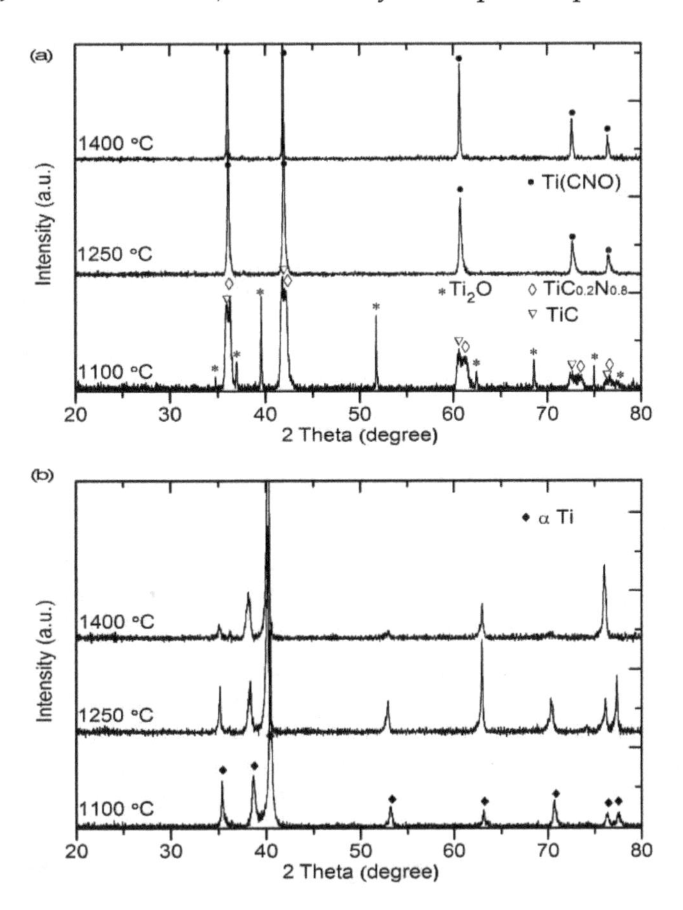

Figure 6. X-ray diffraction (XRD) patterns of the as-sintered small Ti compacts (**a**) on the surface and (**b**) cross-sections of compact interior. The compact was sintered for 4 h at 1100, 1250 and 1400 °C, respectively.

3.5. Mechanical Properties and Fractography

Typical stress–strain curves of the tensile specimens taken from three different positions in the sintered cylinder are shown in Figure 7. The contaminated surface scale was removed from the large cylinder. It is clear that specimen S3, which was taken 2 mm underneath the outermost surface, failed prior to yielding. Specimen S4 sliced 4 mm underneath the outermost surface attained a minor plastic strain of 1.5%. Specimens taken from further inside demonstrated much greater plastic strain—up to 10% for specimen S5. The average tensile properties of the specimens taken from the different positions are presented in Figure 7b. An average of ~6% plastic strain was obtained in the tensile specimens that were taken from the compact interior while a negligible plastic strain for the specimens was taken close to the contaminated surface. For the specimens taken from the interior, the yield strength (YS) was in the range of 385 to 500 MPa and ultimate tensile strength (UTS) varied from 470 to 580 MPa. In addition, the Young's moduli of all the specimens were similar (100 to 110 GPa), which correspond to the sintered densities (95.3% to 95.9%, Table 1).

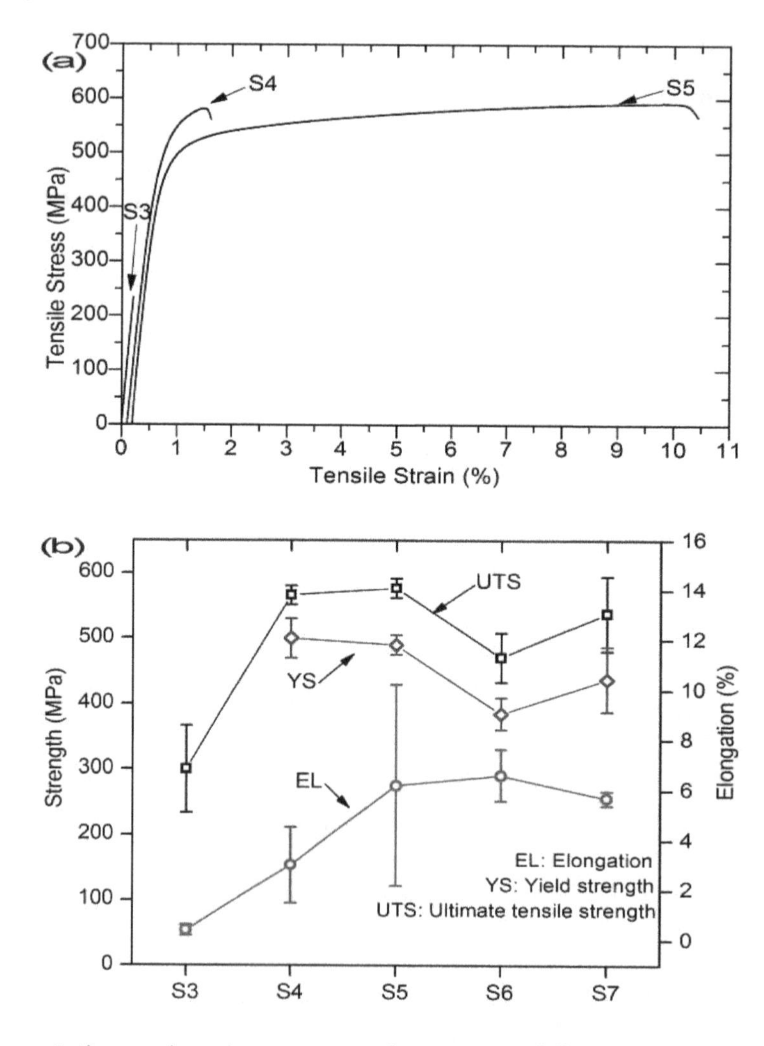

Figure 7. (**a**) Representative engineering stress-strain curves and (**b**) a summary of tensile mechanical properties for slices cut from different positions of the sintered cylinder.

Table 1. Sintered density in terms of percentage of theoretical density for slices taken from the large cylinder.

Slices	S3	S4	S5	S6	S7
Sintered density (%)	95.7 ± 0.1	95.6 ± 0.3	95.9 ± 1.0	95.3 ± 0.1	95.8 ± 0.2

Figure 8 shows the fracture surface of the specimens S4 and S5. Smooth facets and river patterns are visible in S4 (Figure 8a), which indicate a brittle fracture mode. On the other hand, many dimples can be found in specimen S5, indicating a ductile fracture mode. Similar fracture features were observed on the fractured surface of specimens S6 and S7.

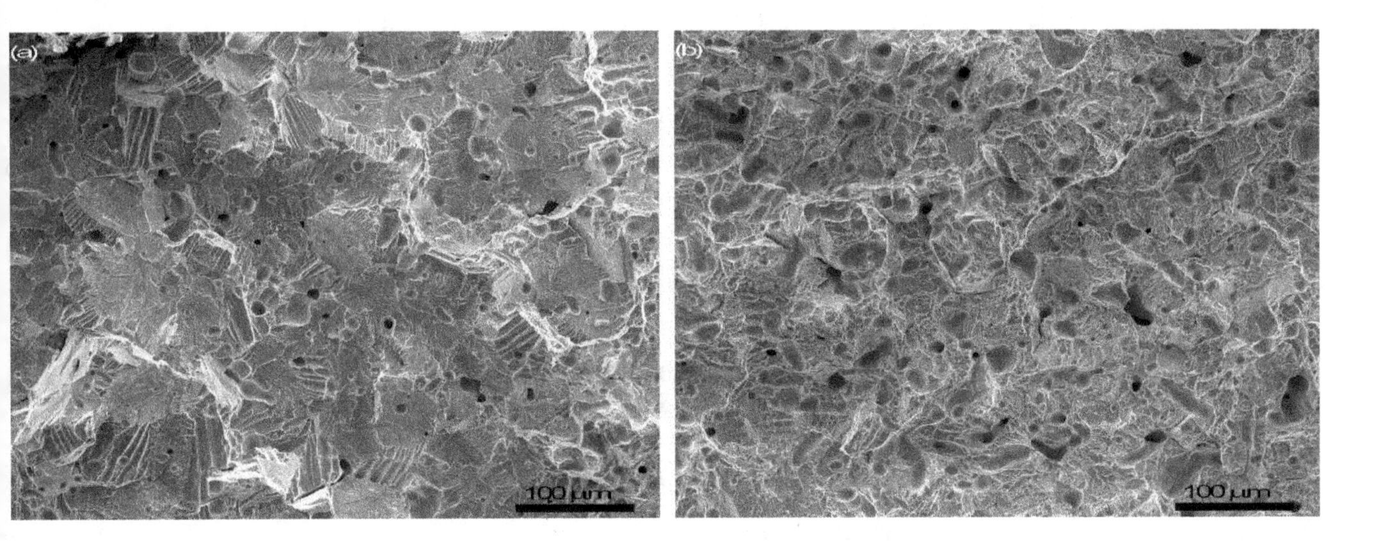

Figure 8. Fracture surfaces of tensile slices from sintered titanium samples: (**a**) S4 and (**b**) S5.

4. Discussion

4.1. Densification

As shown in Figure 2, the sintered density increases by 5%–10% with increasing of the holding time from 4 to 8 h for all three temperatures. This increase is expected because powder densification by sintering is accomplished through diffusion. For the same reason, increasing sintering temperature also results in a higher sintered density. However, as pointed out in Section 3.1, the measured sintered density does not reflect the densification level because of the presence of the porous contaminated scale. In order to evaluate the effect of temperature and holding time on densification, the contaminated surface scale should be removed before the density is measured. After removing the contaminated layer, the density of the sintered compacts is presented in Figure 9, as a function of sintering temperature and holding time. The favoring effect of increasing temperature or holding time is obvious. This is in accordance with the microstructural observations (Figure 4). Although low densities were observed in the small sintered compacts (Figure 9), the densification data of the large cylinder indicates a 96% relative density achieved after sintering at 1250 °C for 4 h (Table 2). It is noted that a sintered density ranging from 95% to 99% is common in PM Ti-6Al-4V if the blended elemental approach is used [24]. However, it needs to point out that a sintered density >98% is necessary for PM Ti products to achieve similar static property levels to wrought Ti. The use of fine particles, high compaction pressure and proper sintering conditions could readily achieve such a densification level. Such a level of densification in Ti powder products is acceptable for non-fatigue applications [4,14].

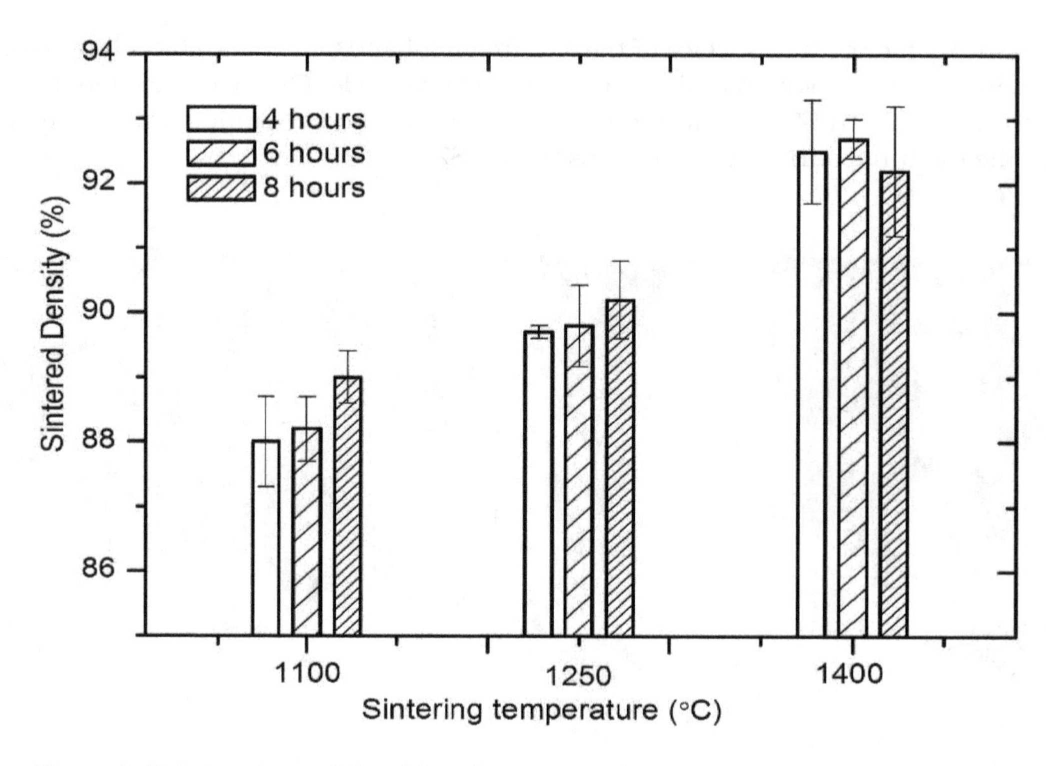

Figure 9. Relative sintered densities of compacts after removing contaminated scale.

Table 2. Densities and densification level for large cylinder prepared for tensile testing. CIP: Cold isostatic pressing.

Specimen	Green Density (600 MPa CIP, %)	Sintered Density (%)	Densification (%)
Cylinder	87.0	95.8.	69.2

4.2. The Formation of Contaminated Surface Layer and the Effect of Interstitials on Mechanical Properties

In contrast to the sintered compact interior where only α-Ti phase is present, the surface of the sintered compact is a complex compound Ti(CNO). The carbon is thought to arise from the graphite heating elements and insulation liner material within the furnace, while nitrogen may originate from backfilled argon and oxygen from the raw powder and argon. In order to further understand the formation of the surface scale resulting from the contamination of interstitials, a formation mechanism is schematically shown in Figure 10.

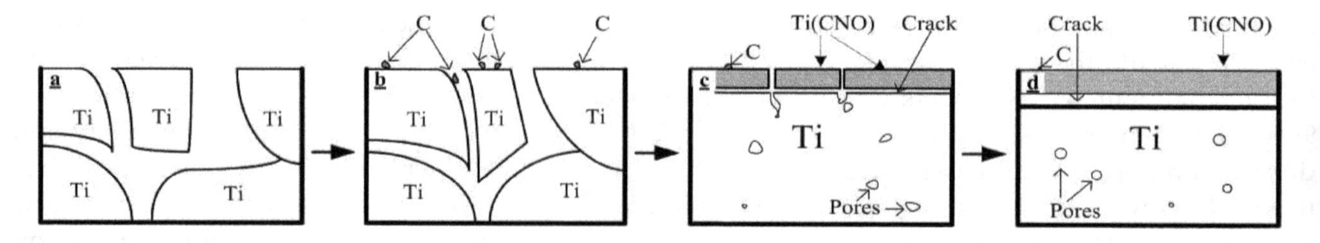

Figure 10. Schematic of the formation process of the contaminated scale. (**a**) Starting powder compact; (**b**) contaminants such as C are deposited on the Ti powder particles; (**c**) Ti reacts with the contaminants forming Ti(CNO). Cracking occurs as a result of different coefficients of thermal expansion; (**d**) larger cracks and more pores are formed at a higher temperature.

Upon heating, the residual carbon in the furnace or newly evaporated carbon from the graphite heating element would deposit on the compact surface (Figure 10b) and react with Ti. Meanwhile, the oxygen and nitrogen in the backfilled argon are also involved in this reaction and, consequently, a complicated Ti(CNO) solution of TiC, TiN and TiO forms (Figure 10c). The formed Ti(CNO) film is porous. During sintering, the powder compact starts to densify and contraction occurs, which poses a significant stress between the Ti(CNO) film and the underlying Ti. The interfacial stress can also be caused by the difference in the coefficients of thermal expansion between Ti(CNO) and Ti. Such interfacial stress is the cause of the many cracks and porous structure of the Ti(CNO) compound (Figure 10d).

The formation of the Ti(CNO) compound would prevent the interior from interstitial contamination if the film were dense and intact, which is supported by a study of Lefebvre et al., who observed that the O content in Ti solid solution remained constant although the thickness of the surface oxide layer increased [25]. Unfortunately, this does not seem true in our study. Only when the residual interstitials are completely consumed or the sintering procedure is terminated does the Ti(CNO) layer stop growing. This explains a constant thickness of Ti(CNO) of approximately 300 μm regardless of sintering temperature (for the same sintering time). In other words, the outermost layer on the titanium powder compact acts as a scavenger of interstitials.

The analyses of interstitials on the transverse of the sintered compact indicate that C and O diffuse inwards, as shown in Figure 11. The observed O content at position S2 to S7 (the center) ranges from 0.7% to 0.4%. The carbon content varies from 1000 to 300 ppm. In spite of a possible scavenging effect of the outermost layer on the Ti compact, it seems impossible to avoid the inward diffusion of the interstitials. Only the very interior of the compact appears to not have been contaminated—for instance, specimens S5 and S6 in the large cylindrical compact (Figure 11).

Apparently, the measured tensile properties largely depend on the interstitials, on top of porosity, present in the specimens. Interstitials, usually represented by oxygen, can increase strength but reduce ductility dramatically [26]. It has been extensively reported that oxygen levels ~0.3 wt % significantly reduce the ductility of PM C.P. Ti to 11% from 37% at 0.07% O (with the same sintered density) [7]. A similar finding was also reported in C.P. Ti produced by metal injection moulding and sintering [27]. At the same densification level, tensile specimen S3 does not show any plastic deformation and its UTS value is also much lower than other specimens. Such poor properties are ascribable to the extremely high C (~0.1 wt %) and O content (0.6 wt %). Specimen S4, which contains 0.5% O and 0.05% C, does show a small plastic strain (average 3%, Figure 7b). On the other hand, specimen S5 contains 0.4% O and 0.04% C and therefore demonstrates much greater plastic elongation. When the other impurities are taken into account, an oxygen equivalent $O_{eq} = O + 2N + 0.75C$ is usually used to reflect the effect of the interstitials [28,29]. An oxygen equivalent level higher than 0.4% for C.P. Ti is likely unacceptable. The same oxygen threshold applies for Ti-6Al-4V alloy [30]. In this present study, specimens S5 and S6 demonstrated the highest ductility, which is in accordance with their oxygen equivalents. Specimen S5 had an O_{eq} level of 0.47% while S4 had 0.60% oxygen equivalent. The ductility obtained in this study is in agreement with an early study by Kusaka et al. [27] who observed an elongation of ~7.0% in the as-sintered C.P. Ti that had a density of 99%. Such ductility is slightly lower than that specified in the newest ASTM B988-13 for Grade 5 PM (i.e., 9% elongation).

Although the high levels of impurities in the starting powder did not affect the densification of C.P. Ti, the resultant mechanical properties suffer significantly from the presence of these impurities. In order to mitigate the adverse effect of interstitials, not only reasonably low impurity levels are required, but also the sintering atmosphere should be controlled. It is expected that if some purifying media are placed surrounding the powder compact to be sintered, the contaminated surface could be minimized. Our additional experiments show that a contamination-free sintering could be accomplished by a proper wrapping of the Ti compact. In this separate study, we used HDH Ti powder (O: 0.29%, C: 0.04% and particle size <75 μm) as raw material and a Ti sponge as the scavenger to sinter Ti in the same graphite furnace with identical sintering parameters. The sintered Ti demonstrated an elongation of ~16%.

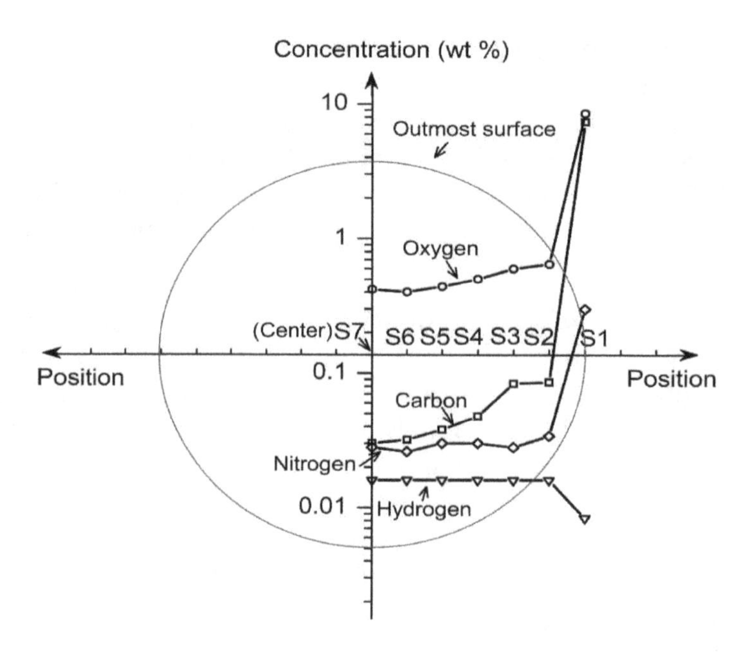

Figure 11. The distribution of interstitial contents at varied slices taken from different positions. Superimposed is the contour of the cylindrical compact.

The fractography results (Figure 8) suggest that the fracture mode is also strongly related to the interstitials, particularly O and C. At a low level of C and O, the sintered Ti-6Al-4V fails through microvoid nucleation, growth and coalescence, which results in a typical dimple rupture. However, at a high level of C and O, the sintered Ti-6Al-4V fails via cleavage and an intergranular fracture mode. Macroscopically, a fracture at 45° with respect to the tension axis was observed in a relatively low-interstitial specimen (S7, Figure 12b), whereas a flat (0°) rupture was observed in a high-interstitial specimen (S3, Figure 12a). The critical levels seem to be 0.05% C and 0.3% to 0.4% O for the ductile-to-brittle transition. This follows the general consensus that an oxygen level >0.3% is detrimental to the ductility of the PM Ti-6Al-4V [26]. The same speculation could be also applicable to other Ti alloys, while the critical values for C and O contents might be different. It is postulated that, at different interstitial levels, the dislocation mechanisms might be different as well. Further transmission electron microscopic (TEM) investigations are necessary. In addition to the interstitial contents, the morphology and size of the pores also affect the fracture behavior. A small pore size and spherical pore shape would favor the ductile fracture mode from a fracture mechanics point of view.

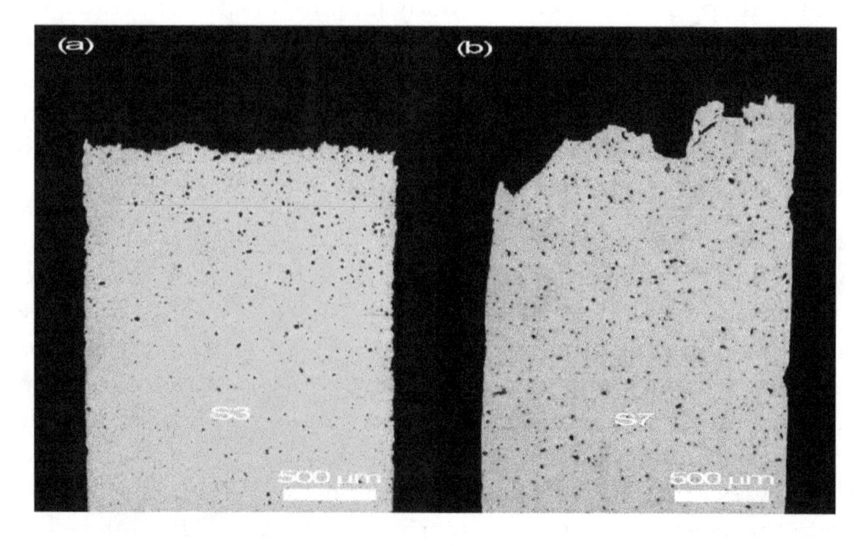

Figure 12. Macroscopic fracture surface of the tensile specimens (**a**) S3 and (**b**) S7.

5. Conclusions

This work investigated the sintering of commercially pure titanium powders in a graphite furnace backfilled with argon. A complex compound Ti(CNO) was formed on the surface with a thickness of 300–500 μm. The interiors of sintered specimens were pure single α-Ti phase. The formation of Ti(CNO) is caused by the residual interstitials in the furnace from the raw powder and in the argon gas used for backfilling. Densification was enhanced by increasing sintering temperature and/or holding time. Tensile properties were largely determined by the interstitial contents for samples where the sintered density reached at least 96%. With 0.04% C and 0.4% O contents, the sintered Ti shows an average of 6% elongation with yield strength of 385 to 490 MPa. On the other hand, for the sintered specimens containing higher C and O contents, the ductility was significantly reduced. The critical level of C and O content for ductile-to-brittle transition appeared to be ~0.04% C and 0.3% to 0.4% O. This work indicates that it is feasible to realize isothermal atmospheric sintering of C.P. HDH titanium powders in argon atmosphere in a graphite furnace and satisfactory mechanical properties can be achieved.

Acknowledgments: This work was financially supported by the Ministry of Business, Innovation and Employment (MBIE), New Zealand.

Author Contributions: Changzhou Yu conducted experimental work as part of his Ph.D project. He also analyzed the data. Peng Cao conceived and co-designed the experiments, analyzed the data and revised the manuscript. Mark Ian Jones discussed the results and revised the manuscript. Mark Ian Jones and Peng Cao were the supervisors of Changzhou Yu.

References

1. Donachie, J.M.J. *Titanium—A Technical Guide*, 2nd ed.; ASM International: Materials Park, OH, USA, 2000.
2. Froes, F.H. *Titanium—Physical Metallurgy, Processing, and Applications*; ASM International: Materials Park, OH, USA, 2015.
3. Zhang, L.-C.; Attar, H. Selective Laser Melting of Titanium Alloys and Titanium Matrix Composites for Biomedical Applications: A Review. *Adv. Eng. Mater.* **2016**, *18*, 463–475. [CrossRef]
4. Abkowitz, S.; Rowell, D. Superior Fatigue Properties for Blended Elemental P/M Ti-6Al-4V. *JOM* **1986**, *38*, 36–39. [CrossRef]
5. Henriques, V.A.R.; Galvani, E.T.; Petroni, S.L.G.; Paula, M.S.M.; Lemos, T.G. Production of Ti-13Nb-13Zr alloy for surgical implants by powder metallurgy. *J. Mater. Sci.* **2010**, *45*, 5844–5850. [CrossRef]
6. Whittaker, P. Dynamet Technology Approved by Boeing as Qualified Supplier for Powder Metallurgy Titanium Alloy Products. Available online: http://www.pm-review.com/dynamet-technology-approved-by-boeing-as-qualified-supplier-for-powder-metallurgy-titanium-alloy-products/ (accessed on 22 January 2017).
7. Qian, M. Cold compaction and sintering of titanium and its alloys for near-net-shape or preform fabrication. *Int. J. Powder Metall.* **2010**, *46*, 29–44.
8. Bolzoni, L.; Esteban, P.G.; Ruiz-Navas, E.M.; Gordo, E. Mechanical behaviour of pressed and sintered titanium alloys obtained from prealloyed and blended elemental powders. *J. Mech. Behav. Biomed. Mater.* **2012**, *14*, 29–38. [CrossRef] [PubMed]
9. Fujita, T.; Ogawa, A.; Ouchi, C.; Tajima, H. Microstructure and properties of titanium alloy produced in the newly developed blended elemental powder metallurgy process. *Mater. Sci. Eng. A* **1996**, *213*, 148–153. [CrossRef]
10. Yang, Y.F.; Luo, S.D.; Bettles, C.J.; Schaffer, G.B.; Qian, M. The effect of Si additions on the sintering and sintered microstructure and mechanical properties of Ti-3Ni alloy. *Mater. Sci. Eng. A* **2011**, *528*, 7381–7387. [CrossRef]
11. Yang, Y.F.; Luo, S.D.; Schaffer, G.B.; Qian, M. The Sintering, Sintered Microstructure and Mechanical Properties of Ti-Fe-Si Alloys. *Metall. Mater. Trans. A* **2012**, *43*, 4896–4906. [CrossRef]
12. Xu, Q.; Gabbitas, B.; Matthews, S.; Zhang, D. The development of porous titanium products using slip casting. *J. Mater. Process. Technol.* **2013** *213*, 1440–1446. [CrossRef]

13. Kroll, W. Malleable Alloys of Titanium. *Z. Metall.* **1937**, *29*, 189–192.
14. Qian, M.; Schaffer, G.B.; Bettles, C.J. *Sintering of Titanium and its Alloys, in Sintering of Advanced Materials: Fundamentals and Processes*; Fang, Z.Z., Ed.; Woodhead Publishing: Philadelphia, PA, USA, 2010; pp. 324–355
15. Dean, R.S.; Wartman, F.S.; Hayes, E.T. Ductile Titanium—Its Fabrication and Physical Properties. *Trans. Am. Inst. Mining Met. Eng.* **1946**, *166*, 381–389.
16. Limberg, W.; Ebel, T.; Pyczak, F.; Schimansky, F.P. Influence of the sintering atmosphere on the tensile properties of MIM-processed Ti 45Al 5Nb 0.2B 0.2C. *Mater. Sci. Eng. A* **2012**, *552*, 323–329. [CrossRef]
17. Arensburger, D.S.; Pugin, V.S.; Fedorchenko, I.M. Properties of electrolytic and reduced titanium powders and sinterability of porous compacts from such powders. *Sov. Powder Metall. Met. Ceram.* **1968**, *7*, 362–367 [CrossRef]
18. Heaney, D.F.; German, R.M. *Proceedings of PM 2004 Powder Metallurgy World Congress*; European Powder Metallurgy Association: Vienna, Austria, 2004; pp. 222–227.
19. Kanto Yakin Kogyo Co. Available online: http://www.k-y-k.co.jp/en/product01.html (accessed on 22 January 2017).
20. Metal Powder Reports. Furnace Masters Difficult Metals. *Met. Powder Rep.* **2004**, *59*, 12.
21. Yu, C.; Cao, P.; Jones, M.I. Effect of Contaminants on Sintering of Ti and Ti-6Al-4V Alloy Powders in an Argon-Back-Filled Graphite Furnace. *Key Eng. Mater.* **2012**, *520*, 139–144. [CrossRef]
22. Yu, C.Z.; Jones, M.I. Investigation of chloride impurities in hydrogenated–dehydrogenated Kroll processed titanium powders. *Powder Metall.* **2013**, *56*, 304–309. [CrossRef]
23. German, R.M. *Sintering Theory and Practice*; Wiley: New York, NY, USA, 1996.
24. Hausner, H.H.; Smith, G.D.; Antes, H.W. *Modern Development in Powder Metallurgy, Volume 13: Ferrous and Nonferrrous Materials*; Metal Powder Industries Federation: Prinston, NJ, USA, 1981; pp. 537–549.
25. Lefebvre, L.P.; Baril, E. Effect of Oxygen Concentration and Distribution on the Compression Properties on Titanium Foams. *Adv. Eng. Mater.* **2008**, *10*, 868–876. [CrossRef]
26. Conrad, H. Effect of interstitial solutes on the strength and ductility of titanium. *Prog. Mater Sci.* **1981**, *26*, 123–403. [CrossRef]
27. Kusaka, K.; Kohno, T.; Kondo, T.; Horata, A. Tensile Behavior of Sintered Titanium by MIM Process. *J. Jpn. Soc. Powder Powder Metall.* **1995**, *42*, 383–387. [CrossRef]
28. Conrad, H. The rate controlling mechanism during yielding and flow of α-titanium at temperatures below 0.4 TM. *Acta Metall.* **1966**, *14*, 1631–1633. [CrossRef]
29. Okazaki, K.; Conrad, H. Effects of interstitial content and grain size on the strength of titanium at low temperatures. *Acta Metall.* **1973**, *21*, 1117–1129. [CrossRef]
30. Wang, H.; Fang, Z.Z.; Sun, P. A critical review of mechanical properties of powder metallurgy titanium. *Int. J. Powder Metall.* **2010**, *46*, 45–57.

9

The Microstructure Evolution, Mechanical Properties and Densification Mechanism of TiAl-Based Alloys Prepared by Spark Plasma Sintering

Dongjun Wang [1,2,*], Hao Yuan [3] and Jianming Qiang [3]

1 National Key Laboratory for Precision Hot Processing of Metals, Harbin Institute of Technology, Harbin 150001, China
2 Key Laboratory of Micro-Systems and Micro-Structures Manufacturing, Ministry of Education, Harbin 150001, China
3 School of Materials Science and Engineering, Harbin Institute of Technology, Harbin 150001, China; yh18345168977@gmail.com (H.Y.); q2564108815@gmail.com (J.Q.)
* Correspondence: dongjunwang@hit.edu.cn

Academic Editor: Mark T. Whittaker

Abstract: The microstructure evolution and mechanical properties of a Ti-Al-Cr-Nb alloy prepared by spark plasma sintering (SPS) at different temperatures and stresses were investigated in detail. Sintering temperature plays a key role in the densification process and phase transformation, which determines the microstructure. The mechanical properties of the sintered alloys depend on the microstructure caused by the sintering. Furthermore, the densification process and mechanism of TiAl-based metallic powders during SPS were studied based on experimental results and theoretical analysis, the results of which will help fabricate these kinds of intermetallic alloys using a powder metallurgy technique and accelerate their industrial applications.

Keywords: intermetallics; aerospace; powder metallurgy; microstructure evolution; mechanical property; densification

1. Introduction

Alloys based on the intermetallic phase γ-TiAl are increasingly used as potential replacements for nickel-based superalloys in different application fields, e.g., turbine blades, space vehicles, and stationary turbines [1,2]. TiAl-based alloys have attracted this attention due to their low density (about 4 g/cm^3), high yield strength at high temperature, good oxidation resistance, and corrosion resistance [3,4].

TiAl-based alloys can usually be produced by conventional casting or ingot metallurgy, etc. However, microstructural defects such as porosity, coarse grain and composition heterogeneity, and low material utilization ratio of TiAl-based alloys hinder their actual engineering applications. The mechanical properties of TiAl-based alloys mainly depend on their microstructure [5,6], and thus the alloys are regularly forcibly treated using hot isostatic processing (HIP), or hot processing [7,8] to eliminate porosity or refine grains, which can improve their performance, but inversely gives rise to a longer manufacturing duration and higher cost of investment.

In recent years, powder metallurgy (PM) has been considered as an alternative processing technique for the preparation of TiAl-based alloys as the near-net-shape forming method [9,10]. Furthermore, TiAl-based metallic powders with fine grains and homogeneous composition can be obtained during a gas atomization process. Afterwards, the atomized powders usually consolidate into bulk by hot pressing, or the HIP method [11]. Particularly, spark plasma sintering (SPS) can satisfactorily compact powders through high intensity pulsed direct current and stress, and it is currently attracting the attention of the industrial field due to its advantages such as rapidity, cheapness, and simplicity [12–14]. Lin et al. [15] fabricated a high Nb-containing Ti-45Al-8.5Nb-(W, B, Y) alloy using the SPS and HIP methods. Couret et al. [16] obtained a near-lamellar Ti-48Al-2W-0.02B alloy and the effects of B addition were investigated. Liu et al. [17] conducted spark plasma sintering of a beta phase-containing Ti-44Al-3Nb-1Mo-1V-0.2Y alloy with potentially good hot deformability To fabricate TiAl-based alloys with considerable properties through the PM route, it is necessary to carry out a densification process by which the alloys with high density are obtained. Moreover, the sintering parameters can affect the phase morphology and thus influence the mechanical properties. To attempt the near-net fabrication of TiAl alloys and parts utilized at a temperature of ~800 °C, it was of importance to understand the relationship between the sintering process, densification, microstructure, and properties of TiAl-based alloys during sintering, which can provide and supply more information about this alloy using the PM technique.

In this paper, based on the study of the microstructure evolution and mechanical properties of TiAl-based alloys prepared by SPS, the effect of parameters on the densification process was analyzed, leading to a more basic understanding of intermetallic alloys for fabrication using the PM technique.

2. Materials and Methods

Gas atomized TiAl powders (Ti-46.5Al-2.15Cr-1.90 Nb, atomic percent) with sizes between 20 and 80 μm were prepared and sintered using 3.20-MK-V SPS equipment for 7 min at different temperatures (900, 1050, 1100, 1150, and 1250 °C) under a pressure of 50 MPa. For comparison, more samples were sintered under pressures of 10 and 30 MPa at 1150 °C, as well as under pressure of 50 MPa at 1000 °C. The sintering temperatures were measured with a pyrometer on the external surface of the graphite molds. The pressure and temperature started simultaneously at the time of zero and the pressure was kept constant during sintering. The heating rate was initially 120 K/min, and was reduced to 20 K/min for the last 100 K to mitigate temperature overshooting. The temperatures given in this paper were the monitored temperatures. After holding at the sintering temperature, the heating current and the pressure were released, followed by sample cooling in the SPS chamber. Two typical samples were fabricated with cylindrical shapes of Φ 20 mm × 10 mm and Φ 45 mm × 15 mm, respectively. The structural characteristics were investigated by X-ray diffraction (XRD, D/Max-RA diffractometer (Rigaku Corporation, Tokyo, Japan), operated with Cu Kα), differential thermal analysis (DTA, NETZSCH STA 449C, NETZSCH company, Selb, Germany), and scanning electron microscopy (SEM, Quanta200FEG, FEI company, Hillsboro, OR, USA), equipped with energy dispersive spectroscopy (EDS, FEI company, Hillsboro, OR, USA). For the SEM analysis, the back-scattered electron (BSE) detector (FEI company, Hillsboro, OR, USA) was used. The densities of the sintered samples were determined using the "Archimedes" method. Micro-hardness was measured using a Matsuzawa SeikiMHT-1 micro-hardness tester ((MATSUZAWA SEIKI Co. LTD, Tokyo, Japan) under a load of 50 g. The mechanical responses with a dimension of Φ 3 mm × 4.5 mm (aspect ratio of 1.5) were evaluated by the quasi-static compression test at room temperature with a strain rate of 5×10^{-4} s^{-1}. The two sides of each specimen were carefully ground and polished until they were parallel to each other in order to mitigate the effect of friction during tests. The fracture surfaces of the samples were also examined by SEM (secondary electron detector).

3. Results and Discussion

3.1. Microstructure Evolution

Figure 1 shows the XRD patterns of the TiAl powders and the samples sintered at different temperatures. For the atomized powders, they were mainly composed of a single α_2 phase due to the rapid cooling rate. Similar phase constitution has been reported in small size TiAl powders prepared by gas atomization in Reference [18]. These results also suggested that the atomized powders were in a non-equilibrium state caused by rapid solidification. The diffraction patterns of samples after sintering were similar. As can be seen, the sharp diffraction peaks belonging to TiAl and Ti_3Al confirmed that the TiAl-based alloys after SPS contained substantive amounts of γ (e.g., dark phase in Figure 2D,E), and α_2 phases (e.g., bright phase in Figure 2D,E). Moreover, the bulk TiAl-based alloy with a dimension of Φ 45 mm × 15 mm and cylinder-shaped part with the dimension of Φ 30 (internal diameter 10) mm × 30 mm prepared by SPS are shown in the inset of Figure 1. One can see that the spark plasma sintered TiAl-based alloy exhibited a shining metallic luster in appearance.

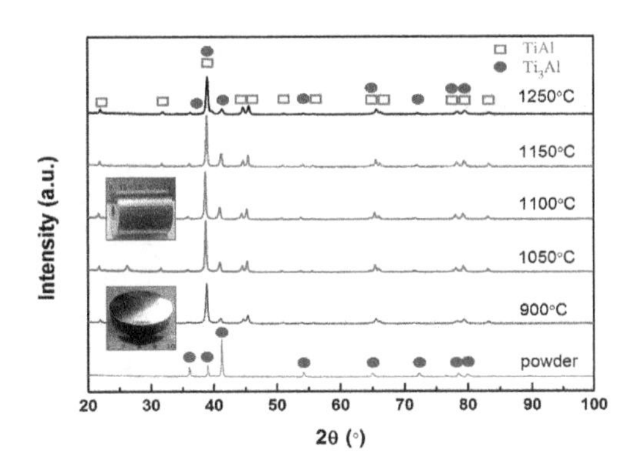

Figure 1. X-ray diffraction patterns of the samples.

The SEM observations of the powders and samples sintered at different temperatures are shown in Figure 2. As can be seen, the powder exhibited a dendritic-like microstructure containing a large amount of out-of-equilibrium α_2 phase (Figure 2A). Based on EDS analysis, the composition of metallic powders was $Ti_{50.46}Al_{45.61}Cr_{1.99}Nb_{1.94}$ (at %), which was close to the nominal composition of this alloy.

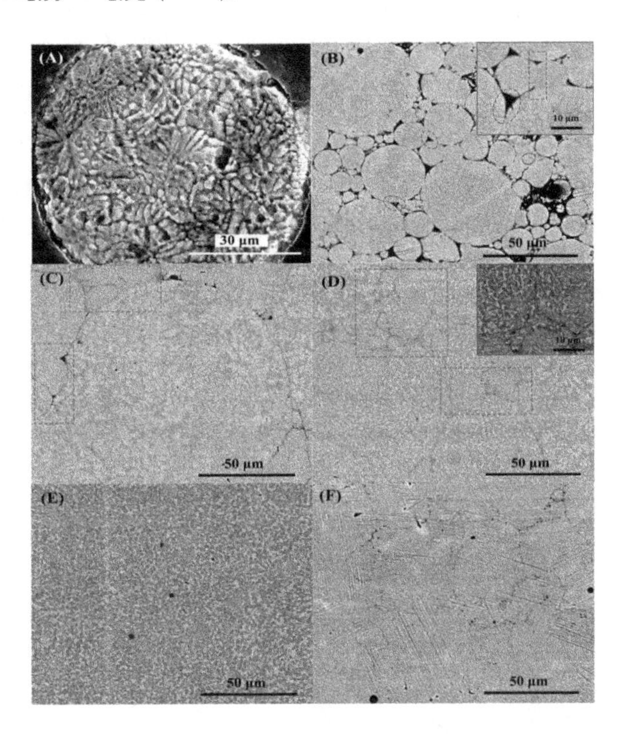

Figure 2. Scanning electron microscopy (SEM) images of the atomized powder (**A**) and the sintered samples at (**B**) 900 °C; (**C**) 1050 °C; (**D**) 1100 °C; (**E**) 1150 °C; and (**F**) 1250 °C.

For the sample consolidated at 900 °C, the microstructure had significant porosity, especially at triple contact areas of powder particles (blue dashed region of Figure 2B). It was also noted that the plastic deformation of some small powders occurred at this temperature (red dashed region of inset of

Figure 2B), suggesting that densification initially took place. When the sintering temperature increased to 1050 °C (Figure 2C), a few pores could still be seen, although a more compact microstructure was obtained. Moreover, more deformed features of small powders were observed (red dashed region in Figure 2C). With an increase in temperature to 1100 °C (Figure 2D), macroscopic pores disappeared post-sintering. However, the grain boundaries (GBs) were still visible at the contact areas among the deformed powders (e.g., red dashed region), which indicates that plastic deformation plays a key role in the densification process. In addition, compared with the alloys at lower sintering temperatures (Figure 2B,C), the non-equilibrium dendritic microstructure (Figure 2A) of the powders disappeared for the sample sintered at 1100 °C (inset of Figure 2D). With further increases in temperature (Figure 2E), one could see a highly dense microstructure of the alloy sintered at 1150 °C without GBs observed, revealing a high sintering density and sufficient densification at this temperature. It is also of importance to note that a double-phase microstructure was obtained. From a chemical composition perspective, the composition of the bright α_2 phase was $Ti_{53.98}Al_{41.85}Cr_{2.24}Nb_{1.94}$ (at %, relatively rich in Ti), and that of the dark γ phase was $Ti_{47.48}Al_{48.76}Cr_{1.90}Nb_{1.86}$ (at %, relatively rich in Al). Aside from the compact microstructure, the sample sintered at 1250 °C exhibited lamella morphology (Figure 2F). As shown, this lamella microstructure contained homogeneous lamellar colonies α_2/γ. Moreover, the composition of these lamellas was $Ti_{49.63}Al_{46.40}Cr_{2.15}Nb_{1.82}$ (at %) based on EDS, which is close to the nominal composition of this alloy.

To verify the phase transformation point of TiAl-based powders, a DTA analysis was conducted, and the experimental curve is shown in Figure 3. In this curve, two endothermal peaks based on phase transformation were observed. Based on this, the eutectoid temperature (T_e) and the α transus temperature (T_α) of the powders were estimated to be approximately 1260 and 1315 °C, respectively. During the SPS, the temperature can induce phase transformation and thus affect the microstructure and densification. For a given experiment, the actual temperature could even be 160 °C higher than the monitored SPS temperature [19]. When the sintering temperature was 1150 °C, it was deduced that the actual temperature of the powders could be higher than 1260 °C (T_e). In this case, the sample was composed of γ phase and α phase ($\alpha + \gamma$ phase region) at this sintering temperature. Upon cooling, the α phase transformed into an ordered α_2 phase, while the γ phase remained and a double-phase microstructure formed for the sintered sample (Figure 2E). When the powders were sintered at 1250 °C, the actual temperature could be above T_α, thus the microstructure consisted of α grains in this situation. During cooling, the formation of a lamellar microstructure (Figure 2F) took place following the evolution of $\alpha \rightarrow \alpha + \gamma \rightarrow \alpha_2 + \gamma$. Meanwhile, it was also of interest to note that the dendritic-like microstructure of rapid atomized powders was not seen after densification. To understand this, it is well documented that recrystallization can occur dynamically during densification of the powders [12]. Since a clear tendency to recrystallize can occur in the deformed zones [20], the deformation of metallic powders (Figure 2B–D) during densification will result in recrystallization, and in turn an equilibrium microstructure, due to a large amount of stored deformation energy.

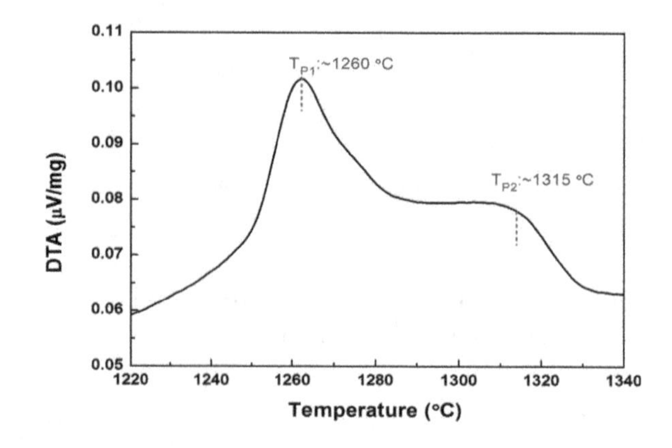

Figure 3. Differential thermal analysis curve of the powder sample.

3.2. *Mechanical Property*

The density and micro-hardness of the samples sintered at different temperatures were investigated and are shown in Table 1. One can clearly see that the density and hardness of the sample with a lower sintering temperature exhibited relatively smaller values due to numerous pores after sintering, e.g., 900 °C. When the sintering temperature increased to 1050, 1100, 1150, and 1250 °C, the density increased sharply and then nearly kept at constant. Furthermore, the hardness of the samples sintered at higher temperatures was also similar, which indicates that hardness is mainly dependent on density.

Table 1. Mechanical properties of TiAl-based alloys sintered at different temperatures.

Temperature (°C)	**900**	**1050**	**1100**	**1150**	**1250**
Density (g/cm^3)	3.392	3.944	3.966	3.967	3.965
Hardness (HV)	278.6	413.0	420.0	417.0	430.0

The relationship between fracture true strength, plastic strain, and sintering temperature is shown in Figure 4, as are the true stress-true strain curves of the sintered samples. When the sintering temperature was low (900 °C), the sample broke with no plastic strain and a very low fracture strength of ~350 MPa. Many pores in this sample (Figure 2B) resulted in its weak compressive response. When the sintering temperature rose to 1050 and 1100 °C, the densification process gradually took place and more compact microstructures were achieved (Figure 2C,D). Therefore, the mechanical performances of these two samples were significantly improved compared with samples sintered at lower temperatures, e.g., the fracture true strength and plastic strain of the samples sintered at 1050 and 1100 °C were 1795 MPa and 24.8%, and 1754 MPa and 26.2%, respectively. As shown, the sample sintered at 1150 °C with high density (Figure 2E) had optimal compressive properties at room temperature. The fracture true strength was 1820 MPa and the plastic true strain could also be as high as 32.6%. Although the density was like that of the sample sintered at 1150 °C, both the fracture strength and the plastic strain of the sample sintered at 1250 °C dramatically decreased. Based on the above composition results, it was noted that the change of composition before and after sintering was slight, suggesting that a homogeneous composition was obtained during SPS. Further to composition, it is well known that the mechanical properties of alloys are determined by the microstructure, namely the synergetic effect of density and phase transformation for the sintered TiAl-based alloys. For the sample sintered at 1150 °C, the high strength and considerable plasticity was attributed to the high density and small grain size. However, the poor strength and limited plasticity of the lamellar microstructure (1250 °C) was caused by the lack of texture [19]. Therefore, the dislocations could propagate more easily and thus initiate the crack, which eventually led to the rapid failure of the sample. It can be concluded from the abovementioned results that the temperature of SPS has two main roles: (1) giving rise to densification; (2) changing the microstructure through phase transformation. Thus, to improve the mechanical properties of TiAl-based alloys, the PM technique can provide a simplified route, i.e., the combined processing for fabrication and microstructural optimization for the required composition.

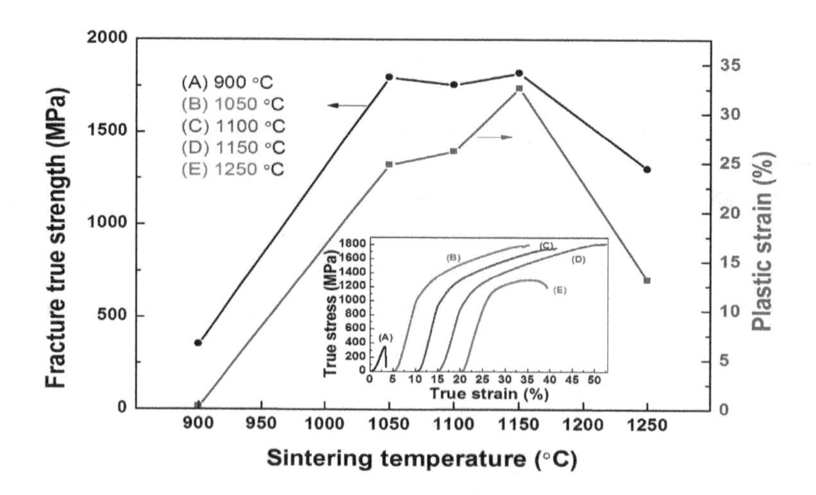

Figure 4. Room temperature compressive properties of the sintered TiAl alloys. The true stress-strain curves have been off-set for better visualization.

The typical fracture morphologies of the sintered samples are presented in Figure 5. Based on Figure 5A, one can see melt-like features at the triple junctions of powder particles, suggesting that spark plasma sintering promotes integration among the contact areas of the particles. Nevertheless, it was evident that a few globate powder particles were removed from the surface by external loading, which agrees with the high porosity and poor mechanical properties of the sample sintered at 900 °C (Figure 2B and Table 1). For the sample sintered at 1150 °C (Figure 5B), intergranular fracture characteristics were observed. This fracture morphology was like that of the as-cast TiAl-based alloys, indicating its high density and good mechanical properties. When the sintering temperature increased to 1250 °C, cracks were seen on the fracture surface. In particular, the lamella-like imprints dominated the fracture morphology and further confirmed the lamellar microstructure of the sample sintered at this temperature.

Figure 5. Fracture morphologies of the sintered samples: (**A**) 900 °C; (**B**) 1150 °C; (**C**) 1250 °C.

3.3. Densification Process

To better understand the effect of sintering parameters on the densification of TiAl-based metallic powders during SPS, a model suggested by Bernard-Granger and Guizard [21] was used to analyze the sintering process. In this model, the flow stress for high-temperature deformation of the alloy is described as a function of strain rate ($\dot{\varepsilon}$) and temperature (T) by Equation (1) [22].

$$\dot{\varepsilon} = \frac{\mathrm{d}\varepsilon}{\mathrm{d}t} = \mathrm{A}\frac{DG_0 b}{kT}\left(\frac{b}{d}\right)^p\left(\frac{\sigma}{G_0}\right)^n \tag{1}$$

where $\dot{\varepsilon}$ is the strain rate; A is the constant; D is the diffusion coefficient; G_0 is the shear modulus; b is the Burgers vector; k is Boltzmann's constant; T is temperature; d is grain size; σ is the macroscopic applied stress; p is the inverse grain size exponent; n is the stress exponent; and t is time.

During sintering, the strain rate is compared to the densification rate as per Equation (2) [23].

$$\dot{\varepsilon} = \frac{1}{\rho}\frac{\mathrm{d}\rho}{\mathrm{d}t} \tag{2}$$

where ρ is the density. By taking the integral transformation of Equation (2), one can obtain Equation (3):

$$\varepsilon = \ln \rho \tag{3}$$

As per Equations (1)–(3), there are three main factors that determine the densification (ρ) during sintering: the diffusion coefficient D, the temperature T, and the applied stress σ. It is well documented that the diffusion coefficient D is mainly dominated by temperature; namely, the higher the temperature, the larger the diffusion coefficient. Therefore, when the temperature rises, the macroscopic sintering density increases and is maintained at nearly constant (Table 1).

In addition, based on the experimental data, the relationships between temperature, punch displacement, and sintering time were recorded, and a typical result of the sample sintered at 1150 °C is shown in Figure 6. During SPS, sintering can be divided into different stages to analyze the densification [24,25]. As shown in Figure 6, there were two clear sintering stages in our study, namely curve A-B and curve B-C-D for the displacement data. For the first stage (curve A-B), the punch displacement decreased due to the powder thermal expansion caused by the increase in temperature. The second stage (curve B-C-D) was important for densification, and the density increased sharply in accordance with the increase in temperature and punch displacement.

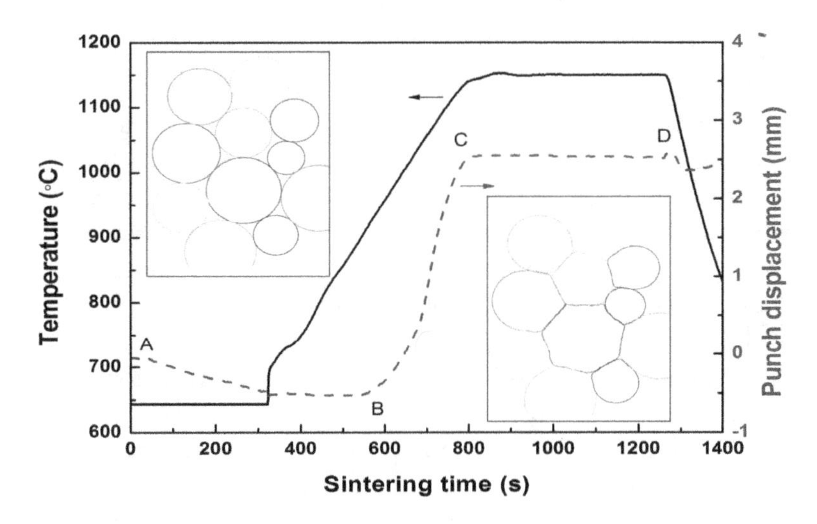

Figure 6. Temperature and punch displacement curves versus time for sintering at 1150 °C.

It was noted that sintering involved several mechanisms microscopically operating simultaneously over these stages. Plastic deformation, e.g., in the form of dislocation movement, can play a role in densification [26]. During sintering, the loose powders initially encounter each other through the combined effect of temperature and sintering stress (upper left inset of Figure 6). When sintered at 900 °C, the particles were in contact and were forced to deform, followed by the formation of sintering necks (inset of Figure 2B). Due to the applied pressure, densification took place through the plastic deformation of the powder particles and the dislocation emitted atoms as it moved close to the neck. However, this densification was not sufficient, and many large pores were visible (Figure 2B). In particular, the pores at the contacts of the powder particles exhibited sharp cusps (blue dashed region of Figure 2B) at this temperature. It is well known that the vacancy concentration that can act as an atomic diffusion path under a curved surface, depends on the curvature of the two perpendicular radii of curvature for the surface. Based on the two-sphere sintering model, the particle surface is convex and the sintering bond is concave, thus there will be a vacancy concentration gradient between these two [27]. Moreover, the solid surface energy—due to a concave curvature at the neck—also generates capillary stress as a driving force for atom diffusion. Therefore, the result is that vacancy flows away from the neck and atomic diffusion moves into the neck, resulting in the blunting of sharp cusps for the pores. Nevertheless, the pore cusps for the sample sintered at 900 °C (blue dashed region of Figure 2B) suggested that the diffusion mechanism played a minor role during this initial stage of densification and the deformation was considered as a dominant mechanism in this stage. By increasing the sintering temperature (1050–1100 °C), densification could be further associated with more plastic deformation of the particles (red dashed regions in Figure 2C,D) due to lower yield stress with increased temperatures. It was obvious that some small rounded powders deformed into elliptic shapes or even irregular shapes to fill the vacancies. The plastic deformation led to the flattening of contact areas and the reduction of porosity (Figure 2C,D), as demonstrated in the lower right inset of Figure 6. For the sample sintered at 1150 °C (Figure 2E), sufficient densification occurred and a nearly full dense microstructure was obtained.

In contrast, at the areas of the powder contacts during SPS there was current through the powder particles. Coupled with the plastic deformation discussed above, electro-thermal and heat transfer can occur at the contact interface when axial pressure applied. It is well known that motion such as thermal diffusion caused by heat transfer can promote mass transfer driven by Gibbs-Thomson driving forces [20], which is also responsible for interface bonding and causes densification from a microscopic perspective. After neck formation through deformation in the initial stage (900 °C), the grain boundaries formed within the neck (Figures 7a and 2B–D) between individual particles as random grain contacts led to misaligned crystals [28]. When the sintering temperature (such as 1000 °C) increased, the diffusion mechanism significantly affected densification in comparison with lower temperatures. The evident sharp cusps (Figure 2B) of the pores were blunted and the pore structure became rounded (Figure 7b). In addition, blunted neck morphology between adjacent particles was observed and typical results are shown in the red dashed regions in Figure 7b. Similar neck features were also seen in the samples sintered at 1050 and 1100 °C. Based on the above discussions, our results indicate that the diffusion mechanism occurred at a relatively higher temperature and the atoms diffusion (caused by capillary stress due to concave curvature) was performed to remove this curvature gradient. During diffusion, the collaborated mass moved from the solid particle to deposit on the pores, i.e., the atoms moved along the particle surfaces (surface diffusion), along grain boundaries (grain boundary diffusion), and through the lattice interior (volume diffusion). A schematic illustration of the microscopic diffusion paths and interface bonding during the densification of TiAl powders is shown in Figure 7a. Although the surface diffusion produced neck growth, it did not lead to a change in particle spacing [29], that is, no densification occurred since the mass flow originated and terminated at the particle surface. Densification took place only by bulk transport, as the mass responsible for growing the sintering neck comes from inside the powder particles, for example, from grain boundary diffusion and volume diffusion [30]. As seen in Figure 7a, the crystalline solid powders joined at the interparticle neck with a misalignment of crystal planes, resulting in a grain boundary where defective atomic bonding enabled rapid atomic diffusion and thus contributed to the densification [31]. Furthermore, effective volume diffusion involved the motion of vacancies along the lattice paths and a counter flow of atoms into the pores, which required relatively higher activation energy. Compared with volume diffusion activated at higher temperatures, there existed a sufficient grain boundary area due to the small particle size of this work (Figures 2B–D and 7b). Therefore, the grain boundary diffusion was considered as the dominant diffusion densification mechanism. Based on the above analysis, the sintering necks would form and gradually grow among adjacent particles through the synergetic mechanism of deformation and thermal diffusion. Thus, the pores were filled and the microstructure became compacted.

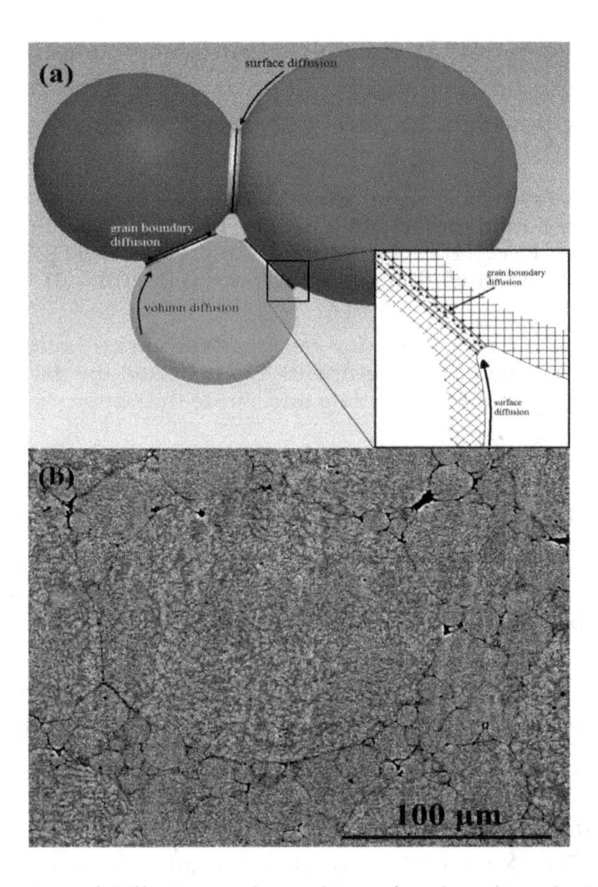

Figure 7. Schematic illustration of diffusion paths and interface bonding during spark plasma sintering (SPS) densification of TiAl powders (**a**); and the microstructure of the sample sintered at 1000 °C (**b**).

Besides temperature, it was also noted from Equations (1)–(3) that the applied stress could affect densification. To verify the influence of sintering stress, more samples were sintered at 1150 °C under different applied stresses, as shown in Table 2. As seen, a higher sintering density was obtained by applying larger sintering stress. Thus, the imposition of higher sintering pressure on powder-powder interfaces led to more severe deformation and accelerated the mass transport of the particles, which could promote the densification as per the abovementioned discussions.

Table 2. Sintering densities of TiAl-based alloys under different stresses at 1150 °C.

Stress (MPa)	**10**	**30**	**50**
Density (g/cm^3)	3.947	3.961	3.967

4. Conclusions

Sintering temperature mainly determines the densification and phase transformation of TiAl-based metallic powders during SPS. By increasing the temperature, the density of sintered alloys increased and then stayed nearly constant. For the alloys sintered at 1150 and 1250 °C, the microstructures exhibited double-phase and lamellar characteristics, respectively, due to the phase transformation, despite having similar densities. The mechanical properties of the sintered alloys depended on the microstructures caused by sintering conditions, such as porosity and phase morphology. At lower temperatures and stress, the micro-hardness, fracture strength, and plastic strain increased with the increase in density. For alloys with similar densities sintered at a high temperature and stress, the phase morphology affected the mechanical properties. In this work, the alloy sintered at 1150 °C with a double-phase microstructure showed optimal properties for a fracture true strength of 1820 MPa and a plastic true strain of 32.6% at a room-temperature compressive test. Thus, sintering parameters including temperature and pressure can influence the densification process and a high density can be achieved through sufficient plastic deformation and thermal diffusion, such as the grain boundary diffusion of metallic powders caused by the effect of temperature and pressure.

Acknowledgments: This work is financially supported by the National Natural Science Foundation of China (No. 51674093), the Natural Science Foundation of Heilongjiang Province (No. E201425), the Fundamental Research Funds for the Central Universities (No. HIT.KLOF.2013021 and HIT.MKSTISP.2016019), and the Postdoctoral Scientific Research Development Fund of Heilongjiang Province (No. LBH-Q15040).

Author Contributions: Dongjun Wang conceived and designed the experiments. Hao Yuan and Jianming Qiang searched literatures. Hao Yuan performed the experiments and collected the data. Hao Yuan and Jianming Qiang interpreted the data. Dongjun Wang analyzed the data and wrote the paper.

References

1. Lagos, M.A.; Agote, I. SPS synthesis and consolidation of TiAl alloys from elemental powder: Microstructure evolution. *Intermetallics* **2013**, *36*, 51–56. [CrossRef]
2. Liss, K.D.; Funakoshi, K.I.; Dippenaar, R.J.; Higo, Y.; Shiro, A.; Reid, M.; Suzuki, H.; Shobu, T.; Akita, K. Hydrostatic compression behavior and high-pressure stabilized β-phase in γ-based titanium aluminide intermetallics. *Metals* **2016**, *6*, 165. [CrossRef]
3. Liu, H.W.; Rong, R.; Gao, F.; Li, Z.X.; Liu, Y.G.; Wang, Q.F. Hot deformation behavior and microstructural evolution characteristics of Ti-44Al-5V-1Cr alloy containing ($\gamma + \alpha_2$ + B2) phases. *Metals* **2016**, *6*, 305. [CrossRef]
4. Clemens, H.; Mayer, S. Design, processing, microstructure, and applications of advanced intermetallic TiAl alloys. *Adv. Eng. Mater.* **2013**, *15*, 191–215. [CrossRef]
5. Niu, H.Z.; Chen, Y.Y.; Xiao, S.L.; Xu, L.J. Microstructure evolution and mechanical properties of a novel beta γ-TiAl alloy. *Intermetallics* **2012**, *31*, 225–231. [CrossRef]
6. Edalati, K.; Toh, S.; Iwaoka, H.; Watanabe, M.; Horita, Z.; Kashioka, D.; Kishida, K.; Inui, H. Ultrahigh strength and high plasticity in TiAl intermetallics with bimodal grain structure and nanotwins. *Scr. Mater.* **2012**, *67*, 814–817. [CrossRef]
7. Miriyev, A.; Levy, A.; Kalabukhov, S.; Frage, N. Interface evolution and shear strength of Al/Ti bi-metals processed by a spark plasma sintering (SPS) apparatus. *J. Alloy Compd.* **2016**, *678*, 329–336. [CrossRef]
8. Sadeghi, E.; Karimzadeh, F.; Abbasi, M.H. Thermodynamic analysis of Ti-Al-C intermetallics formation by mechanical alloying. *J. Alloy Compd.* **2013**, *576*, 317–323. [CrossRef]
9. Guillaume, B.G.; Chrisian, G. Spark plasma sintering of a commercially available granulated zirconia powder: I. Sintering path and hypotheses about the mechanism(s) controlling densification. *Acta Mater.* **2007**, *55*, 3493–3504.
10. Jiang, D.T.; Hulbert, D.M.; Kuntz, J.D.; Anselmi-Tamburini, U.; Mukherjee, A.K. Spark plasma sintering: A high strain rate low temperature forming tool for ceramics. *Mater. Sci. Eng. A* **2013**, *463*, 89–93. [CrossRef]
11. Li, J.; Liu, Y.; Liu, B.; Wang, Y.; Liang, X.; He, Y. Microstructure characterization and mechanical behaviors of a hot forged high Nb containing PM TiAl alloy. *Mater. Charact.* **2014**, *95*, 148–156. [CrossRef]
12. Trzaska, Z.; Couret, A.; Monchoux, J.P. Spark plasma sintering mechanisms at the necks between TiAl powder particles. *Acta Mater.* **2016**, *118*, 100–108. [CrossRef]
13. Ghasali, E.; Pakseresht, A.H.; Alizadeh, M.; Shirvanimoghaddam, K.; Ebadzadeh, T. Vanadium carbide reinforced aluminum matrix composite prepared by conventional, microwave and spark plasma sintering. *J. Alloy Compd.* **2016**, *688*, 527–533. [CrossRef]
14. Shirvanimoghaddam, K.; Hamim, S.U.; Akbari, M.K.; Fakhrhoseini, S.M.; Khayyam, H.; Pakseresht, A.H.; Ghasali, E.; Zabet, M.; Munir, K.S.; Jia, S.; et al. Carbon fiber reinforced metal matrix composites: Fabrication process and properties. *Composites Part A* **2017**, *92*, 70–96. [CrossRef]
15. Wang, Y.H.; Lin, J.P.; He, Y.H.; Wang, Y.L.; Chen, G.L. Microstructures and mechanical properties of Ti-45Al-8.5Nb-(W, B, Y) alloy by SPS-HIP route. *Mater. Sci. Eng. A* **2008**, *489*, 55–61. [CrossRef]
16. Voisin, T.; Monchoux, J.P.; Perrut, M.; Couret, A. Obtaining of a fine near-lamellar microstructure in TiAl alloys by spark plasma sintering. *Intermetallics* **2016**, *71*, 88–97. [CrossRef]

17. Liu, X.W.; Zhang, Z.L.; Sun, R.; Liu, F.C.; Fan, Z.T.; Niu, H.Z. Microstructure and mechanical properties of beta TiAl alloys elaborated by spark plasma sintering. *Intermetallics* **2014**, *55*, 177–183. [CrossRef]
18. Wang, Y.H.; Lin, J.P.; He, Y.H.; Wang, Y.L.; Chen, G.L. Fabrication and SPS microstructure of Ti-45Al-8.5Nb-(W, B, Y) alloying powders. *Intermetallics* **2008**, *16*, 215–224. [CrossRef]
19. Couret, A.; Molenat, G.; Galy, J.; Thomas, M. Microstructures and mechanical properties of TiAl alloys consolidated by spark plasma sintering. *Intermetallics* **2008**, *16*, 1134–1141. [CrossRef]
20. Jabbar, H.; Couret, A.; Durand, L.; Monchoux, J.P. Identification of microstructural mechanisms during densification of a TiAl alloy by spark plasma sintering. *J. Alloy Compd.* **2011**, *509*, 9826–9835. [CrossRef]
21. Lee, G.; Yurlova, M.S.; Giuntini, D.; Grigoryev, E.G.; Khasanov, O.L.; Mckittrick, J.; Olevsky, E.A. Densification of zirconium nitride by spark plasma sintering and high voltage electric discharge consolidation: A comparative analysis. *Ceram. Int.* **2015**, *41*, 14973–14987. [CrossRef]
22. Niraj, C.; Koundinya, N.T.B.N.; Srivastav, A.K.; Kottada, R.S. On correlation between densification kinetics during spark plasma sintering and compressive creep of B2 aluminides. *Scr. Mater.* **2015**, *107*, 63–66.
23. Lodhe, M.; Chawake, N.; Yadav, D.; Balasubramanian, M. On correlation between $\beta \rightarrow \alpha$ transformation and densification mechanisms in SiC during spark plasma sintering. *Scr. Mater.* **2016**, *115*, 137–140. [CrossRef]
24. Voisin, T.; Durand, L.; Karnatak, N.; Gallet, S.L.; Thomas, M.; Berre, Y.L.; Castagné, J.F.; Couret, A. Temperature control during spark plasma sintering and application to up-scaling and complex shaping. *J. Mater. Process. Technol.* **2013**, *213*, 269–278. [CrossRef]
25. Ghasali, E.; Shirvanimoghaddam, K.; Pakseresht, A.H.; Alizadeh, M.; Ebadzadeh, T. Evaluation of microstructure and mechanical properties of Al-TaC composites prepared by spark plasma sintering process. *J. Alloy Compd.* **2017**, *705*, 283–289. [CrossRef]
26. Wang, J.W.; Wang, Y.; Liu, Y.; Li, J.B.; He, L.Z.; Zhang, C. Densification and microstructural evolution of a high niobium containing TiAl alloy consolidated by spark plasma sintering. *Intermetallics* **2015**, *64*, 70–77. [CrossRef]
27. German, R. *Sintering from Empirical Observations to Scientific Principles*, 1st ed.; Butterworth-Heinemann: Oxford, UK, 2014; pp. 197–198.
28. Zhang, W.; Gladwell, I. Sintering of two particles by surface and grain boundary diffusion—A three dimensional model and numerical study. *Comp. Mater. Sci.* **1998**, *12*, 84–104. [CrossRef]
29. Wang, J.C. Analysis of early stage sintering with simultaneous surface and volume diffusion. *Metall. Mater. Trans. A* **1990**, *21*, 305–312. [CrossRef]
30. Chng, H.N.; Pan, J. Cubic spline elements for modeling microstructural evolution of materials controlled by solid-state diffusion and grain boundary migration. *J. Comp. Phys.* **2004**, *196*, 724–750. [CrossRef]
31. Svoboda, J.; Riedel, H. Quasi-equilibrium sintering for coupled grain boundary and surface diffusion. *Acta Metall. Mater.* **1995**, *43*, 499–506. [CrossRef]

Effects of Trace Si Addition on the Microstructures and Tensile Properties of Ti-3Al-8V-6Cr-4Mo-4Zr Alloy

Hongbo Ba, Limin Dong *, Zhiqiang Zhang and Xiaofei Lei

Institute of Metal Research, Chinese Academy of Sciences, 72 Wenhua Road, Shenyang 110016, China; bahongbo@sina.com (H.B.); zqzhang@imr.ac.cn (Z.Z.); xflei13b@imr.ac.cn (X.L.)
* Correspondence: lmdong@imr.ac.cn

Abstract: The microstructural evolution and tensile properties of Ti-3Al-8V-6Cr-4Mo-4Zr titanium alloys with various Si contents were investigated. The results revealed that the addition of trace Si and the presence of Zr induced the formation of $(TiZr)_6Si_3$ silicides, in the size range from 100 nm to 300 nm. The fine silicide precipitates refined β grains. The tensile strength increased about 40 MPa due to precipitation strengthening and grain refinement, and the ductility of the two alloys was similar. The tensile fracture mode of the alloys was dimple ductile fracture.

Keywords: titanium alloy; microstructure; silicide; tensile properties

1. Introduction

Titanium alloys are widely used in aerospace applications due to their high strength to weight ratio and excellent corrosion resistance [1–3]. Except for compressor disks and blades, a considerable fraction of fasteners and springs are fabricated from α + β and β titanium alloys, because they exhibit a favorable strength/toughness combination and high fatigue strength [4,5]. Among these titanium alloys, the β type Ti-3Al-8V-6Cr-4Mo-4Zr (known as Beta C) alloy has attracted the attention of researchers in recent years [6,7]. This alloy is developed from Ti-13V-11Cr-3Al, and it is easier to melt and exhibits less segregation due to a low Cr content [8]. In addition, the alloy can be hardened by solution plus aging treatment, and its ultimate tensile strength can reach 1380 MPa [8,9]. It can be utilized in either solution-treated or solution-aging-treated conditions; thus, a wide variation in mechanical properties can be obtained for different applications. However, the rapid β grain coarsening is still an open problem when they are solution-treated above the β transus, especially for β type titanium alloys.

The changes in microstructures and phase structures influence the mechanical properties of titanium alloys significantly. For β type titanium alloys, the body-centered cubic β phases possess good ductility, and precipitated α phases enhance their strength after aging treatment. For high-temperature titanium alloys, such as Ti-5.8Al-4Sn-3.5Zr-0.7Nb-0.35Si-0.06C [10] or Ti-6.3Al-1.6Zr-3.4Mo-0.3Si [11], the addition of Si can improve their creep performances due to the precipitation of silicides. Up to now, two different types of Ti-Zr-Si ternary silicides have been recognized in titanium alloys, S1 (a = 0.780 nm, c = 0.544 nm) and S2 (a = 0.701 nm, c = 0.368 nm), with stoichiometries $(TiZr)_5Si_3$ and $(TiZr)_6Si_3$ [12–14], respectively. Flower et al. [15] reported that they have a hexagonal crystal structure in aged Ti-5Zr-1Si alloy, using X-ray diffraction.

Apart from improving the creep properties, the addition of Si also plays an important role in inhibiting β grain growth. Bermingham et al. [16] showed that a small amount of Si addition to commercial purity Ti produced fine prior-β grains. Tavares et al. [17] presented that the Si addition to the β type Ti-35Nb alloy made beta phases more stable and achieved grain refinement. However, the addition of high amounts of Si will decrease the ductility of titanium alloys at room temperature. Ramachandra and Singh [18,19] showed that the presence of 0.25 wt % Si in Ti-6Al-5Zr-0.5Mo-0.25Si alloy led to a drastic reduction in tensile ductility at room temperature. In terms of Ti-3Al-8V-6Cr-4Mo-4Zr alloy, Morito et al. [20] also investigated the effect of higher than 0.2 wt % Si addition on the microstructure and aging behavior, but high contents of Si was found to possibly embrittle this alloy. In order to obtain the balance of strength and ductility for industrial applications, it is necessary to gain more insight into the effect of trace Si addition on the microstructures and properties of Ti-3Al-8V-6Cr-4Mo-4Zr alloy.

Based on the above, trace Si as an alloying element was added to Ti-3Al-8V-6Cr-4Mo-4Zr alloy, and the objective of the experiment was to investigate the influence of Si either in solution or in the form of silicides on its microstructures and performances.

2. Materials and Methods

In this study, Ti-3Al-8V-6Cr-4Mo-4Zr (designated C1) and Ti-3Al-8V-6Cr-4Mo-4Zr-0.05Si (designated C2) alloys were fabricated from pure Ti, Al, V, Cr, Mo, Zr and Si (≥99.9% purity) by three times vacuum arc melting and conventional rolling. The chemical compositions of the two alloys are listed in Table 1. The β transus for the alloys was ~740 °C, as measured by a metallographic technique. The rolled bars, 12 mm in diameter, for both alloys were cut and subjected to solution treatment at 800 °C for 1 h, and were then water quenched to room temperature.

Table 1. Chemical composition of the alloys.

Alloy	Composition (wt %)						
	Al	V	Cr	Mo	Zr	Si	Ti
C1	3.54	8.00	6.02	4.05	4.00	0.02	Bal.
C2	3.48	8.00	5.95	4.05	4.00	0.06	Bal.

The microstructures were characterized using an optical microscope (OM, ZEISS Axiovert 200MAT, Carl Zeiss Shanghai Co., Ltd, Shanghai, China) and scanning electron microscope (SEM, ShimadzuSSX-550, Shimadzu Corporation, Tokyo, Japan) equipped with an Energy Dispersive X-ray (EDX) detector, after samples were manually grounded, polished and chemically etched for 30 s in a solution of 5% HF, 10% HNO_3, and 85% water at room temperature. The back-scattered electron (BSE) mode was utilized for the observation of precipitates, and the polished samples were not be etched to avoid the dissolution of silicides into the HF etchant. Image-Pro Plus software was utilized for grain size analysis, and the grain size distributions were obtained. The average grain size and standard deviation were calculated on the basis of the grain diameter. TEM specimens were manually ground down to a thickness of 50 μm, punched into discs of Φ 3 mm, and then ion-milled to electron transparency in a Gatan ion miller. TEM observation was performed on a Tecnai G2 20 transmission electron microscope (TEM) (FEI Company, Hillsboro, OR, USA) with an Energy Dispersive Spectroscopy (EDS).

Tensile tests at room temperature were carried out on an Instron 5582 testing machine (Instron, Chicago, IL, USA) at a constant cross-head speed of 1 mm/min. Two specimens were tested for each alloy. Tensile specimens with a gage diameter of 5 mm and a gage length of 30 mm were used. The fracture morphologies were observed by SEM.

3. Results and Discussion

3.1. Microstructure Characteristics

Figure 1 presents the optical microstructures of the alloys after solution treatment at 800 °C for 1 h. The microstructures were composed of equi-axed β grains, and the grain size of alloy C2 was finer than that of alloy C1. The grain size distribution histograms, average grain size, and standard deviation of both alloys are shown in Figure 2. It can be observed that the grain size distribution is markedly different for the two alloys. In C1, the average grain size was 58 μm, the grain size ranged from 10 μm to 140 μm, and over 44% of the grains were larger than 60 μm. In C2, the average grain size was 31 μm, and almost all grains were in the grain size range between 10 μm and 60 μm. The standard deviation of the alloys was 25 μm and 9 μm, respectively. These results indicate that the grain size uniformity of alloy C2 is superior to that of C1.

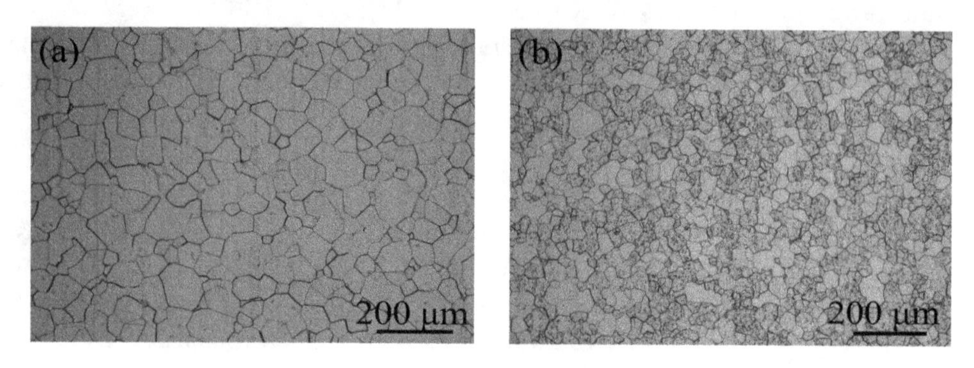

Figure 1. Optical microstructures of the alloys after solution treatment at 800 °C for 1 h: (**a**) alloy C1; (**b**) alloy C2.

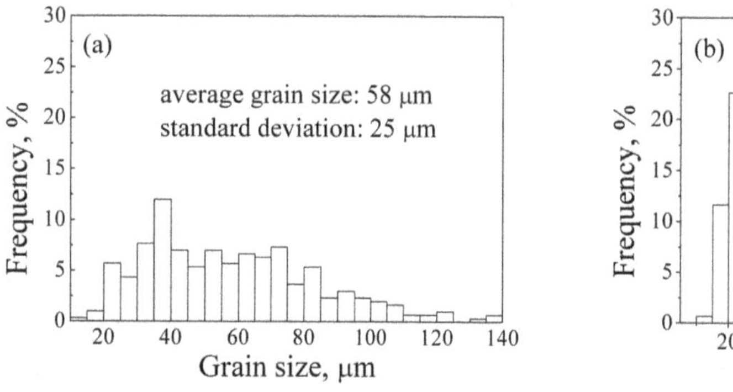

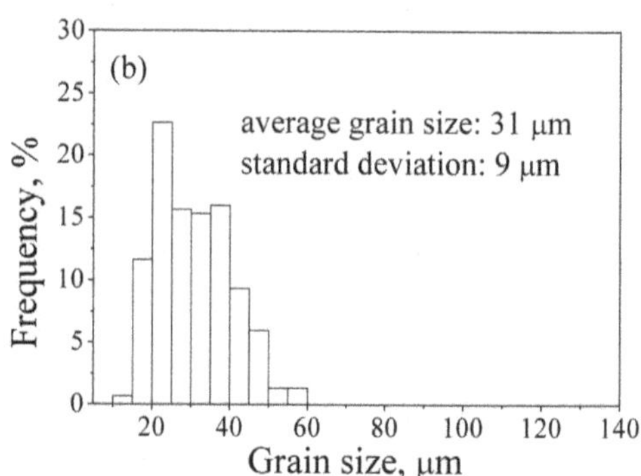

Figure 2. Grain size distributions of alloys: (**a**) alloy C1; (**b**) alloy C2.

Figure 3 shows the BSE images and EDX spectra of the alloys in solution-treated condition before etching. No resolvable precipitates are observed in C1 (Figure 3a), while white particles are clearly observed in C2 (Figure 3b). Although the distribution and morphologies of particles are not exactly confirmed in polished C2 samples, the size is very tiny. By comparison, no Si peak is found in the EDX spectra for the matrix (Figure 3c), but the white precipitates are rich in Zr and Si (Figure 3d). This confirms that the white particles are Ti-Zr-Si silicides.

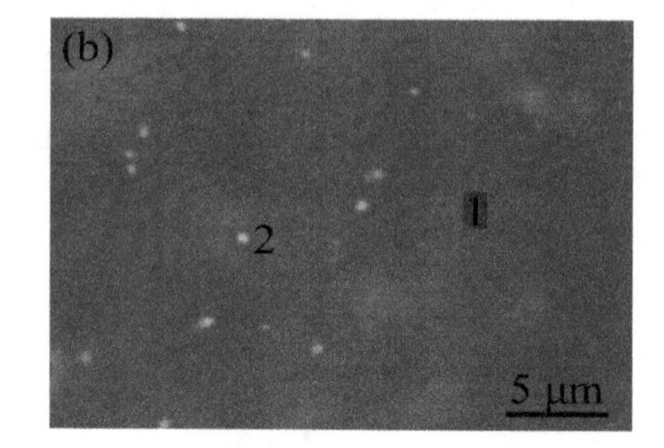

Figure 3. *Cont.*

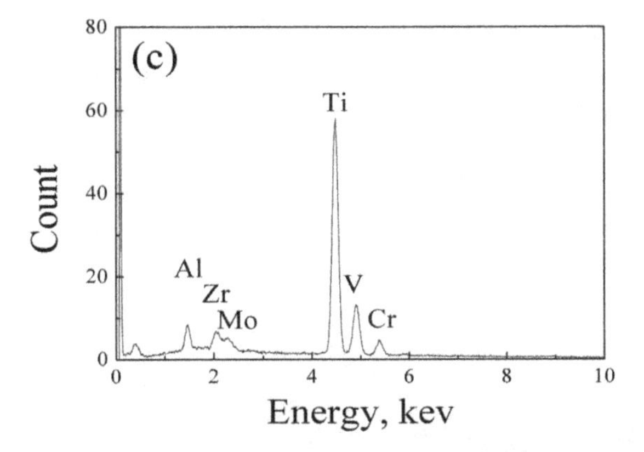

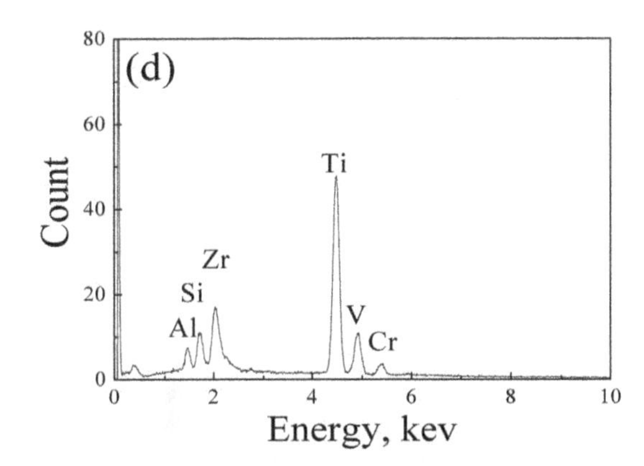

Figure 3. BSE (Back-Scattered Electron) images of microstructures of C1 (**a**) and C2 (**b**) before etching; EDX (Energy Dispersive X-ray) spectra of matrix marked by 1 (**c**) and white precipitates marked by 2 (**d**).

The selected area electron diffraction (SAED) and EDS analysis were carried out on TEM, and the precipitate was particle shaped, about 200 nm in size (Figure 4a). The precipitate is rich in Zr and Si, identified as the $(TiZr)_6Si_3$ silicides through the electron diffraction pattern (Figure 4b) and composition information (Figure 4c,d).

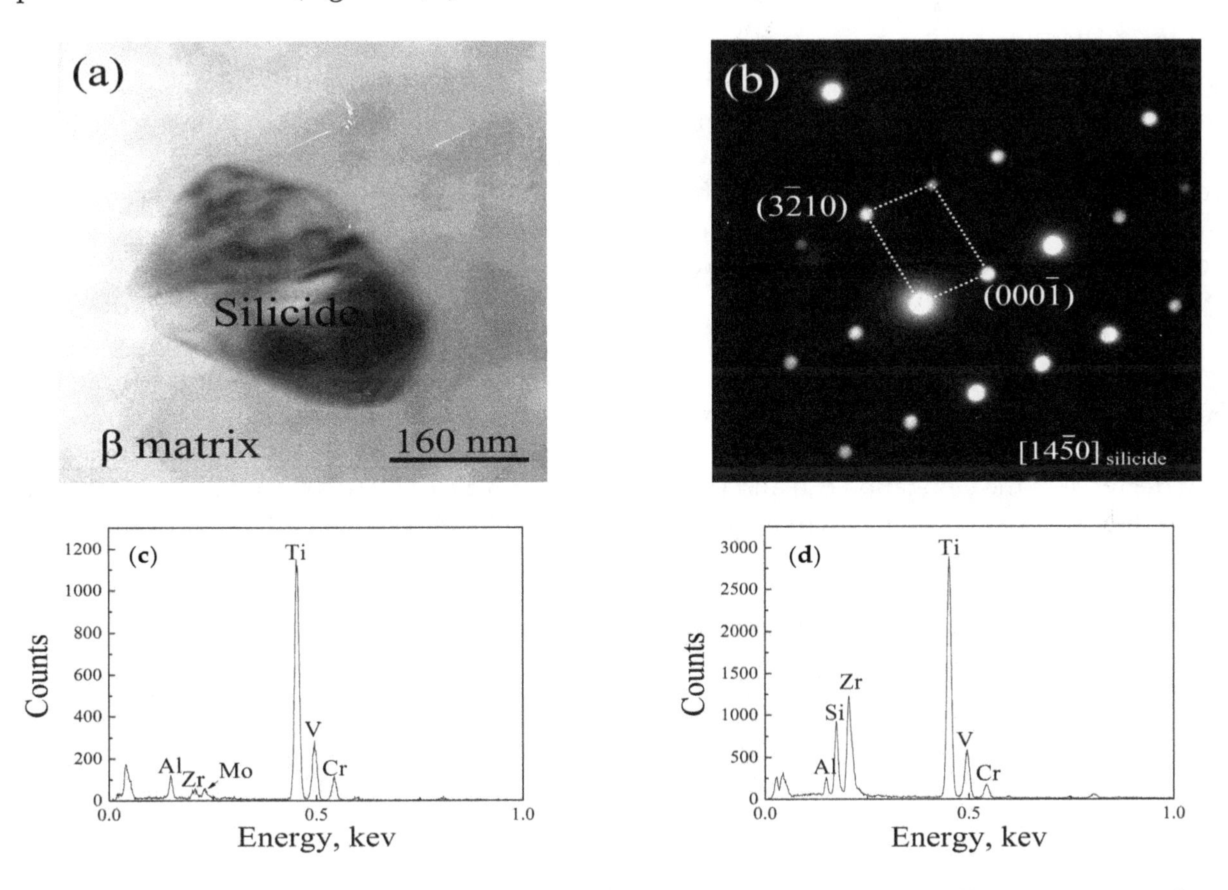

Figure 4. TEM image (**a**); corresponding SAED (Selected Area Electron Diffraction) pattern for the silicide (**b**); EDS (Energy Dispersive Spectroscopy) spectra of β matrix (**c**) and silicide (**d**) for solution-treated alloy C2.

Earlier studies demonstrated that Zr addition will promote the formation of silicides in titanium alloys [11,12,21]. Therefore, the precipitation of silicides resulted from the addition of Si and the presence of Zr in alloy C2. However, the Si content in C1 is only ~0.02 wt % from raw materials, which is below the saturation concentration of Si in the alloy, therefore Si dissolves in the β matrix.

Figure 5 shows the SEM morphologies of microstructures of both alloys after etching. There are no etch pits in C1, but a large number of etch pits are observed in alloy C2, marked by the arrow. It is a common chemical phenomenon that silicides are dissolved by the HF etchant, hence the formation of etch pits should result from the dissolution of silicides. Meanwhile, it is reasonable that the quantity and sites of etch pits correspond to those of silicides in alloy C2. According to the distribution of etch pits, it is deduced that silicides precipitate both on the β grain boundaries and in the interior of grains (Figure 5b).

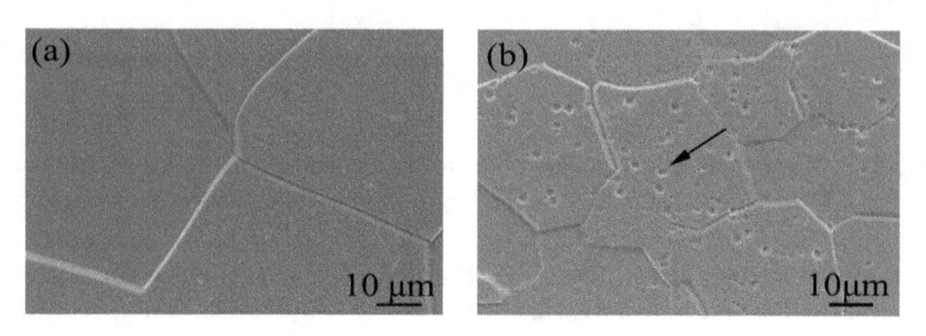

Figure 5. SEM morphologies of microstructures of the alloys after etching: (**a**) alloy C1; (**b**) alloy C2.

Singh et al. [22] reported that the kinetics of silicide precipitation in Ti-6Al-1.6Zr-3.3Mo-0.3Si (VT9) alloy was faster than in Ti-6Al-5Zr-0.5Mo-0.25Si (IMI685), which was attributed to the higher content of strong β-stabilizing element Mo in the former alloy. Therefore, it is understandable that the kinetics of the precipitation of silicides is fast in C2. On one hand, fine dispersed silicide precipitates promote the recrystallization nuclei during solution treatment at 800 °C. On the other hand, fine silicides play an important role in hindering the movement of the grain boundaries during the grain growth. This is the reason why alloy C2 has a much finer grain size.

3.2. Tensile Properties

The tensile stress-strain curves of both alloys are presented in Figure 6. A summary of the data is given in Table 2. Based on these data, it can be seen that the yield strength (YS) and ultimate tensile strength (UTS) of alloy C2 are about 40 MPa higher than those of C1, respectively. Meanwhile, there is a slight decrease in the elongation (El) and the reduction in area (RA). These results are similar to those found in the literature [23–25], where the silicide precipitates are considered to lead to an increase in strength and a slight drop in ductility for near α titanium alloys.

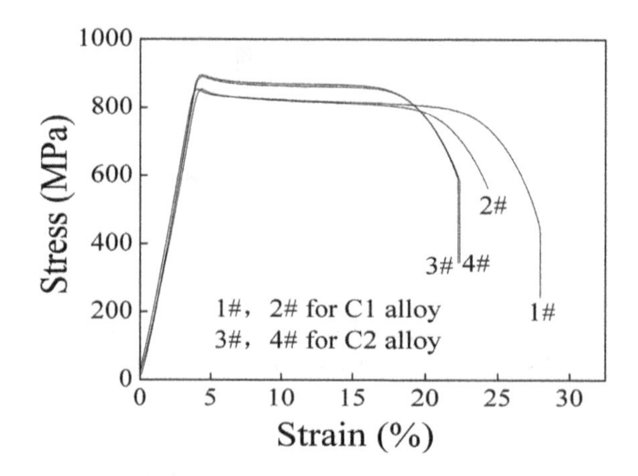

Figure 6. Tensile stress-strain curves of C1 and C2 alloys solution-treated at 800 °C for 1 h.

Table 2. Tensile properties of both alloys solution treated at 800 °C for 1 h.

Alloy	YS, MPa	UTS, MPa	El, %	RA, %
C1	852	853	29.0	73
	852	854	27.5	76
C2	888	891	23.5	62
	893	896	24.0	61

The differences in the tensile properties of the alloys are largely due to their microstructures. Firstly, the β grain size of alloy C2 was finer than that of C1. According to the Hall-Petch law, grain refinement enhances yield strength [26,27]. Secondly, a large number of fine silicides dispersed in alloy C2, which have an effect of precipitation strengthening. These explain the difference in the tensile properties of the alloys.

The changes in ductility also result from the precipitation of dispersed silicides, which would hinder the dislocation motion during tensile deformation. The precipitates tend to induce local stress concentration and crack initiation, resulting in the slightly low plasticity of alloy C2.

Figure 7 shows the fracture morphologies of both alloys. The fracture consists of fiberous and shear lip zones (Figure 7a,b). A large number of dimples spread across the fracture surfaces, which are the characteristic of ductile fracture. Dimples are more uniform and smaller in alloy C2 than those in C1 (Figure 7c,d). This is related to the presence of the dispersed silicides in alloy C2. Compared with C2, alloy C1 experiences a larger plastic deformation before fracture, resulting from its single β phase microstructure.

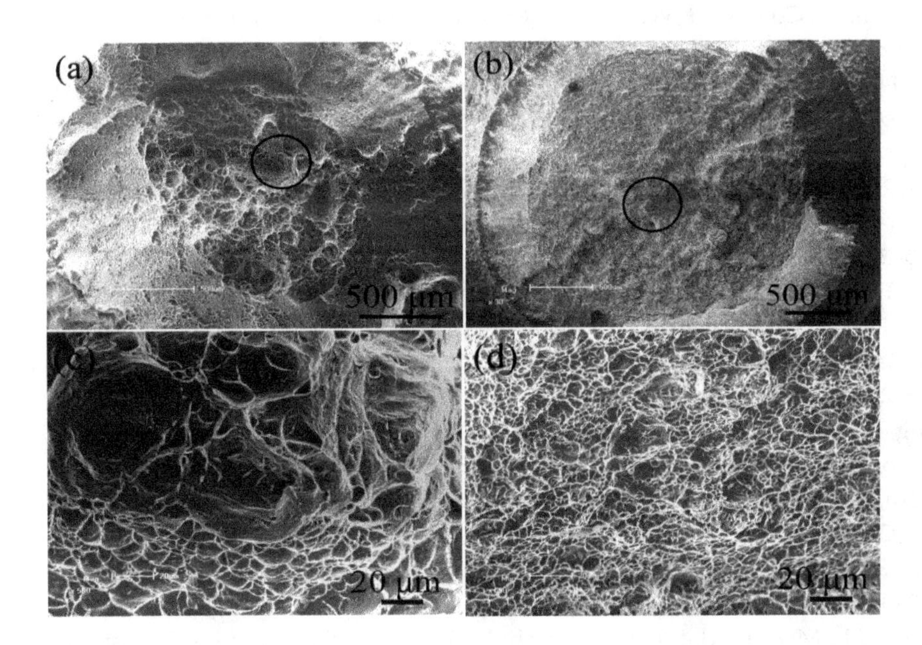

Figure 7. Tensile fractographies of the alloys: (**a**,**c**): alloy C1; (**b**,**d**): alloy C2.

4. Conclusions

(1) The addition of trace amount of Si (~0.05 wt %) and the presence of 4 wt % Zr induced the formation of $(TiZr)_6Si_3$ silicides in Ti-3Al-8V-6Cr-4Mo-4Zr-0.05Si alloy.

(2) The dispersed silicides refined β grains, and the average grain size of Ti-3Al-8V-6Cr-4Mo-4Zr-0.05Si alloy was more uniform than that of Ti-3Al-8V-6Cr-4Mo-4Zr alloy. Adding Si was an effective way to refine β grains for the Ti-3Al-8V-6Cr-4Mo-4Zr alloy.

(3) The silicide precipitates enhanced the tensile strength of Ti-3Al-8V-6Cr-4Mo-4Zr-0.05Si alloy, and the change in ductility was slight. The increase in strength was attributed to both precipitation strengthening and grain refinement. The fracture mode of both alloys was ductile.

Author Contributions: Hongbo Ba and Limin Dong conceived and designed the experiments; Hongbo Ba, Zhiqiang Zhang, and Xiaofei Lei performed the experiments and analyzed the data; Hongbo Ba wrote the paper

References

1. Williams, J.C.; Starke, E.A. Progress in structural materials or aerospace systems. *Acta Mater.* **2003**, *51*, 5775–5799. [CrossRef]
2. Banerjee, D.; Williams, J.C. Perspectives on titanium science and technology. *Acta Mater.* **2013**, *61*, 844–879 [CrossRef]
3. Lei, X.F.; Dong, L.M.; Zhang, Z.Q.; Liu, Y.J.; Hao, Y.L.; Yang, R.; Zhang, L.C. Microstructure, texture evolution and mechanical properties of VT3-1 titanium alloy processed by multi-pass drawing and subsequent isothermal annealing. *Metals* **2017**, *7*, 131. [CrossRef]
4. Kim, J.H.; Lee, C.H.; Hong, J.K.; Kim, J.H.; Yeom, J.T. Effect of surface treatment on the hot forming of the high strength Ti-6Al-4V fastener. *Mater. Trans.* **2009**, *50*, 2050–2056. [CrossRef]
5. Kume, K.; Furui, M.; Ikeno, S.; Ishisaka, Y.; Yamamoto, M. Screw form rolling of beta type titanium alloy preliminary worked by torsion. *Mater. Sci. Forum* **2010**, *654*, 906–909. [CrossRef]
6. Solek, A.L.; Krawczyk, J. The analysis of the hot deformation behaviour of the Ti-3Al-8V-6Cr-4Zr-4Mo alloy, using processing maps, a map of microstructure and of hardness. *Mater. Des.* **2015**, *65*, 165–173. [CrossRef]
7. Salam, A.; Hammond, C. Superplasticity and associated activation energy in Ti-3Al-8V-6Cr-4Mo-4Zr alloy. *J. Mater. Sci.* **2005**, *40*, 5475–5482. [CrossRef]
8. Boyer, R.; Welsch, G.; Collings, E.W. *Materials Properties Handbook: Titanium Alloys*, 1st ed.; ASM International: Materials Park, OH, USA, 1994; pp. 797–828.
9. Scmidt, P.; El-chaikl, A.; Christ, H.J. Effect of duplex aging on the initiation and propagation of fatigue cracks in the solute-rich metastable β titanium alloy Ti 38–644. *Metall. Mater. Trans. A* **2011**, *42*, 2652–2667. [CrossRef]
10. Es-souni, M. Creep behaviour and creep microstructures of a high-temperature titanium alloy Ti-5.8Al-4.0Sn-3.5Zr-0.7Nb-0.35Si-0.06C (Timetal 834): Part I. Primary and steady-state creep. *Mater. Charact.* **2001**, *46*, 365–379. [CrossRef]
11. Gu, Y.; Zeng, F.H.; Qi, Y.L.; Xia, X.Q.; Xiong, X. Tensile creep behavior of heat-treated TC11 titanium alloy at 450–550 °C. *Mater. Sci. Eng. A* **2013**, *575*, 74–85. [CrossRef]
12. Singh, A.K.; Ramachandra, C. Characterization of silicides in high-temperature titanium alloys. *J. Mater. Sci.* **1996**, *32*, 229–234. [CrossRef]
13. Li, J.X.; Wang, L.Q.; Qin, J.N.; Chen, Y.F.; Lu, W.J.; Zhang, D. The effect of heat treatment on thermal stability of Ti matrix composite. *J. Alloys Compd.* **2011**, *509*, 52–56. [CrossRef]
14. Ramachandra, C.; Singh, V. Silicide precipitation in alloy Ti-6Al-5Zr-0.5Mo-0.25Si. *Metall. Trans. A* **1982**, *13*, 771–775. [CrossRef]
15. Flower, H.M.; Swann, P.R.; West, D.R.F. Silicide precipitation in the Ti-Zr-Al-Si system. *Metall. Mater. Trans. B* **1971**, *2*, 3289–3297. [CrossRef]
16. Bermingham, M.J.; Mcdonald, S.D.; Dargusch, M.S.; Stjohn, D.H. The mechanism of grain refinement of titanium by silicon. *Scr. Mater.* **2008**, *58*, 1050–1053. [CrossRef]
17. Tavares, A.M.G.; Ramos, W.S.; Blas, J.C.G.; Lopes, E.S.N.; Caram, R.; Batista, W.W.; Souza, S.A. Influence of Si addition on the microstructure and mechanical properties of Ti-35Nb alloy for applications in orthopedic implants. *J. Mech. Behav. Biomed. Mater.* **2015**, *51*, 74–87. [CrossRef] [PubMed]
18. Ramachandra, C.; Singh, V. Effect of silicide precipitation on tensile properties and fracture of alloy Ti-6Al-5Zr-0.5Mo-0.25Si. *Metall. Trans. A* **1985**, *16*, 227–231. [CrossRef]
19. Ramachandra, C.; Singh, V. Effect of silicides on tensile properties and fracture of alloy Ti-6Al-5Zr-0.5Mo-0.25Si from 300 to 823 K. *J. Mater. Sci.* **1988**, *23*, 835–841. [CrossRef]
20. Morito, F.; Muneki, S.; Takahashi, J.; Kainuma, T. The effect of Silicon Addition on the Microstructure and the

Aging Behavior of Ti-3Al-8V-6Cr-4Mo-4Zr Alloy. In Proceedings of the Titanium 95: Eighth World Congress on Titanium, Birmingham, UK, 22–26 October 1995; pp. 2494–2501.

21. Singh, A.K.; Roy, T.; Ramachandra, C. Microstructural stability on aging of an $\alpha + \beta$ titanium alloy: Ti-6Al-1.6Zr-3.3Mo9-0.30Si. *Metall. Mater. Trans. A* **1996**, *27*, 1167–1173. [CrossRef]
22. Singh, A.K.; Ramachandra, C.; Tavafoghi, M.; Singh, V. Microstructure of β-solution-treated, quenched and aged $\alpha + \beta$ titanium alloy Ti-6Al-1.6Zr-3.3Mo9-0.30Si. *J. Alloys Compd.* **1992**, *179*, 125–135. [CrossRef]
23. Jayaprakash, M.; Ping, K.H.; Yamabe-mttaeai, Y. Effect of Zr and Si addition on high temperature mechanical properties of near-α Ti-Al-Zr-Sn based alloys. *Mater. Sci. Eng. A* **2014**, *612*, 456–461. [CrossRef]
24. Popov, A.; Rossina, N.; Popova, M. The effect of alloying on the ordering processes in near-alpha titanium alloys. *Mater. Sci. Eng. A* **2013**, *564*, 284–287. [CrossRef]
25. Jia, W.; Zeng, W.; Yu, H. Effect of aging on the tensile properties and microstructures of a near-alpha titanium alloy. *Mater. Des.* **2014**, *58*, 108–115. [CrossRef]
26. Tsai, Y.L.; Wang, S.F.; Bor, H.Y.; Hsu, Y.F. Effects of Zr addition on the microstructure and mechanical behavior of a fine-grained nickel superalloy at elevated temperatures. *Mater. Sci. Eng. A* **2014**, *607*, 294–301. [CrossRef]
27. Muszka, K.; Majta, J.; Bienias, L. Effect of grain refinement on mechanical properties of microalloyed steels. *Metall. Found. Eng.* **2006**, *32*, 87–97. [CrossRef]

Sintering Behavior and Microstructure of TiC-Me Composite Powder Prepared by SHS

Elena N. Korosteleva [1,2,*], Victoria V. Korzhova [2] and Maksim G. Krinitcyn [1,2]

[1] Department of High Technology Physics in Mechanical Engineering, Tomsk Polytechnic University, 30 Lenin av., 634050 Tomsk, Russia; krinmax@gmail.com

[2] Institute of Strength Physics and Materials Science of Siberian Branch of the Russian Academy of Sciences, 2/4, pr. Akademicheskii, 634055 Tomsk, Russia; vicvic5@mail.ru

* Correspondence: elenak@ispms.tsc.ru

Abstract: Titanium, its alloys, and refractory compounds are often used in the compositions of surfacing materials. In particular, under the conditions of electron-beam surfacing the use of synthesized composite powder based on titanium carbide with a metal binder (TiC-Me) has a positive effect. These powders have been prepared via the self-propagating high-temperature synthesis (SHS) present in a thermally-inert metal binder. The initial carbide particle distribution changes slightly in the surfacing layer in the high-energy rapid process of electron-beam surfacing. However, these methods also have their limitations. The development of technologies and equipment using low-energy sources is assumed. In this case, the question of the structure formation of composite materials based on titanium carbide remains open, if a low-energy and prolonged impact in additive manufacturing will be used. This work reports the investigation of the sintered powders that were previously synthesized by the layerwise combustion mode of a mixture of titanium, carbon black, and metal binders of various types. The problems of structure formation during vacuum sintering of multi-component powder materials obtained as a result of SHS are considered. The microstructure and dependences of the sintered composites densification on the sintering temperature and the composition of the SH-synthesized powder used are presented. It has been shown that under the conditions of the nonstoichiometric synthesized titanium carbide during subsequently vacuum sintering an additional alloy formation occurs that can lead to a consolidation (shrinkage) or volumetric growth of sintered TiC-Me composite depending on the type of metal matrix used.

Keywords: titanium carbide; self-propagating high-temperature synthesis (SHS); metal binder; composite powders; sintering

---

1. Introduction

Titanium-based materials possess many unique properties that enable them to be used in a wide range of applications in various industries [1–6]. Titanium-containing materials are in demand both as casting alloys and as powder composites. Interest in the powder composite materials, in particular, has grown considerably with the development of additive manufacturing, including surfacing processes and 3D-printing of finished parts and products. The desire to create volumetric printing products with complex surface geometry gave an additional incentive for the study of powder materials containing titanium [4–6]. Composites based on titanium carbide occupy a special place among the known groups of titanium materials. The high melting point, high strength, high thermal conductivity, stability in aggressive media and low abrasion make titanium carbide indispensable in a many areas of human activity [1–6]. Titanium carbide is remarkable in that, according to the equilibrium state diagram, it has a wide homogeneity region in the titanium-carbon system [7]. The boundaries of this interval have not yet been fully studied, but it can be assumed that it is wider than 40–50 at % of carbon at room temperature. TiC has a structure similar to that of NaCl (Bi) The carbon is usually located in the lattice between the points of the lattice, forming an interstitial solution. It was found that decreasing carbon content decreases the lattice period and hardness [2,3].

Most studies have been shown that TiC of stoichiometric composition is difficult to produce. A number of researchers note that it is impossible to obtain TiC in which all interstitial sites are occupied by carbon atoms [8–10]. One of the reasons is the ability of oxygen and nitrogen atoms to penetrate into the TiC lattice together with carbon atoms. The TiO and TiN phases also have a lattice similar to the NaCl (Bi) lattice and are isomorphous with TiC. Therefore, oxygen and nitrogen pose great difficulties in obtaining pure TiC. Production of titanium carbide with the self-propagating high-temperature synthesis (SHS) method is characterized by the fact that the combustion reaction in the compact of titanium powder and carbon black completes in just a few seconds [11–18]. The synthesis is most often carried out in a protective inert atmosphere. In addition, raw titanium must contain the least possible amount of impurities, including oxygen. The combination of the structural characteristics of the hard inclusions (titanium carbide) and the metal binder (matrix) provides the functional properties of the composite material. There are many methods of producing metal matrix composites with titanium carbide [19–27]. Classical methods of powder metallurgy (sintering, hot pressing and extrusion) and combination of them with mechanical activation or chemical-thermal processes can be used. In addition, the structure formation of the metal matrix composite can occur both as a result of heat treatment of a powder's mixture containing a metal base with titanium carbide, and as a result of the synthesis of titanium with carbon components and other metals. In the first case, it is difficult to use a large amount of titanium carbide. In the second case, the excess carbon content is desirable in order to achieve an acceptable volume fraction of carbide. The method of producing a metal matrix composite by the SHS method is quite simple when the synthesis takes place directly during the reaction of titanium and carbon in the presence of a thermally-inert metal matrix [28–30]. In this case, a structure with carbides surrounded by a metal binder is formed directly during the synthesis process. This method allows us to achieve the optimum volume fraction, and a homogeneous distribution with a high degree of dispersion of titanium carbide that has a positive effect on the properties of the metal matrix composite. Such characteristics are difficult to achieve with the conventional method of introducing the titanium carbide into a metal matrix. As a rule, such a synthesized compact is quite fragile and easily subjected to grinding. After grinding, the resulting powder product is sieved into fractions. As a result, the finished powder can be obtained with a predetermined structure of a metal matrix composite based on titanium carbide. Subsequently, the synthesized powder can be used as consumables (feedstock) for surfacing processes and other additive production. The behavior of such powders under the conditions of the electron beam surfacing can be assessed by us already, where high-energy fast processes are realized. However, at the moment it is difficult to answer how the synthesized composite powders will behave under low-energy prolonged impact. This paper presents the results of our research, which were carried out in order to explain the behavior of such non-equilibrium heterogeneous structures during subsequent heat treatment.

2. Materials and Methods

2.1. Materials and Experimental Procedure

The powder materials under study were obtained in two stages:

(1). Synthesis of the powder base from the reaction mixture by a layerwise combustion mode in an argon atmosphere, followed by milling of the synthesized material and sieving of the necessary fraction.

(2). Vacuum sintering of the synthesized powders into compacts.

Standardized industrial powders were used in the experiment:

Titanium TPP-8 (POLEMA JSC, Tula, Russia) with dispersion 50–130 μm;

Carbon black P-803 (Omsk Carbon Group, Omsk, Russia) (< 0.1 μm);

High-speed steel R6M5 (POLEMA JSC, Tula, Russia) (50–100 μm);

High-chromium cast iron PG-S27 (POLEMA JSC, Tula, Russia) (50–100 μm);

Nickel-based self-fluxing alloy PR-N77H15S3R2 (POLEMA JSC, Tula, Russia) (50–100 μm).

The titanium powder TPP-8 contains no less than 99.4% of the primary component, and no more than 0.33% iron, 0.12% chlorine, and 0.1% oxygen. The high-speed steel powder contains 1% carbon, alloying dopants (Cr, 4%; W, 6.5%; Mo, 5%; V, 2%) and impurities (Si, 0.5%; Mn, 0.55%; Ni, 0.4%).

Powder mixing was carried out in stationary conditions using axial mixing machine SMU-PB-180 (Agromash, Moscow, Russia). The metal binder content is determined by the maximally possible volume fraction of the thermally-inert metal component that still permits the initiation and realization of the SHS process in a mixture of titanium, carbon black, and the chosen metal matrix powders Calculation of the initial powder mixture was made based on the assumption that the estimated volume fraction of equiatomic titanium carbide is formed during SHS. The following compositions were chosen (Table 1).

Table 1. Compositions of experimental powder mixtures (SHS: Self-propagating high-temperature synthesis).

Composition Number	The Calculated Content of Phases in the SHS Powders (with the Assumption of Equiatomic TiC Formation)
1	TiC + 50 vol % Ti
2	TiC + 60 vol % Ti
3	TiC + 50 vol % high-chromium cast iron
4	TiC + 50 vol % high-speed steel
5	TiC + 20 vol % nickel-based self-fluxing alloy
6	TiC + 30 vol % nickel-based self-fluxing alloy
7	TiC + 40 vol % nickel-based self-fluxing alloy
8	TiC + 50 vol % nickel-based self-fluxing alloy

The powder compositions were pressed into cylindrical compacts 35 mm in diameter to facilitate the passage of the combustion front. Then the compacts were placed in a reactor for synthesis in a layerwise combustion mode using argon gas with an excess pressure of about 0.5 atm. The reaction of layerwise combustion was initiated using short-term heating by means of a molybdenum filament of an igniting tablet pressed from the powder mixture of titanium and silicon corresponding to the stoichiometric composition Ti_5Si_3. The combustion product remains of titanium silicide were removed after synthesis; the reacted compact was milled and sieved into fractions. Using this procedure, synthesized powders of all investigated compositions containing titanium carbide were obtained [28–32]. The typical microstructure and morphology of the studied synthesized powder compositions are shown in Figure 1. The finest fraction (not exceeding 50 μm) was selected for studying the behavior of composite powders under vacuum sintering. Cylindrical specimens pressed using a plasticizer with residual porosity not exceeding 45% were placed in a vacuum furnace and heated in vacuum at 10^{-2} Pa with a heating rate of 3–5 °C·min^{-1} to achieve sintering temperatures of 1300 °C and 1350 °C. The holding time at the sintering temperature was 3 h.

(a) (b)

Figure 1. *Cont.*

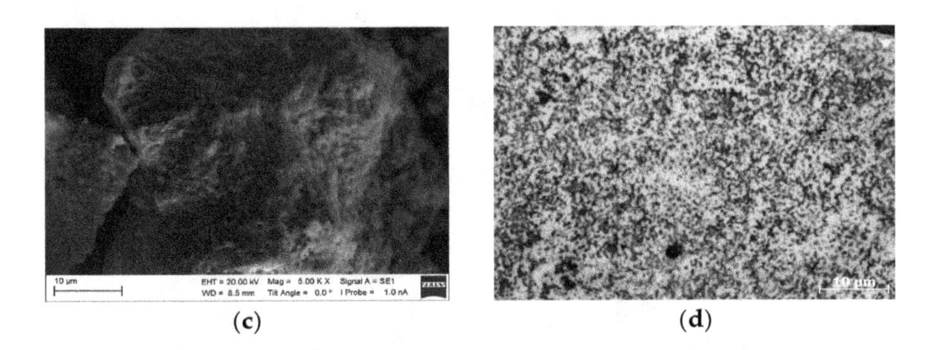

Figure 1. Morphology and microstructure of the following powder composites: (**a**,**b**) TiC-50% binder of high-chromium cast iron; and (**c**,**d**) TiC-50% binder of high-speed steel.

2.2. *Characterisation*

The density of specimens was measured before and after sintering by the Archimedes method according to the ASTM B962-17 standard. Metallographic specimens were prepared using the standard metallographic procedures: grinding with abrasive materials; fine polishing and etching in Keller's reagent (95 mL H_2O; 2.5 mL HNO_3; 1.5 mL HCl; 1 mL HF) for 5–10 s. An AXIOVERT-200MAT metallographic microscope (Carl Zeiss, Oberkochen, Germany) and a LEO EVO 50 scanning electron microscope (Carl Zeiss, Oberkochen, Germany) were used for the analysis of the SHS-powder morphology, the microstructure of the SHS-powder, and the sintered compacts. The phase composition was determined by X-ray analysis using Co K_α. For phase identification and calculation of the lattice parameters of the phases was used the X-ray data of the ASTM data file and the software package PDWin (4.0, NPP Bourevestnik, Saint-Petersburg, Russia, 2015).

3. Results

The results of vacuum sintering of SH-synthesized metal-matrix powders based on titanium carbide are presented in Figure 2. These results are best analyzed by the change in density before and after sintering. The density of the green compacts varies from 3.1 g/cm^2 to 3.7 g/cm^2, depending on the metal binder used. Our estimates show that this density corresponds to approximately 60–70% of the theoretical maximum for the respective system. In our studies, we preferred to use the actual density values of the sintered compacts. The porosity was determined by the metallographic method (quantitative metallography). In multicomponent powder materials, it is difficult to calculate the theoretical density of sintered compacts to estimate their residual porosity, since the real phase relationship after sintering is unknown.

The experimental results show that vacuum sintering at 1300–1350 °C has a significant effect on the structure formation of composite materials sintered from SH-synthesized powders, i.e., from the already-reacted powder composition. The process of diffusion interaction was once again activated during the subsequent sintering. Comparing the values of densification for different compositions in Figure 2, it is evident that the qualitative composition of the metal matrix affects the sinterability of the composite powder compacts.

Titanium, as a metallic binder, will be well-sintered in any temperature range (Figure 3). Its inclusions, which stimulate densification of the compact during sintering, are insignificant (by volume) relative to the dominant titanium carbide (Figure 2a). However, the interaction of free surfaces of "titanium-titanium" is not the only factor leading to shrinkage of the samples. According to microstructure analysis, an additional carbon migration occurs during sintering, which increases the proportion of non-stoichiometric titanium carbide (Figure 3b).

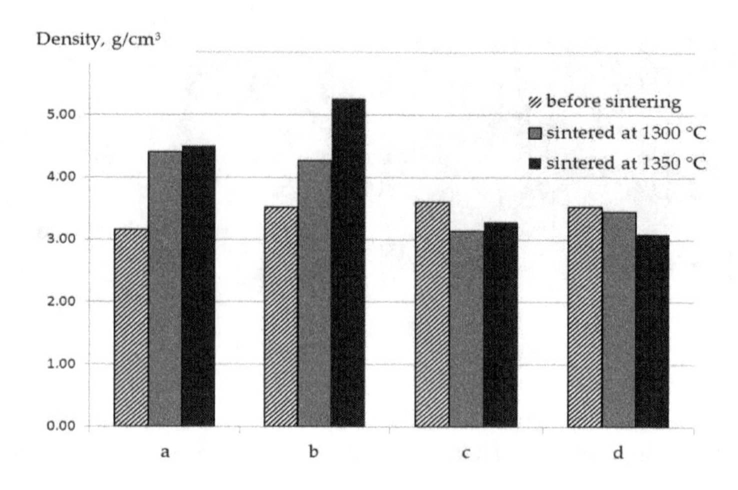

Figure 2. Density change of TiC-Me (titanium carbide with metal binder) compacts sintered at different temperature: (**a**) TiC-50% Ti; (**b**) TiC-50% high-chromium cast iron; (**c**) TiC-50% nickel-based self-fluxing alloy; (**d**) TiC-50% high-speed steel.

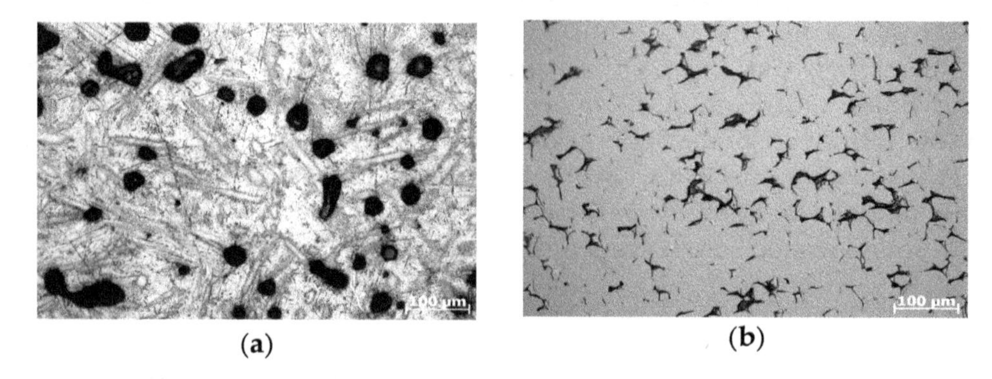

Figure 3. Microstructure of titanium powder: (**a**) TiC-50% Ti synthesized powder before sintering at 1350 °C; (**b**) after sintering at 1350 °C.

Composite powders with a high-chromium cast iron binder also densify well (Figure 4). In this case, it can be assumed that the fixed carbon, which is a component of the cast iron alloy, acts as an activator of sintering. Evidently, its affinity to titanium is stronger than the bond with iron and other alloying components (chrome) that are alloy additives of cast iron.

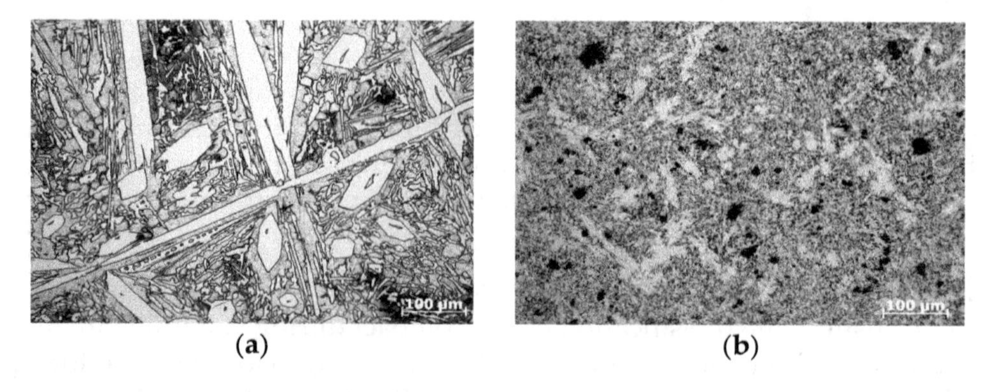

Figure 4. Microstructure of high-chromium cast iron powder: (**a**) TiC-50% high-chromium cast iron binder synthesized powder before sintering at 1300 °C; (**b**) after sintering at 1300 °C.

The use of other alloys (high-speed steel and nickel-based self-fluxing alloy) as a metallic binder demonstrates the opposite effect (Figure 5). Vacuum sintering of these synthesized powders causes the residual porosity increase that leads to a density decrease.

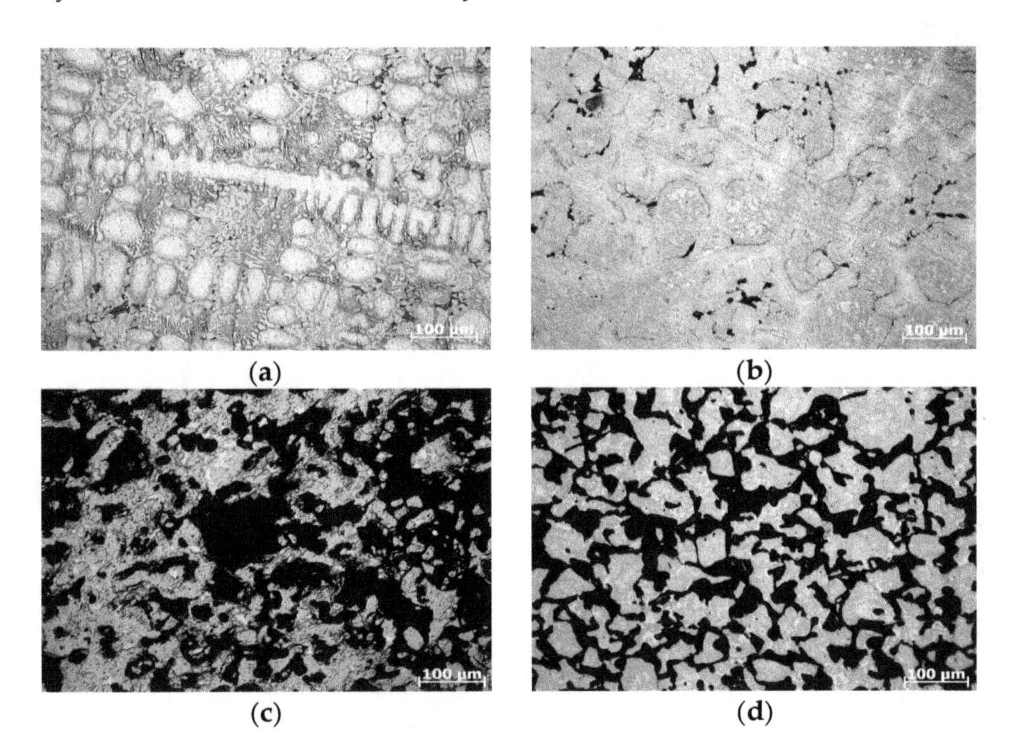

Figure 5. Sintered at 1300 °C: (**a**) microstructure of nickel-based self-fluxing alloy powder; (**b**) high-speed steel powder; (**c**) TiC-50% nickel-based self-fluxing alloy synthesized powder; and (**d**) TiC-50% high-speed steel synthesized powder.

Although the powders of high-speed steel and nickel-based self-fluxing alloys themselves are well-sintered (Figure 5a,b), in the presence of the titanium carbide the "negative" redistribution of the carbon takes place towards the metallic binder. This is assisted by the presence of carbon-active components in the alloys of high-speed steel and nickel-based self-fluxing alloy: Tungsten and molybdenum in high-speed steel and chromium in the nickel-based self-fluxing alloy, where nickel can also have an active influence on the titanium bound in the carbide.

It was found that the non-equilibrium metal-matrix structure of the powders formed as a result of SHS by the layerwise combustion method consisting mainly of non-stoichiometric titanium carbide [30,31]. Many works are devoted to the non-stoichiometricity of titanium carbide [6,8–10]. Unfortunately, there are not enough publications showing the influence of this factor on the various processes of structure formation. Our research shows that subsequent heat treatment in the mode of vacuum sintering activates the tendency of the system to approach a more equilibrated state through the redistribution of fixed carbon. In fact, despite the high affinity of carbon to titanium, some carbon atoms leave the lattice of titanium carbide, which already has a carbon deficit. Some coarsening and coalescence of the carbide particles after vacuum sintering occurs in comparison with the distribution and dispersion of carbide inclusions in the synthesized powders before heat treatment (Figure 6).

Figure 7 shows the effect of temperature on the sinterability of TiC-50% Ti SHS-powders. The metal binder type of synthesized powders have different effects on the sinterability of compacts. In the case of a maximum binder content of titanium, high-chromium cast iron, and a nickel alloy. An increase in temperature from 1300 °C to 1350 °C stimulates densification; in the case of a high-speed steel binder, however, the density at 1350 °C was less than at 1300 °C. It is supposed by us that the temperature of 1350 °C is insufficient to cause the shrinkage process to become dominant. At this temperature, the alloy formation local processes continue, causing the formation of rigid skeletons.

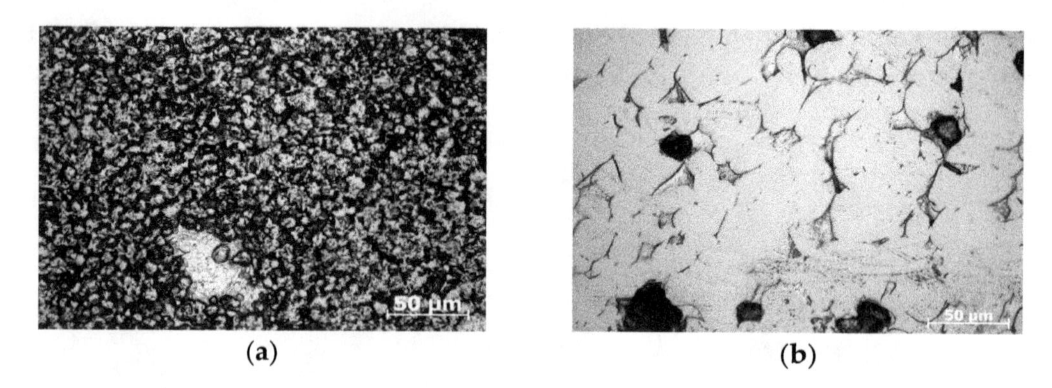

(a) (b)

Figure 6. Microstructure of TiC-50% Ti synthesized powder: (**a**) Before sintering at 1350 °C and (**b**) after sintering at 1350 °C.

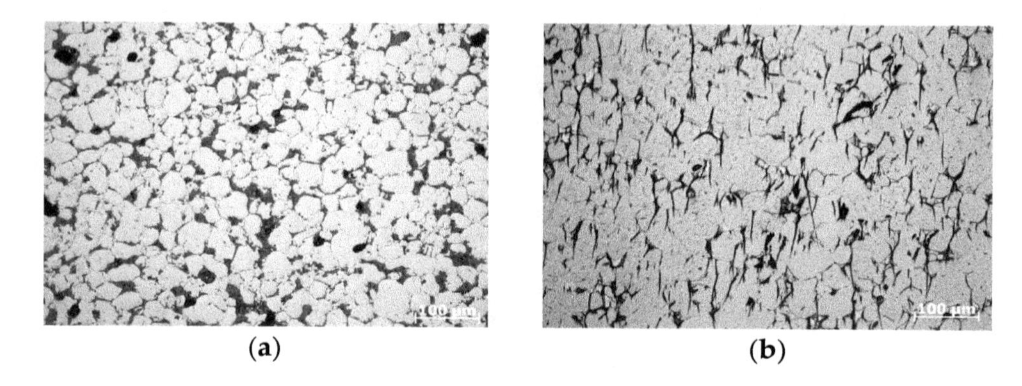

(a) (b)

Figure 7. Microstructure of TiC-50% Ti synthesized powder sintered at: (**a**) 1300 °C and (**b**) 1350 °C.

In the latter compositions, a rigid skeleton is formed from the high-alloy metal binder already at 1300 °C. This rigid skeleton prevents consolidation of the material with an increasing temperature up to the melting point of the binder.

At the same time, there is no noticeable qualitative change in the phase composition of the synthesized powder materials after sintering (Figure 8). The main phases were recognized, and were identical both for synthesized powder and for sintered compacts from them. The formation of any additional phases was not detected. However, a number of the X-ray diffraction (XRD) lines have a superposed view that may indicate the possibility of the existence of additional phases in small volume fractions. It is assumed that additional structural studies are needed in this case.

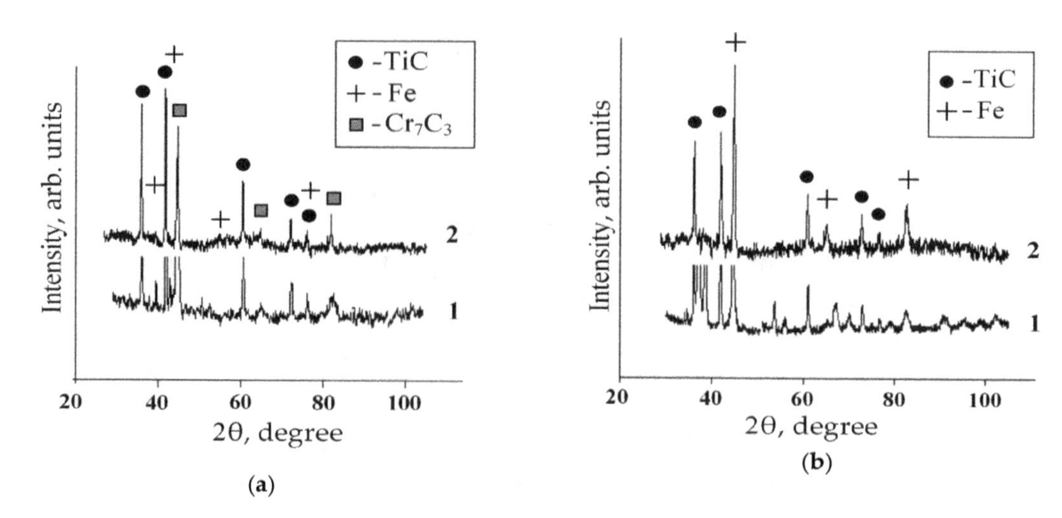

(a) (b)

Figure 8. *Cont.*

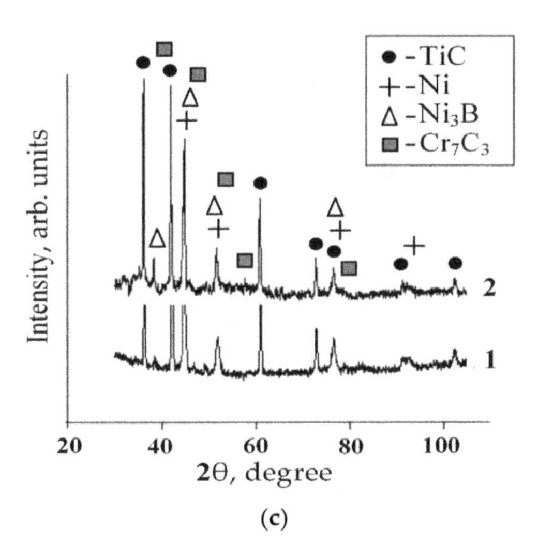

(c)

Figure 8. XRD patterns of the SHS-powder materials before (1) and after (2) sintering at 1350 °C: (**a**) TiC-50% high-chromium cast iron binder; (**b**) TiC-50% high-speed steel binder; (**c**) TiC-50% binder of nickel-based self-fluxing alloy. (SHS: Self-propagating high-temperature synthesis).

The volume fraction of the metal binder affects the density change of the sintered materials depending on its type. In the case of titanium binder, an increase in its bulk content contributes to the shrinkage of the sintered compacts. In the case of a nickel alloy matrix, on the contrary, an increase in the volume fraction of the binder leads to an increased volume growth of the compacts sintered at 1300 °C (Table 2). When the temperature reaches 1350 °C, the densification processes starts to dominate for the compositions with 40–50% nickel-based binder.

Table 2. Relative densification of the SHS compacts with various volume content of metal binder sintered at 1350 °C.

Composition	Relative Densification, %	
	T_{sin} = 1300 °C	T_{sin} = 1350 °C
TiC + 50 vol % Ti	40.1	42.0
TiC + 60 vol % Ti	43.2	50.6
TiC + 20 vol % nickel-based self-fluxing alloy	−10.2	−14.1
TiC + 30 vol % nickel-based self-fluxing alloy	−11.3	−12.5
TiC + 40 vol % nickel-based self-fluxing alloy	−12.9	−9.2
TiC + 50 vol % nickel-based self-fluxing alloy	−13.5	−8.1

4. Discussion

The results of our research show that metal matrix composites "TiC-metal binder" synthesized from powders in a layerwise combustion mode have a non-stoichiometric composition and are in a non-equilibrium state. It is assumed by us that carbon rearrangement occurs not only in the carbide phase, but also at the interface between TiC and the metal matrix upon subsequent heat treatment. Powder-synthesized composite materials of the type "TiC-high-chromium cast iron" and "TiC-Ti" exhibit significant densification during vacuum sintering at temperatures starting from 1250 °C, up to 1300 °C. Conversely, use of high-speed steel and nickel-based self-fluxing alloy as the metal binder leads to an increased pore fraction by volume and, as a result, to an increased volume growth of specimens during sintering. The rise of the temperature up to 1350 °C only increases this expansion. For these multicomponent composites it is necessary to change a composition of synthesized powder by means of metal binder addition. The ability of the two studied compositions ("TiC-high-chromium cast iron" and "TiC-Ti") to significantly densify during sintering at temperatures lower than the melting point of the metal matrix may be of interest for the design of equipment for additive manufacturing, including 3D-printing of composite materials at reduced temperatures.

Acknowledgments: The research was funded from Russian Science Foundation, grant number 17-19-01425. The experimental calculations were carried out at Tomsk Polytechnic University within the framework of Tomsk Polytechnic University Competitiveness Enhancement Program grant.

Author Contributions: Elena N. Korosteleva conceived and designed the experiments, analyzed the data, and planned the manuscript; Victoria V. Korzhova conducted the experimental work for XRD analysis and discussed the results; Maksim G. Krinitcyn carried out the experiments for sintering and microstructure analysis.

References

1. Cui, C.; Hu, B.; Zhao, L.; Liu, S. Titanium alloy production technology, market prospects and industry development. *Mater. Des.* **2011**, *32*, 1684–1691. [CrossRef]
2. Pierson, H.O. *Handbook of Refractory Carbides and Nitrides: Properties, Characteristics, Processing and Applications*, 1st ed.; Noyes publications: Westwood, NJ, USA, 1996.
3. Weimer, A.W. *Carbide, Nitride and Boride Materials Synthesis and Processing*; Springer Science & Business Media: Dordrecht, The Netherlands, 2012.
4. Herderick, E. Additive manufacturing of metals: A review. *Mater. Sci. Technol.* **2011**, 1413–1425.
5. Gong, X.; Anderson, T.; Chou, K. Review on powder-based electron beam additive manufacturing technology. *Manuf. Rev.* **2014**, *1*. [CrossRef]
6. Shabalin, I.L.; Luchka, M.V.; Shabalin, L.I. Vacuum SHS in systems with group IV transition metals for production of ceramic compositions. *Phys. Chem. Solid State* **2007**, *8*, 159–175.
7. Okamoto, H. *Phase Diagrams for Binary Alloys*, 2nd ed.; ASM International: Materials Park, OH, USA, 2010; pp. 176–189.
8. Wanjara, P.; Drew, R.A.L.; Root, J.; Yue, S. Evidence for stable stoichiometric Ti_2C at the interface in TiC particulate reinforced Ti alloy composites. *Acta Mater.* **2000**, *48*, 1443–1450. [CrossRef]
9. Quinn, C.J.; Kohlstedt, D.L. Solid-state reaction between titanium carbide and titanium metal. *J. Am. Ceram. Soc.* **1984**, *67*, 305–310. [CrossRef]
10. Yang, Y.F.; Wang, H.Y.; Zhang, J.; Zhao, R.Y.; Liang, Y.H.; Jiang, Q.C. Lattice parameter and stoichiometry of TiC_x produced in the Ti-C and Ni-Ti-C systems by self-propagating high-temperature synthesis. *J. Am. Ceram. Soc.* **2008**, *91*, 2736–2739. [CrossRef]
11. Merzhanov, A.G. History and recent developments in SHS. *Ceram. Int.* **1995**, *21*, 371–379. [CrossRef]
12. Merzhanov, A.G. Combustion processes that synthesize materials. *J. Mater. Process. Technol.* **1996**, *56*, 222–241. [CrossRef]
13. Munir, Z.A.; Anselmi-Tamburini, U. Self-propagating exothermic reactions: the synthesis of high-temperature materials by combustion. *Mater. Sci. Rep.* **1989**, *3*, 277–365. [CrossRef]
14. LaSalvia, J.C.; Meyer, L.W.; Meyers, M.A. Densification of reaction-synthesized titanium carbide by high-velocity forging. *J. Am. Ceram. Soc.* **1992**, *75*, 592–602. [CrossRef]
15. Strutt, E.R.; Olevsky, E.A.; Meyers, M.A. Combustion synthesis/quasi-isostatic pressing of TiC–NiTi cermets: Processing and mechanical response. *J. Mater. Sci.* **2008**, *43*, 6513–6526. [CrossRef]
16. Samokhin, A.V.; Alekseev, N.V.; Sinayskiy, M.A.; Tsvetkov, J.V. Thermodynamic model of high-temperature synthesis of oxygen-free titanium compounds from titanium tetrachloride. *Contemp. Eng. Sci.* **2015**, *8*, 1449–1460. [CrossRef]
17. Nersisyan, H.H.; Lee, J.H.; Won, C.W. Self-propagating high-temperature synthesis of nano-sized titanium carbide powder. *J. Mater. Res.* **2002**, *17*, 2859–2864. [CrossRef]
18. Kobashi, M.; Ichioka, D.; Kanetake, N. Combustion synthesis of porous TiC/Ti composite by a self-propagating mode. *Materials* **2010**, *3*, 3939–3947. [CrossRef]
19. Gülsoy, H.O.; Gunay, V.; Baykara, T. Influence of TiC, TiN and TiC(N) additions on sintering and mechanical properties of injection moulded titanium based metal matrix composites. *Powder Metall.* **2015**, *58*, 30–35. [CrossRef]
20. Kim, Y.J.; Chung, H.; Kang, S.J. In situ formation of titanium carbide in titanium powder compacts by gas-solid reaction. *Compos. Part A Appl. Sci. Manuf.* **2001**, *32*, 731–738. [CrossRef]

21. Roger, J.; Gardiola, B.; Andrieux, J.; Viala, J.C.; Dezellus, O. Synthesis of Ti matrix composites reinforced with TiC particles: thermodynamic equilibrium and change in microstructure. *J. Mater. Sci.* **2017**, *52*, 4129–4141. [CrossRef]
22. El-Eskandarany, M.S. Structure and properties of nanocrystalline TiC full-density bulk alloy consolidated from mechanically reacted powders. *J. Alloys Compd.* **2000**, *305*, 225–238. [CrossRef]
23. Winkler, B.; Juarez-Arellano, E.A.; Friedrich, A.; Bayarjargal, L.; Yan, J.; Clark, S.M. Reaction of titanium with carbon in a laser heated diamond anvil cell and reevaluation of a proposed pressure-induced structural phase transition of TiC. *J. Alloys Compd.* **2009**, *478*, 392–397. [CrossRef]
24. Yan, Y.; Zheng, Y.; Yu, H.; Bu, H.; Cheng, X.; Zhao, N. Effect of sintering temperature on the microstructure and mechanical properties of Ti(C, N)-based cermets. *Powder Metall. Met. Ceram.* **2007**, *46*, 449–453. [CrossRef]
25. Viljus, M.; Pirso, J.; Juhani, K.; Letunovitš, S. Structure Formation in Ti-C-Ni-Mo Composites during Reactive Sintering. *Mater. Sci.* **2012**, *18*, 62–65. [CrossRef]
26. Luo, S.D.; Li, Q.; Tian, J.; Wang, C.; Yan, M.; Schaffer, G.B.; Qian, M. Self-assembled, aligned TiC nanoplatelet-reinforced titanium composites with outstanding compressive properties. *Scr. Mater.* **2013**, *69*, 29–32. [CrossRef]
27. Li, S.; Sun, B.; Imai, H.; Kondoh, K. Powder metallurgy Ti-TiC metal matrix composites prepared by in situ reactive processing of Ti-VGCFs system. *Carbon* **2013**, *61*, 216–228. [CrossRef]
28. Kalambaeva, S.S.; Korosteleva, E.N.; Pribytkov, G.A. Structure of Composite Powders "TiC-high Chromium Cast Iron Binder" Produced by SHS Method. In Proceedings of the International Conference on Mechanical Engineering, Automation and Control Systems, Tomsk, Russia, 16–18 October 2014. [CrossRef]
29. Krinitcyn, M.G.; Pribytkov, G.A.; Durakov, V.G. Structure and properties of electron beam coatings, overlaid of SHS composite powders "TiC-Ti", synthesized in air. *Key Eng. Mater.* **2016**, *685*, 719–723. [CrossRef]
30. Korosteleva, E.N.; Pribytkov, G.A.; Krinitcyn, M.G.; Baranovskii, A.V.; Korzhova, V.V.; Strelnitskij, V.E. Fabrication of "TiC-HSS steel binder" composite powders by self-propagating high temperature synthesis. *Key Eng. Mater.* **2016**, *712*, 195–199. [CrossRef]
31. Korosteleva, E.N.; Pribytkov, G.A.; Krinitcyn, M.G.; Baranovskii, A.V.; Korzhova, V.V. Problems of Development and Application of Metal Matrix Composite Powders for Additive Technologies. *IOP Conf. Ser. Mater. Sci. Eng.* **2016**, *140*. [CrossRef]
32. Dudina, D.V.; Pribytkov, G.A.; Krinitcyn, M.G.; Korchagin, M.A.; Bulina, N.V.; Bokhonov, B.B.; Batraev, I.S.; Rybin, D.K.; Ulianitsky, V.Y. Detonation spraying behavior of TiC_x-Ti powders and the role of reactive processes in the coating formation. *Ceram. Int.* **2016**, *42*, 690–696. [CrossRef]

Implant Treatment in Atrophic Maxilla by Titanium Hybrid-Plates: A Finite Element Study to Evaluate the Biomechanical Behavior of Plates

María Prados-Privado [1,2,*], Henri Diederich [3] and Juan Carlos Prados-Frutos [4]

1 Department Continuum Mechanics and Structural Analysis, Higher Polytechnic School, Carlos III University, Avenida de la Universidad, 30, 28911 Leganés, Madrid, Spain
2 Research Department, ASISA Dental, Calle José Abascal, 32, 28003 Madrid, Spain
3 Private Practice, 51 Avenue Pasteur, L2311 Luxembourg, Luxembourg; hdidi@pt.lu
4 Department of Medicine and Surgery, Faculty of Health Sciences, Rey Juan Carlos University, Avenida de Atenas s/n, 28922 Alcorcón, Madrid, Spain; juancarlos.prados@urjc.es
* Correspondence: mprados@ing.uc3m.es

Abstract: A severely atrophied maxilla presents serious limitations for rehabilitation with osseointegrated implants. This study evaluated the biomechanical and long-term behavior of titanium hybrid-plates in atrophic maxilla rehabilitation with finite elements and probabilistic methodology A three-dimensional finite element model based on a real clinical case was built to simulate an entirely edentulous maxilla with four plates. Each plate was deformed to become accustomed to the maxilla's curvature. An axial force of 100 N was applied in the area where the prosthesis was adjusted in each plate. The von Mises stresses were obtained on the plates and principal stresses on maxilla The difference in stress between the right and left HENGG-1 plates was 3%, while between the two HENGG-2 plates it was 2%, where HENGG means Highly Efficient No Graft Gear. A mean maximum value of 80 MPa in the plates' region was obtained, which is a lower value than bone resorption stress. A probability cumulative function was computed. Mean fatigue life was 1,819,235 cycles. According to the results of this study, it was possible to conclude that this technique based on titanium hybrid-plates can be considered a viable alternative for atrophic maxilla rehabilitation, although more studies are necessary to corroborate the clinical results.

Keywords: atrophic maxilla; titanium hybrid-plates; finite element analysis; biomechanical analysis

---

1. Introduction

The reconstruction of an atrophic maxilla has always been a challenge [1] because of anatomical and clinical factors due to the serious limitations for conventional implant placement [2]. These limitations are related to the amount of bone, which remains insufficient for the conventional placement of a dental implant [3]. The maxillary bone volume has been classified, among other authors, by Cawood and Howel in five grades (I to V). Grades IV and V are considered as extreme atrophies [4]. The most common alternatives in atrophic maxilla rehabilitation are bone grafting [5], pterygoid [6] or zygomatic implants [7], bone regeneration (with or without mesh) [8,9], and finally, short implants [10].

Bone grafting is the most common technique in the reconstruction of an atrophic maxilla [2]. The goal of hard tissue augmentation is to provide an adequate bone volume for ideal implant placement and to support soft tissue for optimal esthetics and function. Zygomatic implants present a viable alternative for this kind of treatment given their design with self-tapping screw and the appropriate length as this kind of implant can be placed in the bone with very good quality and excellent mechanical behavior [2,7]. Pterygoid implants have the advantage of allowing anchorage in the pterygomaxillary region, eliminating the need for sinus lifts or bone grafts. Additionally, pterygoid implants can eliminate posterior cantilever and improve axial loading [2,11]. Finally, short implants are widely used and have demonstrated their efficiency on implant treatment in atrophic jaw and maxilla [12,13].

The first hybrid plates were introduced by G. Scortecci in July 2000 and were first used in the same year in a patient with a fractured atrophic mandible. They had a large base plate (25, 33, or 43 mm long, 7, 9, or 12 mm wide) [14]. These kinds of plates can be adjusted to the maxilla curvature and put in the best place to maintain occlusal function [15].

There are technical differences between the hybrid plates employed and the sub-periosteal implants. The first difference is that the hybrid plates are made of titanium grade II and machined from a block, and the sub-periosteal implants are made of chrome cobalt and cast individually. The second difference is that hybrid plates are flexible and can be adjusted in situ while sub-periosteal implants are rigid and modeled using an initial impression of the bone site. Sub-periosteal implants can only be cemented to the prosthesis while hybrid plates have a screw connection. To use them, it is necessary to make a groove in the bone to receive the plates so that they may become osseointegrated, which are then fixed with osteosynthesis screws.

Hybrid plates are made of titanium grade II, which is the main cp Ti used for industrial dental implant applications [16]. It is recognized that titanium and its alloys are biomaterials with the best in vivo behavior [17]. Due to the excellent biocompatibility of this material, it is very common to use it for biomedical applications such as dental implants ad hybrid-plates [18]. This biocompatibility is provided by the following properties of titanium: low level of conductivity, high corrosion resistance, thermodynamic state at physiological pH values, and low ion-formation tendency in aqueous environments [19,20].

In that sense, the main differences between hybrid plates and other alternatives to treat atrophic maxilla as pterygoid, zygomatic, or short implants are a titanium alloy. Implants are mostly manufactured by titanium alloy Ti6Al4V.

The protocol employed in this study is called Cortically Fixed @ Once (CF@O). It is an alternative to conventional implant placement for severe atrophied maxilla and mandible. This technique uses plates and pterygoid implants. Plates are fixed to the bone with osteosynthesis screws. There are four types of plates available that differ in size and morphology, although in this study, only two of them were used. This technique has its origins in basal implantology, which was developed by Dr. Scortecci in the early 1980s when he proposed the Diskimplant®, a disc implant system that was inserted laterally, and which he refined over the next few years [21]. Several basal implants were developed during the 1980s and the 2000s with different geometric forms and with perforations over the surface to improve the blood supply around the implant [14].

In the last few decades, the finite element method (FEM) has become very popular in the field of biomechanics as it is a useful tool to numerically calculate aspects such as stresses and strains, and to evaluate the mechanical behavior of biomaterials and human tissues, considering the difficulty in making such an assessment in vivo [22,23].

Dental implants and their components including hybrid-plates are subjected to cyclic loads. Therefore, fatigue of materials is introduced in all dental rehabilitations. Results with a good accuracy are essential in dental studies as fatigue is very sensitive to many parameters [22].

Owing to the variety of techniques for atrophic maxillary implant rehabilitation, a new technique based on innovative hybrid titanium plates is described and numerically analyzed. The aim of this study was to evaluate the biomechanical and long-term behavior of CF@O plates on a completely edentulous and atrophic maxilla by employing a finite element analysis and a probabilistic fatigue approach.

2. Materials and Methods

2.1. Description of the Protocol

The CF@O protocol is an alternative to the existing treatment of the atrophied maxilla and mandible. This technique is less invasive than the conventional procedures, and implants can be loaded with a definitive restoration after 6–10 days. Between the surgery and the definitive prosthesis, the patient has a provisional prosthesis, which is a fixed immediate loading prosthesis made of resin installed on plates.

This protocol does not need a sinus lift for the rehabilitation in cases of atrophied maxilla nor bone graft in maxilla and mandible and is based on both traditional implantology methods, combined with the most modern tools.

The Cortically Fixed @ Once protocol is applicable to edentulous maxillae and mandibles, and to unilateral and bilateral edentulism in maxilla and mandible. This protocol is indicated in the following assumptions [24]:

- Reduced bone volume in the upper and lower jaws such as stage D or E according to the Lekholm and Zarb classification of bone quality.
- Severely reduced vertical bone height over the trajectory of the mandibular canal in the lower jaw where there is insufficient vertical bone height to place conventional implants.
- Where short implants are not deemed justified, or in cases where bone reconstruction is not feasible.
- In very sharp dentoalveolar ridges.

This protocol can be used in patients aged between 35 and 90 years and is performed under local anesthesia. It can be done in nearly all cases of atrophy except in the case of egg shell everywhere in the maxilla and 10 mm of residential bone in the mandible. A stereolithographic model based on CT scan is made. After surgery, a temporary fixed bridge is employed. Amoxicillin is prescribed (2 g a day for 10 days) and in case of pain, ibuprofen 600 (1–3 g a day). Finally, there are no restrictions on food after one month after surgery.

The treatment plan starts with signed patient consent. Then, an open flap in the maxilla, as in this study, is made from the left tuberosity along the crest till the canine region. Two hybrid plates HENGG-1 and HENGG-2 are fixed with osteosynthesis screws and covered. The flap is then closed on the left and right with polytetrafluoroethylene polymer (PTFE) monofilament nonabsorbable suture.

Depending on the atrophy, three types of prosthesis are available: a metal acrylic for a big atrophy and a metal ceramic or zirconia when there is less atrophy with enough bone. Another point to consider is the contribution of the pterygoid implants, which contribute to prosthesis fixation.

From 2013, 155 patients between the ages of 38 and 85 (95 were female and 60 were male and 105 were in the maxilla and 45 in the mandible) were treated with the protocol detailed previously, resulting in three failures of plates in the maxilla and two in the jaw. These lost treatments were related to infection processes in the soft tissues. After a follow-up period of one year, there was a clinical and radiographical check to confirm that the plates were fixed and without complication.

The finite element analysis employed in this study simulated a real case of a 58-year-old female, with patient consistent, who wanted fixed teeth in the maxilla in a compromised bone. The treatment consisted of two pterygoid implants, four hybrid plates fixed with osteosynthesis screws, and a metal acrylic bridge ten days later [25]. In this instance, only the plates were analyzed.

2.2. Plates

The plates used in the CF@O protocol are very thin, lightweight, and highly flexible, and therefore may be adapted to any bone anatomy. In this study, the two plates employed are detailed in Figure 1:

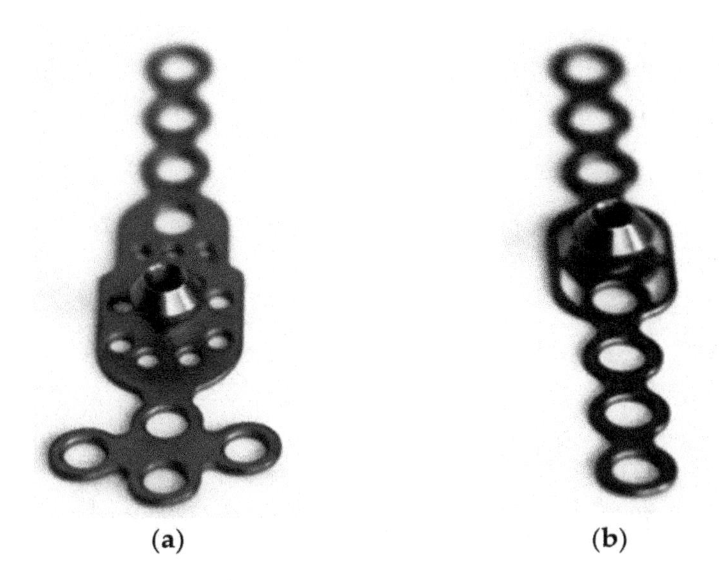

Figure 1. Plates employed in this study: (**a**) HENGG-1; (**b**) HENGG-2.

The HENGG-1 plate is appropriated for atrophied maxilla and is fixed with the zygomatic bone and the palate. The HENGG-2 plate is recommended for premaxilla, and the retromolar region in case of pencil mandible.

The plates are milled in a single piece and may be tilted in two axes to ensure that the implant fits the bone perfectly by manual shaping, making them isoelastic and able to mimic bone. They are minimally invasive and totally adjustable; they can be tilted at 90 degrees and the number of vents needed can be reduced as required, depending on the bone available at the site. They can also be twisted to fit the mandibular anatomy. They are stabilized and fixed by osteosynthesis screws, which give a strong cortical anchorage.

2.3. *Finite Element Reconstruction*

Geometric characteristics of the plates employed in the present study are shown in Figures 2 and 3.

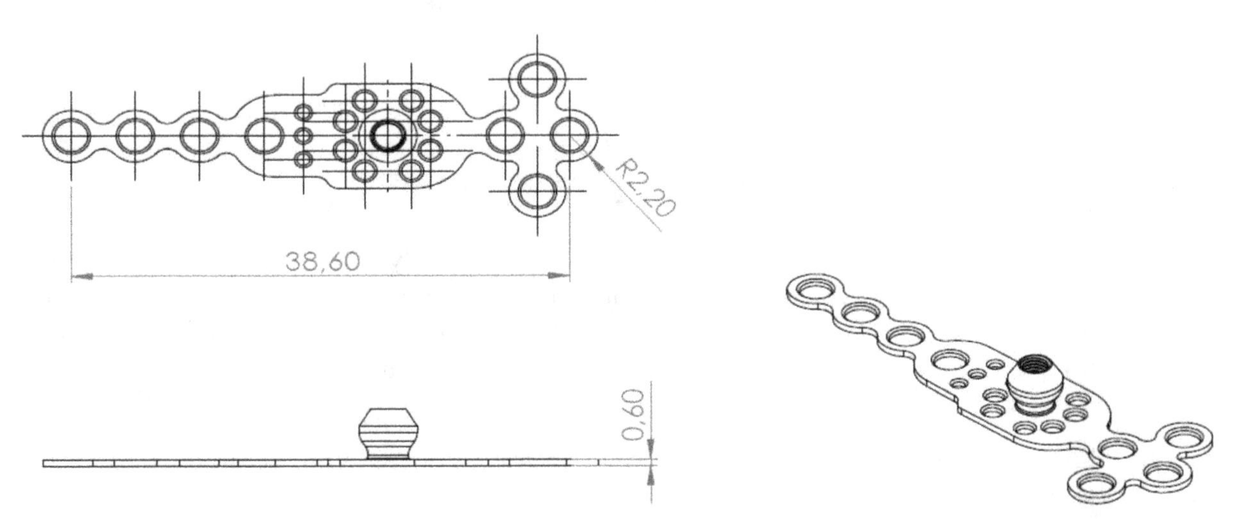

Figure 2. Plate HENGG-1.

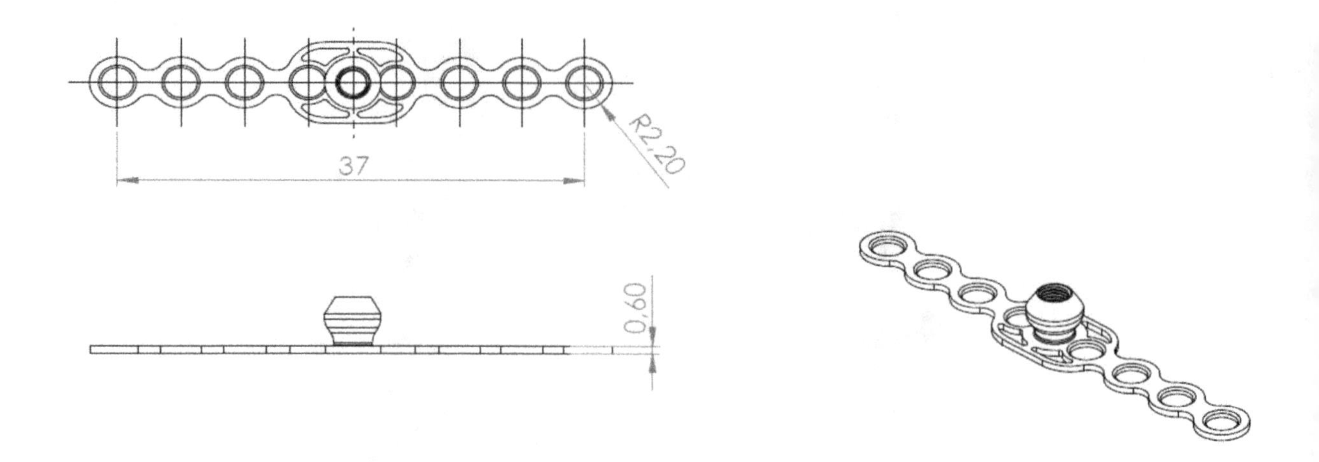

Figure 3. Plate HENGG-2.

All three-dimensional plates were adjusted to the anatomic characteristics of the maxilla (Figure 4b) as the common procedure in a real case (Figure 4a).

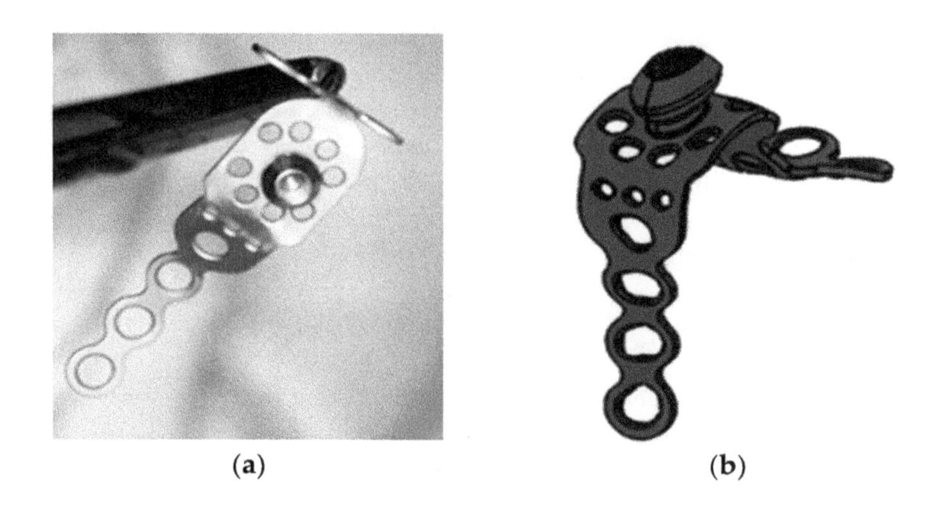

Figure 4. Plate adjusted to the anatomic characteristics of the maxilla: (**a**) Plate deformed before being placed in the patient; (**b**) three-dimensional model of a plate deformed.

The finite element model reproduced the case detailed previously, which is represented in Figure 5. Geometry of the maxilla was obtained using CT and transformed to the STL format. Slice increment was 0.5 mm, according to other studies in the literature. All data in DICOM format were imported into the software package Mimics 10.0 (Materialize, Leuven, Belgium) for the construction of the 3-D model. Plate HENGG-1 was placed in the molar region and HENGG-2 in the premaxilla.

Finally, Figure 6 represents the three-dimensional finite element assembly employed to reproduce the clinical case detailed in this study.

The maxilla in STL format was imported into SolidWorks 2016 (Dassault Systèmes, SolidWorks Corp., Concord, MA, USA) where the assembly with the four plates was done (Figure 6). The trabecular bone was 1 mm thick [26].

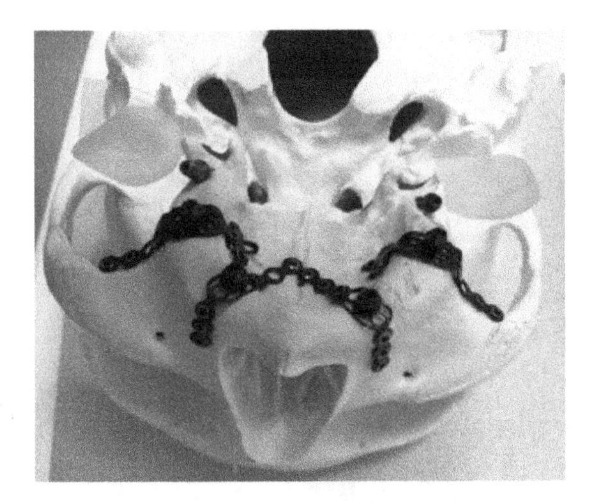

Figure 5. Model employed to reproduce in the finite element model.

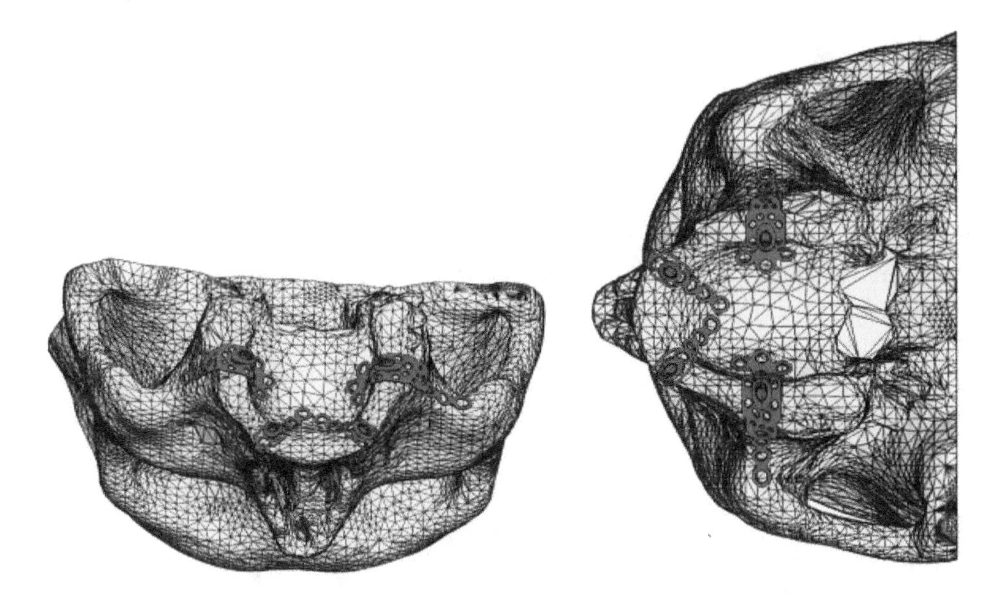

Figure 6. Finite element model employed in this study.

2.4. Material Properties

All materials were considered isotropic, linearly elastic, and homogeneous. The properties of the materials are detailed in Table 1.

Table 1. Material properties employed in this model.

Title	Young's Modulus	Poisson's Ratio
Plates (titanium grade II) [27]	105 GPa	0.37
Maxilla: cortical bone [26]	13.7 GPa	0.3
Maxilla: trabecular bone [26]	1.37 GPa	0.3

2.5. Meshing

Mesh generation was done in SolidWorks 2016 (Dassault Systèmes, SolidWorks Corp., Concord, MA, USA). All components were meshed with a fine mesh and all regions of stress concentration that were of interest were manually refined. The three-dimensional model presented a total of 432,404 nodes and 294,104 elements. The convergence criterion was a change of less than 5% in the von Mises stress in the model [28] (Figure 7).

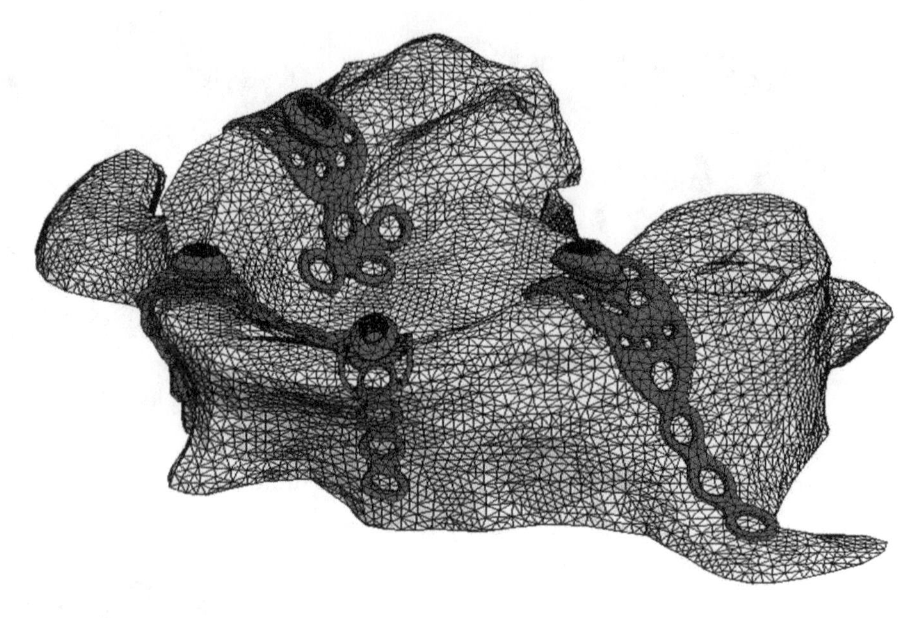

Figure 7. Finite element mesh (isometric view).

2.6. Boundary Conditions and Loading Configuration

The model was subjected to a rigid fixation restriction in the upper and lateral maxilla to prevent displacement in the x, y, and z axes (Figure 8). Plates were in contact with the maxilla and a nonpenetration condition was also added to prevent interferences during the execution process between the plates and the maxilla.

A load of 100 N [29,30] was directly applied perpendicular to the area where the prosthesis was fixed to the plate as detailed in Figure 8.

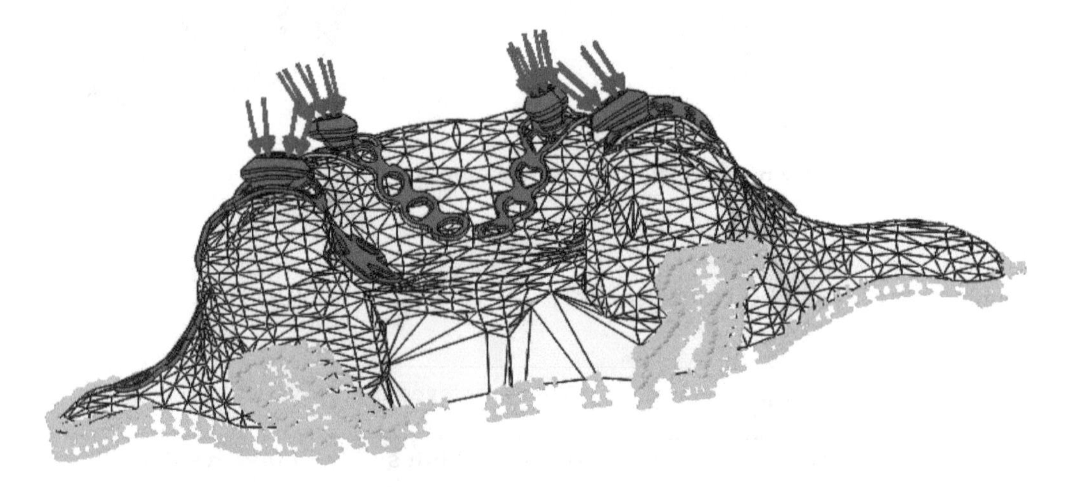

Figure 8. Three-dimensional posterior view of maxilla with load application and boundary conditions. Loads are represented in blue arrows and the rigid fixation restriction is represented in green.

2.7. Probabilistic Fatigue Model

In addition to the previous deterministic finite element analysis, a probabilistic fatigue model at the crack nucleation stage was also implemented. This stage is the most important regarding dental components and, hybrid plates, life, in particular [31]. As Riahi et al. detailed in their study, the probabilistic finite element method is a viable tool to estimate the influence of the stochastic properties of loads, material properties, and geometry on the response [32]. The methodology employed in this study was based on a cumulative damage B-model, which is constructed from the Probabilistic Finite Element Method (PFEM) results and computed for every random variable here considered: the Young's modulus (105 $\pm$ 10 GPa) and the applied loads (100 $\pm$ 10 N) [31]. The input random variables considered in this study were handled via its first order Taylor series expansion. Once all the sensitivities of the random variables are known, it is possible to apply the mean and variance operator. All the sensitivities of the random fields involved, such as displacements field, strain field, and stress field can be obtained.

Bogdanoff and Kozin (B-K) created a number of probabilistic models of cumulative damage based on ideas from Markov chains. This study employed one they called the B-model of unit steps, for its simplicity and suitability to the physical description of the process of fatigue in the crack initiation stage. The hypotheses that serve as a basis for the expansion of the B-K unit step model are [33]:

(1) Damage cycles (DC) are repetitive and of constant severity.
(2) The levels of damage a component will go through until final failure are discrete $(1, 2, \ldots, j, \ldots, b)$, and failure occurs at the last level of damage (b). This hypothesis merely discretizes the total life of the component in b levels.
(3) The accumulation of damage that occurs in each DC depends only on the DC itself and the level of damage of the component at the start of said DC.
(4) The level of damage in each DC can only be increased from the occupied level at the beginning of said DC to the next immediate level.

As damage cycles have been defined as constant severity, the Probability Transition Matrix (P) will be unique and expresses the probability that each DC must be in the same level or the probability will jump to the next DC. This matrix depends on the p_j (probability of remaining in the same DC) and q_j (probability of jumping to the next DC) and is detailed in Equation (1).

$$P = \begin{pmatrix} p_1 & q_1 & 0 & \cdots & 0 & 0 \\ 0 & p_2 & q_2 & 0 & \cdots & 0 \\ 0 & 0 & p_3 & q_3 & \cdots & 0 \\ \vdots & \vdots & \vdots & \vdots & \ddots & \vdots \\ 0 & 0 & 0 & \cdots & p_{b-1} & q_{b-1} \\ 0 & 0 & 0 & \cdots & 0 & 1 \end{pmatrix} \tag{1}$$

The new vector p_x is a vector showing the distribution of damage levels for time $t = x$. Using the results of Markov chains, vector p_x is:

$$p_x = p_{x-1}P = p_o P^x \text{ with } x = 0, 1, 2 \ldots. \tag{2}$$

Finally, to compute the fatigue life estimators, Neuber's rule and a random formulation of the Coffin and Basquin–Manson expressions were employed. Neuber's rule relates the levels of elastic stress and strain obtained by a linear elastic analysis with actual levels of stress and strain, in accordance to the elastic-plastic behavior material [34].

Coffin, for the elastic component of deformations, and Basquin and Manson, for the elastic-plastic component, proposed a nonexplicit relationship between the fatigue life cycles in the nucleation stage of a component and the amplitudes of strain. This relation is shown in Equation (3)

$$\frac{\Delta \varepsilon_{ep}}{2} = \frac{\sigma'_f}{E}\left(2N_f\right)^b + \varepsilon'_f\left(2N_f\right)^c \tag{3}$$

where $\Delta\varepsilon_{ep}$ is the range of elastic-plastic strain suffered by the component at the crack initiation area σ'_f is the fatigue resistance coefficient; ε'_f is the fatigue ductility coefficient; b is the fatigue resistance exponent; c is the fatigue ductility exponent; E is the modulus of elasticity; and N_f is the fatigue life cycles.

The materials properties necessary to solve this probabilistic model are detailed in Table 2:

Table 2. Material properties for the probabilistic model.

Parameter	Value	Parameter	Value
$\Delta\varepsilon_{ep}$	0.352×10^{-2}	σ'_f	0.0140×10^6
b	−0.1203	ε'_f	0.1701
c	−0.349		

3. Results

3.1. Plates

To obtain a correct clinical behavior, loads must be uniformly distributed throughout the four plates and transmit small stresses to the maxilla. Figure 9 shows the von Mises stress on the plates Stress distribution along the plates is different because of the anatomical geometry of the maxilla, however, these differences on stress values are very small, as Figure 10 details.

Figure 9. The von Mises stress on plates in MPa.

In Figure 9, the maximum von Mises stresses appeared around the area where the prosthesis was adjusted to the plate and the body of the plate. The right HENGG-1 plate supported a stress of 185 MPa, while the same plate on the left had a maximum von Mises stress of 179 MPa. Stress on the right HENGG-2 plate was 168 MPa while in the plate placed on the left, it was 165 MPa.

According to Figure 10, the difference between the maximum von Mises stress in the HENGG-1 right and left plates was 3%, while the difference between the HENGG-2 right and left plates was 2%.

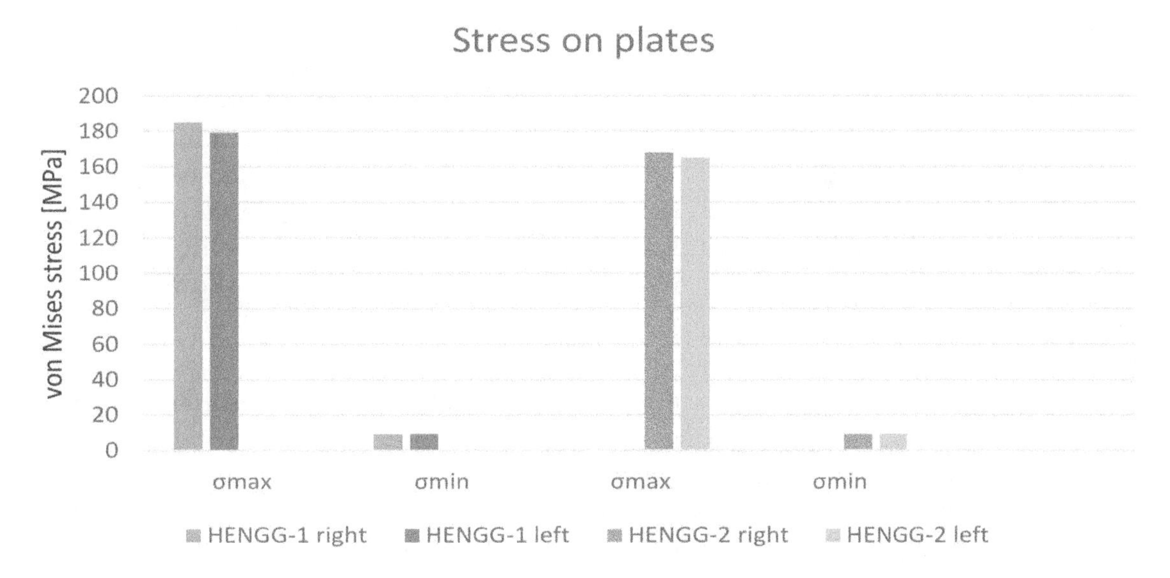

Figure 10. The von Mises stress values (in MPa).

3.2. Maxilla

All plates showed similar distribution patterns of maximum principal stress over the atrophic maxilla. The difference in the principal stress value between the four regions in contact with plates was 5%, with a mean maximum value of 80 MPa in the plates' region.

3.3. Long-Term Behavior

Failure probability of the situation detailed previously was obtained by employing the probabilistic methodology. The expected life computed was 1,819,235 ± 22.6 cycles. Then, the cumulative probability function was computed and represented in Figure 11.

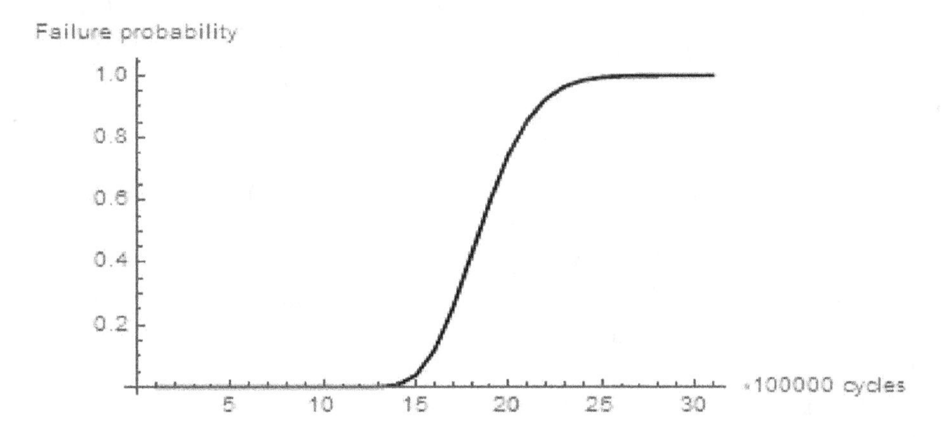

Figure 11. Cumulative probability function.

Figure 11 relates the probability of failure associated with each cycle load. As shown in the previous figure, the probability was equal to zero until 1,300,000 cycles.

4. Discussion

The biomechanical behavior of CF@O plates on a completely edentulous and atrophic maxilla was evaluated in the present study by employing finite element methods.

For a few years, there has been a trend towards minimally invasive implant treatment in very atrophic edentulous jaws and maxillae. The purpose of these concepts is to make an implant treatment with a shorter duration and smaller surgical risks [35]. The existence of insufficient bone can strongly influence the choice of the most appropriate rehabilitation in edentulous patients. There are several studies available in the literature that have employed different techniques to treat edentulous and atrophic maxilla, such as basal disk implants [36] or bone augmentation [37]. This study analyzed a new alternative based on titanium hybrid-plates. The accuracy of the results in numerical simulation studies depends on the precision of the model analyzed, the material properties, and the constraining conditions [38]. CT was used to model the geometry of the atrophic maxilla while the plates were provided by the manufacturer. A real clinical case with four titanium hybrid-plates in an atrophic maxilla was modeled and analyzed with the goal of knowing the biomechanical behavior of those plates.

This study had some assumptions and limitations. All materials were considered homogeneous, isotropic, and linearly elastic. Although these assumptions do not occur in clinical practice, they are common in finite element studies due to the challenges in establishing the properties of living tissues. These assumptions are consistent with other numerical studies [22,23,39]. In addition to these limitations, this work did not analyze the role of pterygoid implants as the goal was to study the biomechanical behavior of the plates. In that sense, ideal load distribution was considered.

The application of loads on the plates were supposed as an ideal transfer of loads from the prosthesis to that plate. If there is a good fit between the plate and the prosthesis, forces will be transmitted uniformly and as it was simulated and assumed in this study.

The ultimate strength in titanium grade II has been described as between 275 and 410 MPa and the ultimate tensile strength as 344 MPa. From these finite element analysis results, the maximum von Mises stress values in plates were lower than the ultimate strength [27]. The difference between the maximum von Mises stress between the distal plates was 3%, while the difference between the mesial plates was 2%. This difference was due to the different geometry of the maxilla in each area.

Küçükkurt et al. [40] compared the biomechanical behavior of different sinus floor elevations for dental implant placement. Under the condition of vertical loadings, von Mises stress in mesial implants were lower than our results in the plates in the case of lateral sinus lifting. However, the plates analyzed in this study obtained lower von Mises stress values than the prosthetic distal cantilever application and short implant placement. Regarding the distal implants, our plates obtained lower stresses than the prosthetic distal cantilever.

Ihde et al. [41] numerically analyzed baseplate implants with a vertical load of 114.6 N and a horizontal load 17.1 N and obtained a maximum von Mises stress of 400 MPa. Ihde et al. [42] detailed the von Mises stress in basal implants depending on the bone interface contact (BIC) degree. In this study, the maximum von Mises stress values were between 649 and 190 MPa. In both cases, the titanium hybrid plates used and analyzed in this study obtained lower stresses.

Kopp et al. [43] calculated the distribution of stress when basal implants in the mandible were loaded at two different stages of bone healing. They applied a load of 450 N located in the middle between the left molar and left canine implant and oriented in a vertical direction. Under these conditions, the von Mises stress in the basal implants was around 565 MPa.

The ultimate stress is an important value to understand the limits of the behavior of a material. According to physiological limits (ultimate stress), overloading in the cortical bone has been described as 170 MPa in compression and 100 MPa in tension [44]. Dos Santos et al. [45] detailed in their study that cortical bone resorption occurred when stress was higher than 167 MPa. Based on these limits, the values observed in this model were lower than those considered physiologic to bone tissue. As bone cannot be considered a ductile material, von Mises stress cannot be calculated in the maxilla. In this case, principal stresses have to be employed and calculated although some published studies have not used this kind of stress [39,45,46].

This is the reason why it is difficult to compare the results obtained in the atrophic maxilla of this study with the results provided in the bone in other published studies.

A good biomechanical behavior of plates is understood when a homogeneous stress is transferred to the bone. In this case, the maximum difference between the region of all four plates was 5%, meaning that the principal stress transferred from the plates to maxilla can be considered homogeneous.

Küçükkurt et al. [40] obtained a similar maximum principal stress in maxilla than our results in the case of short implant placement, and higher principal stress to our results in prosthetic distal cantilever application.

Clinical failures have generally been observed in the posterior maxilla region. Most of those failures were observed in bone types 3 and 4, with a highest probability of failures in bone type 4 [47]. According to the results obtained, the mean expected life in this case was 1,819,235 cycles. As Haug et al. detailed in their study [48], one year of in vivo service corresponds with, approximately, 200,000 cycles. Maló et al. obtained a satisfactory long-term outcome from patients with completely edentulous, severely atrophic maxillae supported by immediately loaded zygomatic implants alone, or in combination with conventional implants [49]. The same satisfactory results were detailed in Migliorança et al. who employed zygomatic implants placed lateral to the maxillary sinus and combined with conventional implants for a rehabilitation of the edentulous maxilla [50].

Further studies simulating these titanium hybrid-plates alternatives for atrophic maxilla and jaw that include dynamic forces that occur during chewing and consider the anisotropic and regenerative properties of bone are needed. Furthermore, some in vivo clinical trials are necessary to validate the model and to confirm the efficiency of this protocol. A numerical study of the combination of prosthesis-plates-implant under different functional conditions (bruxism and other parafunctions) just like the antagonist arcade is also necessary. Finally, a simulation of blood flow and bone regeneration around the plates is also necessary.

5. Conclusions

Based on the study results, it is possible to conclude that in terms of clinical application, the results indicate that titanium hybrid-plates proposed as an alternative to severe atrophic maxilla seems to have, from a mechanical point of view, a better behavior than conventional treatments such as prosthetic distal cantilever application and short implant placement. Titanium hybrid-plates distributed load to the maxilla with similar values as short implants but, with higher values than the prosthetic distal cantilever application. In any case, the resistance limits of bone and titanium were not exceeded. Long-term outcomes also seemed to be better than those clinical cases to treat atrophic maxilla. Finally, this technique can be considered as a viable alternative for atrophic maxilla rehabilitation, although more studies are necessary to corroborate the clinical results. As a clinical implication, this treatment allows the patient to be provided with functionally adequate prosthetic rehabilitation, which implies the recovery of their quality of life in a patient with a severe atrophy and, therefore, with an important challenge to a conventional implant treatment.

Author Contributions: M.P.-P. conceived, designed, and performed the analyses, evaluated the results, and wrote part of the paper. H.D. provided critical analysis. J.C.P.-F. provided critical analysis, interpretation of data, reviewed the literature, and wrote part of the paper.

References

1. Van der Mark, E.L.; Bierenbroodspot, F.; Baas, E.M.; de Lange, J. Reconstruction of an atrophic maxilla: Comparison of two methods. *Br. J. Oral Maxillofac. Surg.* **2011**, *49*, 198–202. [CrossRef] [PubMed]
2. Ali, S.; Karthigeyan, S.; Deivanai, M.; Kumar, A. Implant Rehabilitation For Atrophic Maxilla: A Review. *J. Indian Prosthodont. Soc.* **2014**, *13*, 196–207. [CrossRef] [PubMed]
3. Spencer, K. Implant based rehabilitation options for the atrophic edentulous jaw. *Aust. Dent. J.* **2018**, *63*, S100–S107. [CrossRef] [PubMed]

4. Cawood, J.I.; Howell, R.A. A classification of the edentulous jaws. *Int. J. Oral Maxillofac. Surg.* **1988**, *17*, 232–236. [CrossRef]
5. Chiapasco, M.; Zaniboni, M. Methods to Treat the Edentulous Posterior Maxilla: Implants With Sinus Grafting. *J. Oral Maxillofac. Surg.* **2009**, *67*, 867–871. [CrossRef] [PubMed]
6. Cucchi, A.; Vignudelli, E.; Franco, S.; Corinaldesi, G. Minimally Invasive Approach Based on Pterygoid and Short Implants for Rehabilitation of an Extremely Atrophic Maxilla. *Implant Dent.* **2017**, *26*, 639–644. [CrossRef] [PubMed]
7. Aparicio, C.; Manresa, C.; Francisco, K.; Claros, P.; Alández, J.; González-Martín, O.; Albrektsson, T. Zygomatic implants: Indications, techniques and outcomes, and the Zygomatic Success Code. *Periodontol. 2000* **2014**, *66*, 41–58. [CrossRef] [PubMed]
8. Gultekin, B.A.; Cansiz, E.; Borahan, M.O. Clinical and 3-Dimensional Radiographic Evaluation of Autogenous Iliac Block Bone Grafting and Guided Bone Regeneration in Patients With Atrophic Maxilla. *J. Oral Maxillofac. Surg.* **2017**, *75*, 709–722. [CrossRef] [PubMed]
9. Kaneko, T.; Nakamura, S.; Hino, S.; Norio, H.; Shimoyama, T. Continuous intra-sinus bone regeneration after nongrafted sinus lift with a PLLA mesh plate device and dental implant placement in an atrophic posterior maxilla: A case report. *Int. J. Implant Dent.* **2016**, *2*, 16. [CrossRef] [PubMed]
10. Alqutaibi, A.Y.; Altaib, F. Short dental implant is considered as a reliable treatment option for patients with atrophic posterior maxilla. *J. Evid. Based Dent. Pract.* **2016**, *16*, 173–175. [CrossRef] [PubMed]
11. Candel, E.; Peñarrocha, D.; Peñarrocha, M. Rehabilitation of the Atrophic Posterior Maxilla With Pterygoid Implants: A Review. *J. Oral Implantol.* **2012**, *38*, 461–466. [CrossRef] [PubMed]
12. Anitua, E.; Orive, G.; Aguirre, J.J.; Andía, I. Five-Year Clinical Evaluation of Short Dental Implants Placed in Posterior Areas: A Retrospective Study. *J. Periodontol.* **2008**, *79*, 42–48. [CrossRef] [PubMed]
13. Anitua, E.; Piñas, L.; Begoña, L.; Orive, G. Long-term retrospective evaluation of short implants in the posterior areas: Clinical results after 10–12 years. *J. Clin. Periodontol.* **2014**, *41*, 404–411. [CrossRef] [PubMed]
14. Scortecci, G.; Misch, C.; Benner, K. *Implants and Restorative Dentistry*; Martin Dunitz: London, UK, 2001.
15. Wu, C.; Lin, Y.; Liu, Y.; Lin, C. Biomechanical evaluation of a novel hybrid reconstruction plate for mandible segmental defects: A finite element analysis and fatigue testing. *J. Cranio-Maxillofac. Surg.* **2017**. [CrossRef] [PubMed]
16. Sidambe, A. Biocompatibility of Advanced Manufactured Titanium Implants—A Review. *Materials* **2014**, *7*, 8168–8188. [CrossRef] [PubMed]
17. Torres, Y.; Lascano, S.; Bris, J.; Pavón, J.; Rodriguez, J.A. Development of porous titanium for biomedical applications: A comparison between loose sintering and space-holder techniques. *Mater. Sci. Eng. C* **2014**, *37*, 148–155. [CrossRef] [PubMed]
18. Suzuki, K.; Takano, T.; Takemoto, S.; Ueda, T.; Yoshiari, M.; Sakurai, K. Influence of grade and surface topography of commercially pure titanium on fatigue properties. *Dent. Mater. J.* **2018**, *37*, 308–316. [CrossRef] [PubMed]
19. Elias, C.N.; Lima, J.H.C.; Valiev, R.; Meyers, M.A. Biomedical applications of titanium and its alloys. *JOM* **2008**, *60*, 46–49. [CrossRef]
20. Natali, A.N. *Dental Biomechanics*, 1st ed.; Taylor & Francis Group: London, UK, 2003; ISBN 0-415-30666-3.
21. Scortecci, G.; Bourbon, B. Dentures on the Diskimplant. *Rev. Fr. Prothes. Dent.* **1990**, *13*, 31–48.
22. Prados-Privado, M.; Bea, J.A.; Rojo, R.; Gehrke, S.A.; Calvo-Guirado, J.L.; Prados-Frutos, J.C. A New Model to Study Fatigue in Dental Implants Based on Probabilistic Finite Elements and Cumulative Damage Model. *Appl. Bionics Biomech.* **2017**, *2017*, 3726361. [CrossRef] [PubMed]
23. Ferreira, M.B.; Barão, V.A.; Faverani, L.P.; Hipólito, A.C.; Assunção, W.G. The role of superstructure material on the stress distribution in mandibular full-arch implant-supported fixed dentures. A CT-based 3D-FEA. *Mater. Sci. Eng. C* **2014**, *35*, 92–99. [CrossRef] [PubMed]
24. Agbaje, J.O.; Vrielinck, L.; Diederich, H. Rehabilitation of Edentulous Jaw Using Cortically Fixed at Once (Cf@O) Protocol: Proof of Principle. *Biomed. J. Sci. Tech. Res.* **2018**, *5*, 1–5. [CrossRef]
25. Diederich, H.; Junqueira, M.A.; Guimarães, S.L. Immediate Loading of an Atrophied Maxilla Using the Principles of Cortically Fixed Titanium Hybrid Plates. *Adv. Dent. Oral Health* **2017**, *3*, 1–3. [CrossRef]
26. Bhering, C.L.B.; Mesquita, M.F.; Kemmoku, D.T.; Noritomi, P.Y.; Consani, R.L.X.; Barão, V.A.R. Comparison between all-on-four and all-on-six treatment concepts and framework material on stress

distribution in atrophic maxilla: A prototyping guided 3D-FEA study. *Mater. Sci. Eng. C* **2016**, *69*, 715–725. [CrossRef] [PubMed]

27. Boyer, R.; Welsch, G.; Collings, E.W. *Materials Properties Handbook: Titanium Alloys*; ASM International: Materials Park, OH, USA, 1994.
28. Peixoto, H.E.; Camati, P.R.; Faot, F.; Sotto-Maior, B.; Martinez, E.F.; Peruzzo, D.C. Rehabilitation of the atrophic mandible with short implants in different positions: A finite elements study. *Mater. Sci. Eng. C* **2017**, *80*, 122–128. [CrossRef] [PubMed]
29. Shimura, Y.; Sato, Y.; Kitagawa, N.; Omori, M. Biomechanical effects of offset placement of dental implants in the edentulous posterior mandible. *Int. J. Implant Dent.* **2016**, *2*, 17. [CrossRef] [PubMed]
30. Arat Bilhan, S.; Baykasoglu, C.; Bilhan, H.; Kutay, O.; Mugan, A. Effect of attachment types and number of implants supporting mandibular overdentures on stress distribution: A computed tomography-based 3D finite element analysis. *J. Biomech.* **2015**, *48*, 130–137. [CrossRef] [PubMed]
31. Prados-Privado, M.; Prados-Frutos, J.C.; Calvo-Guirado, J.L.; Bea, J.A. A random fatigue of mechanize titanium abutment studied with Markoff chain and stochastic finite element formulation. *Comput. Methods Biomech. Biomed. Eng.* **2016**, *19*, 1583–1591. [CrossRef] [PubMed]
32. Riahi, H.; Bressolette, P.; Chateauneuf, A. Random fatigue crack growth in mixed mode by stochastic collocation method. *Eng. Fract. Mech.* **2010**, *77*, 3292–3309. [CrossRef]
33. Bogdanoff, J.; Kozin, F. *Probabilistic Models of Cumulative Damage*; Wiley: New York, NY, USA, 1985.
34. Neuber, H. Theory of Stress Concentration for Shear-Strained Prismatical Bodies With Arbitrary Nonlinear Stress-Strain Law. *J. Appl. Mech.* **1961**, *28*, 544–550. [CrossRef]
35. Wentaschek, S.; Hartmann, S.; Walter, C.; Wagner, W. Six-implant-supported immediate fixed rehabilitation of atrophic edentulous maxillae with tilted distal implants. *Int. J. Implant Dent.* **2017**, *3*, 35. [CrossRef] [PubMed]
36. Odin, G.; Misch, C.E.; Binderman, I.; Scortecci, G. Fixed Rehabilitation of Severely Atrophic Jaws Using Immediately Loaded Basal Disk Implants After In Situ Bone Activation. *J. Oral Implantol.* **2012**, *38*, 611–616. [CrossRef] [PubMed]
37. Del Fabbro, M.; Rosano, G.; Taschieri, S. Implant survival rates after maxillary sinus augmentation. *Eur. J. Oral Sci.* **2008**, *116*, 497–506. [CrossRef] [PubMed]
38. Van Staden, R.C.; Guan, H.; Loo, Y.C. Application of the finite element method in dental implant research. *Comput. Methods Biomech. Biomed. Eng.* **2006**, *9*, 257–270. [CrossRef] [PubMed]
39. Almeida, E.O.; Rocha, E.P.; Júnior, A.C.F.; Anchieta, R.B.; Poveda, R.; Gupta, N.; Coelho, P.G. Tilted and Short Implants Supporting Fixed Prosthesis in an Atrophic Maxilla: A 3D-FEA Biomechanical Evaluation. *Clin. Implant Dent. Relat. Res.* **2015**, *17*, e332–e342. [CrossRef] [PubMed]
40. Küçükkurt, S.; Alpaslan, G.; Kurt, A. Biomechanical comparison of sinus floor elevation and alternative treatment methods for dental implant placement. *Comput. Methods Biomech. Biomed. Eng.* **2017**, *20*, 284–293. [CrossRef] [PubMed]
41. Ihde, S.; Goldmann, T.; Himmlova, L.; Aleksic, Z.; Kuzelka, J. Implementation of contact definitions calculated by FEA to describe the healing process of basal implants. *Biomed. Pap. Med. Fac. Univ. Palacky Univ. Olomouc* **2008**, *152*, 169–73. [CrossRef]
42. Ihde, S.; Goldmann, T.; Himmlova, L.; Aleksic, Z. The use of finite element analysis to model bone-implant contact with basal implants. *Oral Surg. Oral Med. Oral Pathol. Oral Radiol. Endodontol.* **2008**, *106*, 39–48. [CrossRef] [PubMed]
43. Kopp, S.; Kuzelka, J.; Goldmann, T.; Himmlova, L.; Ihde, S. Modeling of load transmission and distribution of deformation energy before and after healing of basal dental implants in the human mandible. *Biomed. Tech. Eng.* **2011**, *56*, 53–58. [CrossRef] [PubMed]
44. Pérez, M.A.; Prados-Frutos, J.C.; Bea, J.A.; Doblaré, M. Stress transfer properties of different commercial dental implants: A finite element study. *Comput. Methods Biomech. Biomed. Eng.* **2012**, *15*, 263–273. [CrossRef] [PubMed]
45. Dos Santos Marsico, V.; Lehmann, R.B.; de Assis Claro, C.A.; Amaral, M.; Vitti, R.P.; Neves, A.C.C.; da Silva Concilio, L.R. Three-dimensional finite element analysis of occlusal splint and implant connection on stress distribution in implant–supported fixed dental prosthesis and peri-implantal bone. *Mater. Sci. Eng. C* **2017**, *80*, 141–148. [CrossRef] [PubMed]

46. Gümrükçü, Z.; Korkmaz, Y.T.; Korkmaz, F.M. Biomechanical evaluation of implant-supported prosthesis with various tilting implant angles and bone types in atrophic maxilla: A finite element study. *Comput. Biol. Med.* **2017**, *86*, 47–54. [CrossRef] [PubMed]
47. Sevimay, M.; Turhan, F.; Kiliçarslan, M.A.; Eskitascioglu, G. Three-dimensional finite element analysis of the effect of different bone quality on stress distribution in an implant-supported crown. *J. Prosthet. Dent.* **2005**, *93*, 227–234. [CrossRef] [PubMed]
48. Haug, R.; Fattahi, T.; Goltz, M. A biomechanical evaluation of mandibular angle fracture plating techniques. *J. Oral Maxillofac. Surg.* **2001**, *59*, 1199–1210. [CrossRef] [PubMed]
49. Maló, P.; Nobre Mde, A.; Lopes, A.; Ferro, A.; Moss, S. Five-year outcome of a retrospective cohort study on the rehabilitation of completely edentulous atrophic maxillae with immediately loaded zygomatic implants placed extra-maxillary. *Eur. J. Oral Implantol.* **2014**, *7*, 267–281. [PubMed]
50. Migliorança, R.; Coppedê, A.; Dias Rezende, R.; de Mayo, T. Restoration of the edentulous maxilla using extrasinus zygomatic implants combined with anterior conventional implants: A retrospective study. *Int. J. Oral Maxillofac. Implants* **2011**, *26*, 665–672. [PubMed]

13

Microstructural Characterization of Friction-Stir Processed Ti-6Al-4V

Sergey Mironov [1,*], Yutaka S. Sato [2], Hiroyuki Kokawa [2,3], Satoshi Hirano [4], Adam L. Pilchak [5] and Sheldon Lee Semiatin [5]

1. Laboratory of Mechanical Properties of Nanoscale Materials and Superalloys, Belgorod National Research University, Pobeda 85, 308015 Belgorod, Russia
2. Department of Materials Processing, Tohoku University, 6-6-02 Aramaki-aza-Aoba, Sendai 980-8579, Japan; ytksato@material.tohoku.ac.jp (Y.S.S.); kokawa@tohoku.ac.jp (H.K.)
3. Shanghai Key Laboratory of Materials Laser Processing and Modification, School of Materials Science and Engineering, Shanghai Jiao Tong University, 800 Dongchuan Road, Minhang District, Shanghai 200240, China
4. Hitachi Research Laboratory, Hitachi Ltd., 7-1-1 Omika-cho, Hitachi 319-1291, Japan; satoshi.hirano.xv@hitachi.com
5. Air Force Research Laboratory, Materials and Manufacturing Directorate, Wright-Patterson AFB, OH 45433-7817, USA; adam.pilchak.1@us.af.mil or adam.pilchak.1@afresearchlab.com (A.L.P.); sheldon.semiatin.1@us.af.mil or slsemiatin@gmail.com (S.L.S.)

* Correspondence: mironov@bsu.edu.ru

Abstract: The present work was undertaken to shed additional light on the globular-α microstructure produced during FSP of Ti-6Al-4V. To this end, the electron backscatter diffraction (EBSD) technique was employed to characterize the crystallographic aspects of such microstructure. In contrast to the previous reports in the literature, neither the texture nor the misorientation distribution in the α phase were random. Although the texture was weak, it showed a clear prevalence of the P_1 and C-fiber simple-shear orientations, thus providing evidence for an increased activity of the prism-<a> and pyramidal <c+a> slip systems. In addition, the misorientation distribution exhibited a crystallographic preference of 60° and 90° boundaries. This observation was attributed to a partial $\alpha \rightarrow \beta \rightarrow \alpha$ phase transformation during/following high-temperature deformation and the possible activation of mechanical twinning.

Keywords: Ti-6Al-4V; friction stir processing; electron backscatter diffraction; microstructure; texture

---

1. Introduction

Ti-6Al-4V is an important structural material which is widely used in aerospace, marine, and a number of other industries. The optimal balance of service properties in this alloy is often achieved by the formation of a "globular" microstructure. To this end, the material is subjected to a complex thermo-mechanical treatment, the final step of which involves processing below the β-transus temperature (at which $\alpha + \beta \rightarrow \beta$). Unfortunately, conventional industrial techniques (e.g., rolling, extrusion, forging) do not impose sufficient deformation to achieve the desired fully globular microstructure. Accordingly, the development of novel techniques is necessary to improve the performance of Ti-6Al-4V. One of these approaches is friction-stir processing (FSP) [1].

FSP involves a combination of very large strains and high temperatures and thus often leads to a substantial refinement in microstructure. Because of relatively high technological requirements (e.g., tool materials, working loads, etc.,), FSP of titanium alloys is relatively difficult, and until recently, microstructural observations of FSPed Ti-6Al-4V were limited. However, rapid progress in this technology during the past few years has promoted growing interest.

The first investigations of the FSP of Ti-6Al-4V revealed a relatively narrow processing window for this technique [2–18]. This is believed to be due to the relatively low ductility of Ti-6Al-4V even at elevated temperatures as well as its low thermal conductivity which gives rise to considerable temperature gradients during FSP. Moreover, a significant portion of the processing window often lies above β transus. FSP in such regimes does not provide the globular microstructure thus being not practical for material processing. Another important challenge is extensive wear of the processing tool [5,9,19–22]. This effect shortens tool life (thereby increasing processing cost) and results in a substantial contamination of the processed material by tool debris. Despite these shortcomings, careful control of process parameters may provide a desirable, fully globular microstructure which consists of fine-grain (or even ultrafine-grain) α phase [6,10,23–35]. In such instances, the processed material exhibits excellent superplastic properties [36–41].

Until recently, relatively little attention was given to examination of the crystallographic characteristics of such microstructures. It has sometimes been reported that the texture and misorientation distribution after FSP of Ti-6Al-4V are nearly random [12,24,27,28,34,35] Despite valuable prior efforts to describe the underlying microstructural mechanisms [42], it is still not clear whether such results are isolated cases or represent reproducible, general trends. Hence, the present work was undertaken to shed additional light on this issue. To this end, the electron backscatter diffraction (EBSD) technique was employed to characterize the globular-α microstructure produced during FSP of Ti-6Al-4V.

2. Materials and Methods

The material used in the present effort was Ti-6Al-4V, which was supplied as 7-mm-thick plate in the mill-annealed condition. Microstructural details of the base material are shown in supplementary Figures S1–S3. This material was friction-stir processed using a cobalt-based FSP tool. Based on previous experience, a relatively-low spindle (rotation) rate of 120 rpm was used to keep the processing temperature below the β transus. To investigate two points in a possible processing window, trials using feed rates of 15 and 30 mm/min were conducted. The welding tool had a diameter of 17.5 mm and a probe of 5.8 mm in length. Additional processing details were proprietary to Hitachi Ltd. To maintain consistency with the terminology in the literature, the principal directions of the process geometry are denoted herein as the welding direction (WD), transverse direction (TD), and normal direction (ND).

Microstructural observations were performed on a transverse cross-section (TD $\times$ ND plane) using optical microscopy (OM), scanning electron microscopy (SEM), electron probe microanalysis (EPMA), and EBSD. For the OM, SEM, and EPMA examinations, sectioned samples were first ground with silicon carbide abrasive papers, mechanically polished with 1-μm diamond, and then chemically etched with Kroll's reagent (3% HNO_3, 2% HF, and 95% H_2O). A suitable surface finish for EBSD was obtained by mechanical polishing in a similar fashion followed by long-term (24 h) vibratory polishing with colloidal silica.

SEM and EBSD analysis were conducted using a Hitachi S-4300SE field-emission-gun SEM (FEG-SEM) (Hitachi, Tokyo, Japan) equipped with a TSL OIM^{TM} EBSD system (EDAX, Mahwah, NJ, USA) and operated at an accelerating voltage of 25 kV. All SEM examinations were made in the secondary-electron mode. To reveal the microstructure at different length scales, EBSD maps were acquired with scan step sizes of either 0.5 or 0.2 μm. For each diffraction pattern, nine Kikuchi bands were used to minimize mis-indexing errors. To ensure reliability of the EBSD data, small grains comprising one or two pixels were automatically removed from the maps by applying the grain-dilation feature of the EBSD software. To eliminate spurious boundaries caused by orientation noise, a lower-limit boundary misorientation cut-off of 2° was employed. Furthermore, a 15° criterion was applied to differentiate low-angle boundaries (LABs) and high-angle boundaries (HABs). The grain size was quantified from the EBSD data by applying either the grain-reconstruction approach [43] (i.e., converting each grain to a circle with equivalent area and calculating the associated circle-equivalent diameter) or the conventional grain-boundary intercept method. EPMA measurements were conducted with a JEOL XM-85300FBU FEG-SEM (JEOL, Tokyo, Japan) operated at an accelerating voltage of 15 kV.

3. Results and Discussion

3.1. Preliminary Analysis

Low-magnification optical images of the FSP'ed samples (Figure 1) revealed a distinct stir zone for both feed rates. In the sample friction-stirred at the higher rate, a tunnel-type defect was found (indicated by the arrow in Figure 1b). Considering the relatively small change in process variables for the two cases, this observation suggested a rather-narrow processing window, a finding consistent with prior work [12–18]. To avoid the possible influence of this defect on the interpretation of microstructure evolution and material flow, subsequent examination was thus focused on the defect-free material produced using a feed rate of 15 mm/min.

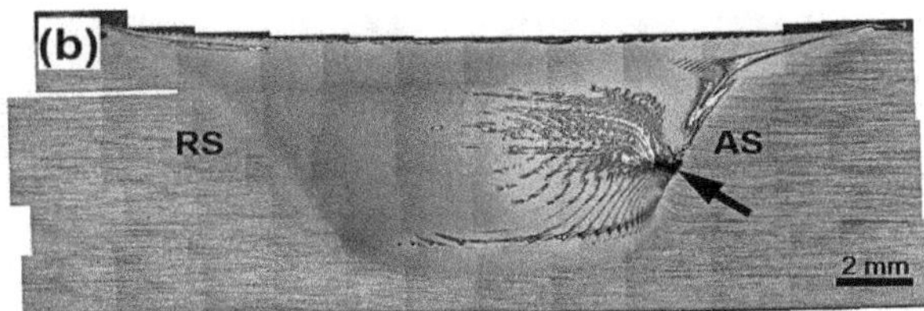

Figure 1. Low-magnification optical images showing a transverse cross section of samples processed using a feed rate of (**a**) 15 mm/min or (**b**) 30 mm/min. RS and AS denote the retreating side and the advancing side, respectively. The reference frame for both cases is shown in the bottom left corner of (**a**). In (**b**), the arrow indicates a tunnel-type defect.

One of the most striking features of the processed material was the very specific structural pattern which evolved at the advancing side and (Figure 1), to a lesser extent, at the upper surface of the stir zone (not shown). Such structures are often observed in friction-stirred Ti-6Al-4V and are usually attributed to tool wear [9,19–22]. SEM and EPMA of the material at the advancing side revealed the origin of this macrostructure (Figure 2). In the SEM image (Figure 2a), the α phase is dark, and the β phase is light. From this micrograph, it is appeared that the lighter features were β (or a β-rich) phase. The conclusion that some regions had transformed to β was rationalized by EPMA chemical maps (Figure 2b–e). These maps did indeed indicate a relatively-low concentration of titanium (Figure 2b) and an enrichment in cobalt, tungsten, and nickel (Figure 2c–e, respectively). Each of these elements are included in the specification for the processing-tool material. Therefore, it was likely that the observed clustering of these elements in the stir zone was associated with tool wear during FSP. The appropriate binary phase diagrams (not shown) indicated that additions of these elements to titanium *decrease* the β transus. Accordingly, a local inter-alloying of the stir zone material by these elements should result in stabilization of the β phase, as was observed (Figure 2a). Moreover, these observations are in good agreement with prior findings [9,19–22].

3.2. Microstructure Distribution within Stir Zone

In addition to the tool wear discussed in the previous section, noticeable variations in microstructure were found in the thickness direction of the stir zone (Figure 3). As indicated by the SEM micrographs, all of the microstructures were dominated by a relatively-fine globular-α phase. This indicates that the processing temperature was indeed below the β transus. In the upper and mid-thickness sections of the stir zone, in particular, a significant amount of secondary α (α_s) was also noted (Figure 3a,b). These observations suggested that the temperature in these areas likely exceeded ~900 °C, and that the material had experienced a *partial* $\alpha \rightarrow \beta \rightarrow \alpha_s$ phase transformation sequence.

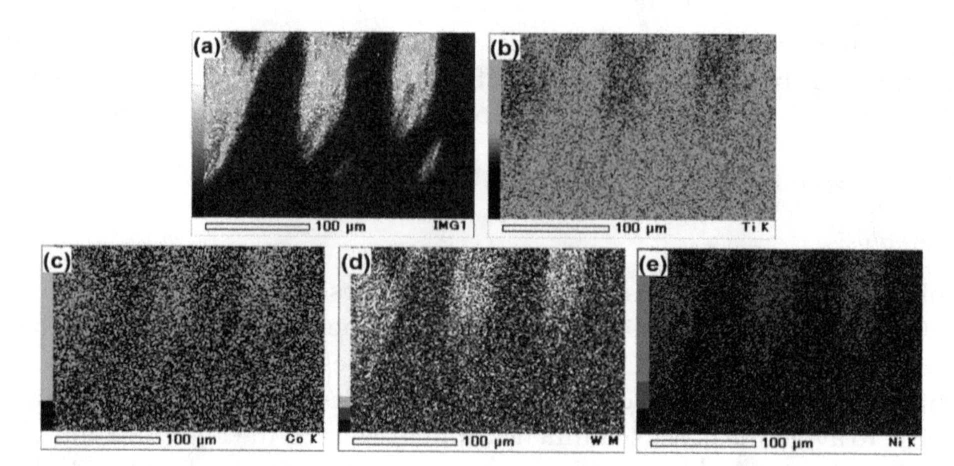

Figure 2. (**a**) SEM image and electron probe microanalysis (EPMA) (composition) maps for (**b**) titanium, (**c**) cobalt, (**d**) tungsten, and (**e**) nickel taken from the advancing side of the stir zone.

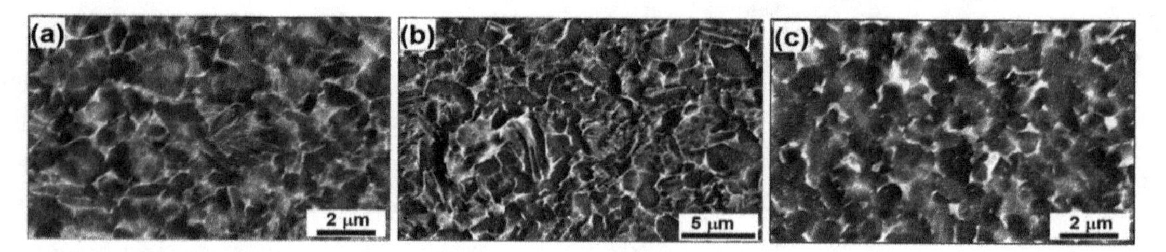

Figure 3. Typical SEM micrographs showing the microstructure distribution in the thickness direction of the stir zone: (**a**) Upper section, (**b**) mid-thickness, and (**c**) bottom section. Note the difference in magnification.

As expected, the finest microstructure was observed at the stir-zone root (Figure 3c). This is consistent with previous work, and is usually explained in terms of the relatively-low thermal conductivity of titanium alloys [13,15,18,24,44–46] and the fact that the main source of heating during FSP is friction between the workpiece surface and tool shoulder [1]. Accordingly, the local temperature is typically believed to decrease in the downward direction, thereby enhancing the microstructure-refinement effect at the stir zone root. Nevertheless, it was surprising to find a fine grain structure in the *upper* region of the stir zone as well (Figure 3a). The source of this effect is less clear, but is sometimes attributed to the relatively-large strain induced by the tool shoulder during FSP. Thus, the coarsest microstructure was observed at the *mid-thickness* of the stir zone (Figure 3b).

3.3. Microstructure Morphology and Grain Size

EBSD maps taken from the central section of the stir zone (Figure 4) provided deeper insight into microstructure evolution. The corresponding microstructure statistics derived from the EBSD maps are summarized in Figures 5–7. Due to difficulties in indexing the β phase, however, only limited data on this microstructural constituent could be obtained. Thus, the analysis in the present investigation was focused on the α phase, and the fine β particles (Figure 3b) were neglected during the evaluation of grain size and misorientation distribution for the α phase.

The *low-resolution* EBSD maps (Figure 4a) showed that the microstructure was very uniform. Equally important, there was no pronounced clustering of α grains sharing a common crystallographic axis (i.e., so-called "microtexture") in contrast to conventional processing of Ti-6Al-4V. This observation has been also highlighted by Pilchak, et al. [42]. In that work, the effect was attributed to the specific nature of deformation during FSP, which is close to simple shear and therefore has no true "end"/stable crystallographic orientation(s). The effect may also be enhanced by the relatively-complex strain path inherent to FSP in general.

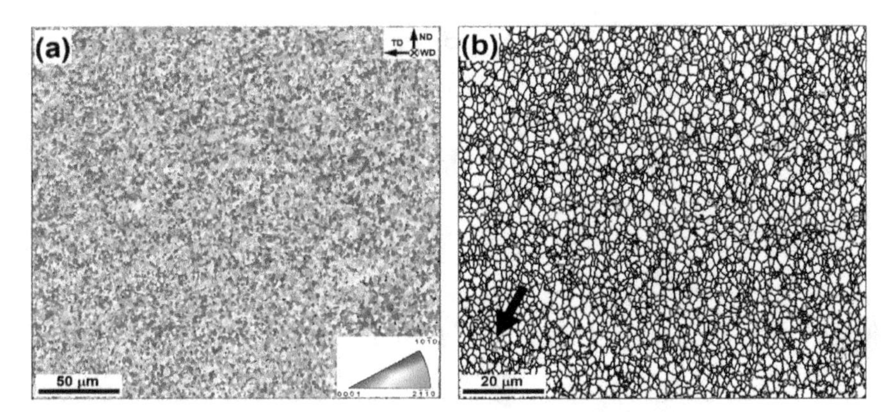

Figure 4. (**a**) Low-resolution EBSD inverse-pole-figure map and (**b**) higher-resolution EBSD grain-boundary map for the α phase taken from the central region of the stir zone. The reference frame for both maps is given in the top right corner of (**a**). In (**b**), low-angle boundaries (LABs) and high-angle boundaries (HABs) are depicted by red and black lines, respectively. Note: The black clusters denote the β phase.

As expected, the microstructure was dominated by the fine grains (Figure 4b) with a mean circle-equivalent diameter of 1.4 μm (Figure 5a) and mean HAB intercept length of 1 μm (Figure 5b). The HAB fraction was 86%. Not surprisingly, the remnant LABs were clustered primarily within relatively-coarse α grains which tended to subdivide them into smaller-scale fragments; an example is indicated by the arrow in Figure 4b. These observations suggested that microstructure evolution within the α phase during FSP occurred by a process of *continuous* dynamic recrystallization (CDRX), thus being consistent with similar reports in the literature [25,35,42]. Also in agreement with previous work, the α grains exhibited a nearly-globular shape, thus confirming the formation of a globular structure (Figure 4b).

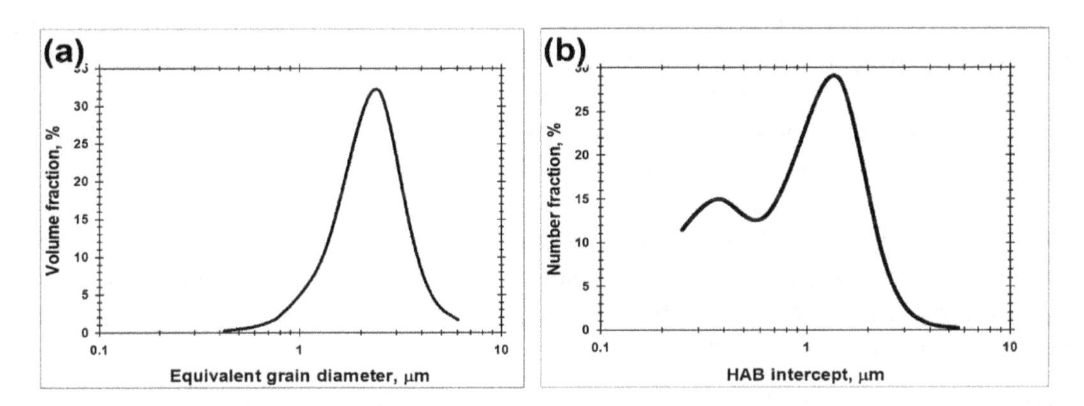

Figure 5. Grain-size distributions for the α phase measured by (**a**) the grain-reconstruction approach and (**b**) the grain-boundary intercept method from the EBSD map in Figure 4b.

3.4. Texture

It is commonly accepted that the deformation imposed during FSP is close to a mode of simple-shear [47,48]. However, according to Pilchak et al. [42], the strain path is simple shear only at limiting points around the tool; elsewhere, it is a mixture of simple and pure shear. The ideal orientations expected during simple shear of hcp crystals (such as α titanium) are given in Figure 6a and Table 1. For FSP, the shearing direction is tangential to the tool-rotation direction [48]. However, the orientation of the shear plane is usually less evident. Often, it is thought to be tangential to the tool-workpiece interface [49] or oriented nearly-parallel to the boundary between the stir zone and the thermo-mechanically affected zone (TMAZ) as in the truncated-cone model [50].

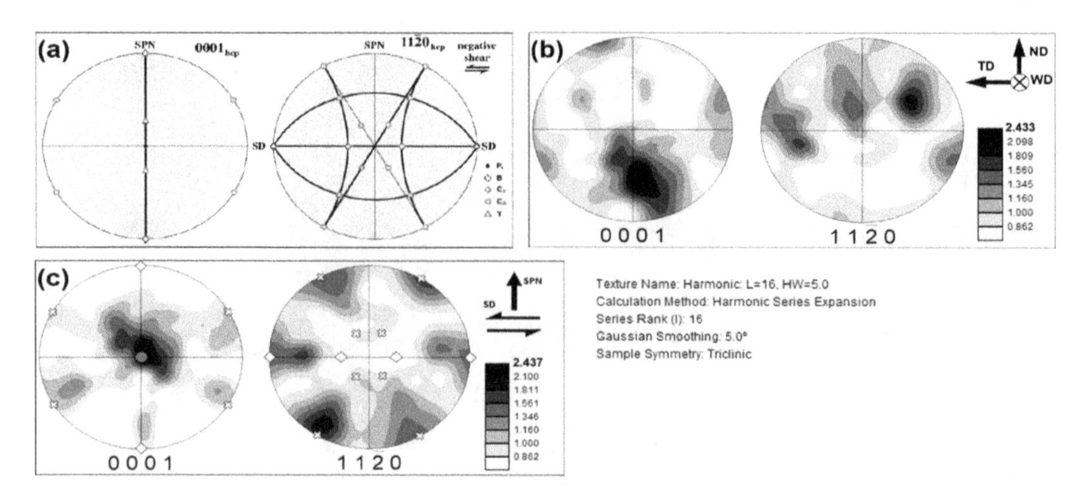

Figure 6. {0001} and $\{11\bar{2}0\}$ pole figures showing (**a**) ideal simple-shear textures expected for hexagonal close packed metals (after Fonda, et al. [48]), (**b**) those derived from Figure 4a, and (**c**) those derived from Figure 4a which were rotated 50° about the TD and then 15° about the ND to align with the presumed geometry of simple shear. For comparison purposes, several ideal simple-shear orientations are indicated in (**c**).

The textures developed during FSP were interpreted in terms of {0001} and $\{11\bar{2}0\}$ pole figures (Figure 6b) derived from the low-resolution EBSD map (Figure 4a). Assuming that the shear plane in the present work was parallel to the stir zone-TMAZ boundary, the measured pole figures were rotated 50° about the TD to align them with the presumed geometry of simple shear. They were also given an additional rotation of 15° about the ND in order to adjust the experimental data with the ideal textures. The rotated pole figures are shown in Figure 6c.

From the comparison of the rotated pole figures with the ideal textures in Figure 6a, it appeared that the FSP material was characterized by a preference for the P_1 $\{\bar{1}100\} < \bar{2}110 >$ and C-fiber components. Based on literature findings (Table 1), such results implied a prevalence of prism- <*a*> and pyramidal-<*c*+*a*> slip during FSP. In addition, there was some evidence for the development of a B-fiber component (Figure 6c), thus implying the possible activation of basal <a> slip as well (Table 1). However, these observations were not clear-cut, and thus further verification may be helpful. While the dominance of prism slip (with its low critical resolved shear stress) was expected, the activation of the pyramidal slip appeared somewhat surprising. It was likely necessitated by strain-compatibility requirements and perhaps the presence of the β phase. Despite these observations, the measured texture was weak with a maximum peak intensity of only ~2.5 times random (Figure 6c). This observation was in the line with previous texture studies [24,27,28,34,35].

It is also worth noting that the final microstructure was produced partially by a local $\alpha \rightarrow \beta \rightarrow \alpha_s$ phase transformation, as mentioned in Section 3.2. This process could influence texture evolution, and thus the analysis given in the present section may be oversimplified. Therefore, orientation measurements within the prior-β phase (and associated the secondary α) are needed to provide further insight into the evolution of texture during FSP.

Table 1. Ideal simple-shear textures in hexagonal close-packed metals (after Li [51], and Beausir et al. [52]).

Notation	Euler Angles (φ1, Φ, φ2)	Miller-Bravais Indices {*hkil*}<*uvtw*>	Primary Slip Mode
P-fiber	(0;0–90;0)	$\{hkil\} < \bar{2}110 >$	Prism <*a*> slip
P_1	(0;0;0)	$\{\bar{1}100\} < \bar{2}110 >$	Prism <*a*> slip
B-fiber	(0;90;0–60)	$\{0001\} < uvtw >$	Basal <*a*> slip
Y-fiber	(0;30;30–60)	-	Pyramidal <*a*> slip
C-fiber	(60;90;0–60)	-	Pyramidal <*c*+*a*> slip

3.5. Misorientation Distribution

Misorientation data extracted from the *high-resolution* EBSD map (Figure 4b) were arranged as distributions of misorientation angles and misorientation (rotation) axes (Figure 7). To assist in the interpretation of these experimental results, a so-called uncorrelated (or texture-derived) misorientation distribution was also calculated. For this approach, 1000 pixels were arbitrarily selected from the EBSD map and all possible misorientations between them were determined using one of the standard options in the EBSD software. Results from this texture-derived distribution were broadly similar to the distribution for a texture comprising randomly oriented grains (Figure 7a). This observation was likely due to the very weak texture that was developed within the stir zone, as discussed in the previous section. On the other hand, the measured grain-boundary misorientation distribution was noticeably different from both the texture-derived and random ones. Specifically, it was characterized by a pronounced low-angle maximum, additional misorientation peaks near ~60° and ~90° (Figure 7a), and the clustering of misorientation axes around several preferred crystallographic directions (Figure 7b).

The low-angle peak was likely attributable to the very large strain experienced by the material during FSP and the continuous nature of CDRX. Surprisingly, the rotation axes of the LABs were distributed in a near-random fashion (Figure 7b). This behavior contrasted with the preferential clustering of LAB axes near the <0001> pole often observed in heavily deformed α titanium [53–55]. The present finding may thus be a result of the complex character of slip involving the activation of both prism- *<a>* and pyramidal *<c+a>* modes, as discussed previously.

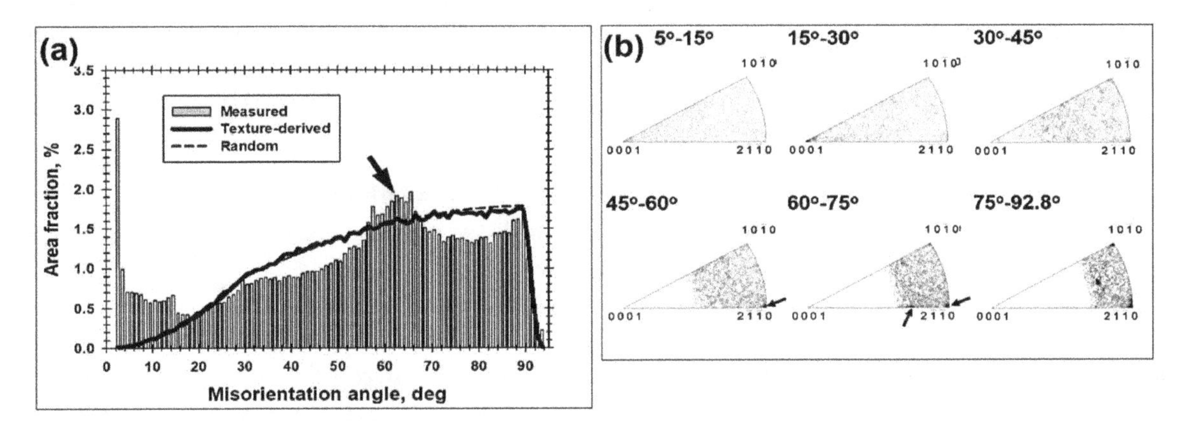

Figure 7. Distributions of (**a**) misorientation angle and (**b**) rotation axis for the α phase derived from the EBSD data in Figure 4b. The arrows show the misorientations which presumably originated from an $\alpha \rightarrow \beta \rightarrow \alpha_s$ phase transformation.

The noticeable proportion of ~60° boundaries was likely associated with the partial $\alpha \rightarrow \beta \rightarrow \alpha_s$ phase transformation, as discussed in Section 3.2. In titanium alloys, this transformation is normally governed by the Burgers orientation relationship, viz., $\{0001\}_\alpha // \{110\}_\beta$ and $< 11\bar{2}0 >_\alpha \ // < 111 >_\beta$. Because of the crystallographic symmetry of the α and β phases, there are 12 crystallographic variants of this relationship. The possible misorientations between the variants inherited from the same prior-β grain are shown in Table 2. From the table, it is seen that 3 of 5 such variants provide a peak near 60° in the misorientation-angle distribution and a clustering of rotation axes near the $< 2\bar{1}\bar{1}0 >$ pole, in broad agreement with the experimental data (indicated by the arrows in Figure 7). This provides an explanation for the origin of the 60° peak therefore. In addition, it is worth noting that the measured fraction of "inter-variant" boundaries constituted only ~6.5% of the total grain-boundary area. The relatively low fraction of such misorientations was probably associated with very fine-grain nature of the microstructure (Figure 3b) which typically results in the nucleation of only a single α variant within some prior β grains. On the other hand, an example of the prior-β grain structure containing several secondary-α colonies (which are responsible for the inter-variant misorientations) is shown in supplementary Figure S4.

Table 2. Predicted misorientations between α variants inherited from the same parent β grain (after Gey et al. [56] and Wang et al. [57]).

Misorientation (Angle-Axis Pair)	Symbol	Location of Misorientation Axes on Stereographic Triangle
$10.5° < 0001 >$	▲	$10\bar{1}0$, 0001, $2\bar{1}\bar{1}0$
$60° < 11\bar{2}0 >$	●	
$60.8° < \overline{1.377}; \bar{1}; 2.377; 0.359 >$	○	
$63.3° < \overline{10}; 5; 5; \bar{3} >$	■	
$90° < 1; \overline{2.38}; 1.38; 0 >$	□	

The crystallographic preference for ~$90° < 2\bar{1}\bar{1}0 >$, ~$90° < 10\bar{1}0 >$, and ~$90° < 16; \bar{4}; \overline{12}; 3 >$ misorientations (Figure 7b) is less clear. From a broad perspective, such boundaries in α titanium can be produced by mechanical twinning involving $\{10\bar{1}2\}$, $\{11\bar{2}3\}$, or $\{10\bar{1}2\} \rightarrow \{11\bar{2}3\}$ modes, respectively e.g., [58]. Indeed, evidence of twinning of the α phase has been found during uniaxial compression of Ti-6Al-4V at room temperature [59] and, very recently, during the hot compression of the single-phase-α alloy Ti-7Al [60].

It should be noted that lowering the FSP temperature (and the concomitant development of ultrafine microstructures similar to that in Figure 3c) suppresses the phase transformation, but may enhance mechanical twinning. Such changes would likely result in a different misorientation distribution. On the other hand, considering the relatively narrow processing window as well as the substantial temperature gradient within the stir zone, phase transformation and twinning may always exert an influence on microstructural evolution to some degree.

4. Conclusions

The present work was undertaken to provide insight into the globular microstructure developed during FSP of Ti-6Al-4V. To this end, the advanced capabilities of the EBSD technique were employed. The main results derived from this study were as follows.

The microstructure developed in the stir zone under nominally subtransus processing conditions results from a complex superposition of several processes. In addition to the strain-induced refinement (common to FSP), it is also influenced noticeably by the partial $\alpha \rightarrow \beta \rightarrow \alpha_s$ phase transformation (induced by the FSP thermal cycle) and inter-alloying due to the wear of the FSP tool. The markedly inhomogeneous microstructure distribution within the stir zone shows evidence of considerable variations in thermo-mechanical conditions.

The microstructure produced in the central section of stir zone is dominated by a fully globular α phase with a mean grain size of ~1 μm and HAB fraction of 86%, and an absence of microtexture. The LAB substructure within remnant, relatively coarse α grains suggest a continuous dynamic recrystallization mechanism of α phase refinement.

Although the texture in the α phase is very weak, there is a crystallographic preference for the formation of P_1 $\{\bar{1}100\} < \bar{2}110 >$ and C-fiber simple-shear components. This observation may be attributed to the dominance of prism- <*a*> and pyramidal <*a*+*c*> slip activity during FSP.

The misorientation distribution in the α phase is characterized by a noticeable proportion of 60° and 90° boundaries. The former boundaries are likely associated with a partial $\alpha \rightarrow \beta \rightarrow \alpha_s$ phase transformation, whereas the latter may indicate the possible activation of $\{10\bar{1}2\}$ and/or $\{11\bar{2}3\}$ twinning.

Supplementary Materials: The following are available online at Figure S1: SEM images of the microstructure of the base material at: (a) low magnification and (b) high magnification. Figure S2: EBSD characterization of the base material: (a) low-resolution orientation (inverse-pole-figure) map, and (b) (0001) and {11-20} pole figures showing the texture. In (a), LABs and HABs are depicted as white and black lines, respectively. Figure S3: EBSD characterization of the base material: (a) high-resolution orientation (inverse-pole-figure) map, and (b) misorientation distribution. In (a), LABs and HABs are depicted as white and black lines, respectively. In (b), the insert in the top right corner shows misorientation-axis distribution. Figure S4: SEM micrograph taken from the stir zone that exemplifies several secondary-alpha colonies within a prior-β grain.

Author Contributions: Conceptualization, S.H., Y.S.S., S.M.; methodology, S.H., Y.S.S., S.M.; formal analysis, S.M.; investigation, S.M.; resources, S.H., Y.S.S., H.K.; data curation, S.H., Y.S.S., S.M.; writing—original draft preparation, S.M.; writing—review and editing, Y.S.S., H.K., S.H., A.L.P., S.L.S.; visualization, S.M.; supervision, S.H., Y.S.S., H.K.; project administration, Y.S.S.; funding acquisition, S.H. All authors have read and agreed to the published version of the manuscript.

References

1. Mishra, R.S.; Ma, Z.Y. Friction stir welding and processing. *Mater. Sci. Eng. R* **2005**, *50*, 1–78. [CrossRef]
2. Juhas, M.C.; Viswanathan, G.B.; Fraser, H.L. Microstructural evolution in Ti alloy friction stir welds. In Proceedings of the Second Symposium on Friction Stir Welding, Gothenburg, Sweden, 26–28 June 2000.
3. Juhas, M.C.; Viswanathan, G.B.; Fraser, H.L. Characterization of microstructural evolution in a Ti-6Al-4V friction stir weld. In Proceedings of the Lightweight Alloys for Aerospace Application, TMS, Warrendale, PA, USA, 12–14 February 2001; Jata, K., Lee, E.W., Frazier, W., Kim, N.J., Eds.; TMS: Warrendale, PA, USA, 2001; pp. 209–217.
4. Ramirez, A.J.; Juhas, M.C. Microstructural evolution in Ti-6Al-4V friction stir welds. In *Material Science Forum*; Trans Tech Publications Ltd.: Zurich-Uetikon, Switzerland, 2003; Volume 426, pp. 2999–3004.
5. Pavka, P.A. Microstructural Evolution of Friction Stir Processed Ti-6Al-4V. Ph.D. Thesis, Ohio State University, Columbus, OH, USA, 2006.
6. Pilchak, A.L.; Juhas, M.C.; Williams, J.C. Microstructural changes due to friction stir processing of investment-cast Ti-6Al-4V. *Metall. Mater. Trans. A* **2007**, *38*, 401–408. [CrossRef]
7. Pilchak, A.L.; Li, Z.T.; Fisher, J.J.; Reynolds, A.P.; Juhas, M.C.; Williams, J.C. The relationship between friction stir processing (FSP) parameters and microstructure in investment cast Ti-6Al-4V. In *Friction Stir Welding and Processing IV*; Mishra, R.S., Mahoney, M.W., Lienert, T.J., Jata, K.V., Eds.; TMS: Warrendale, PA, USA, 2007; pp. 419–427.
8. Pilchak, A.L.; Juhas, M.C.; Williams, J.C. The effect of friction stir processing on microstructure and properties of investment cast Ti-6Al-4V. In Proceedings of the 11th World Titanium 2007 Conference, Kyoto, Japan, 3–7 June 2007; Ninomi, M., Akiyama, S., Ikeda, M., Hagiwara, M., Maruyama, K., Eds.; Japan Institute of Metals: Sendai, Japan, 2007; pp. 1723–1726.
9. Pilchak, A.L.; Juhas, M.C.; Williams, J.C. Observations of tool-workpiece interactions during friction stir processing of Ti-6Al-4V. *Metall. Mater. Trans. A* **2007**, *38*, 435–437. [CrossRef]
10. Pilchak, A.L.; Norfleet, D.M.; Juhas, M.C.; Williams, J.C. Friction stir processing of investment-cast Ti-6Al-4V: Microstructure and properties. *Metall. Mater. Trans. A* **2008**, *39*, 1519–1524. [CrossRef]
11. Pilchak, A.L.; Juhas, M.C.; Williams, J.C. A comparison of friction stir processing of investment cast and mill-annealed Ti-6Al-4V. *Weld. World* **2008**, *52*, 60–68. [CrossRef]
12. Lauro, A. Friction stir welding of titanium alloys. *Weld. Int.* **2012**, *26*, 8–21. [CrossRef]
13. Su, J.; Wang, J.; Mishra, R.S.; Xu, R.; Baumann, J.A. Microstructure and mechanical properties of a friction stir processed Ti-6Al-4V alloy. *Mater. Sci. Eng. A* **2013**, *573*, 67–74. [CrossRef]
14. Lippold, J.C.; Livingston, J.J. Microstructure evolution during friction stir processing and hot torsion simulation of Ti-6Al-4V. *Metall. Mater. Trans. A* **2013**, *44*, 3815–3825. [CrossRef]
15. Edwards, P.; Ramulu, M. Identification of process parameters for friction stir welding Ti-6Al-4V. *J. Eng. Mater. Technol.* **2010**, *132*, 031006–1. [CrossRef]
16. Edwards, P.; Ramulu, M. Effect of process conditions on superplastic forming behavior in Ti-6Al-4V friction stir welds. *Sci. Technol. Weld. Join.* **2009**, *14*, 669–680. [CrossRef]
17. Sanders, D.G.; Ramulu, M.; Edwards, P.D.; Cantrell, A. Effect on the surface texture, superplastic forming, and fatigue performance of Titanium 6Al-4V friction stir welds. *J. Mater. Eng. Perform.* **2010**, *19*, 503–509. [CrossRef]
18. Edwards, P.D.; Ramulu, M. Investigation of microstructure, surface and subsurface characteristics in titanium alloy friction stir welds of varied thicknesses. *Sci. Technol. Weld. Join.* **2009**, *14*, 476–483. [CrossRef]
19. Wang, J.; Su, J.; Mishra, R.S.; Xu, R.; Baumann, J.A. Tool wear mechanisms in friction stir welding of Ti-6Al-4V. *Wear* **2014**, *321*, 25–32. [CrossRef]
20. Pilchak, A.L.; Tang, W.; Sahiner, H.; Reynolds, A.P.; Williams, J.C. Microstructure evolution during friction stir welding of mill-annealed Ti-6Al-4V. *Metall. Mater. Trans. A* **2011**, *42*, 745–762. [CrossRef]

21. Fall, A.; Fesharaki, M.H.; Khodabandeh, A.R.; Jahazi, M. Tool wear characteristics and effect on microstructure in Ti-6Al-4V friction stir welded joints. *Metals* **2016**, *6*, 275. [CrossRef]
22. Wu, L.H.; Wang, D.; Xiao, B.L.; Ma, Z.Y. Tool wear and its effect on microstructure and properties of friction stir processed Ti-6Al-4V. *Mater. Chem. Phys.* **2014**, *146*, 512–522. [CrossRef]
23. Kitamura, K.; Fujii, H.; Iwata, Y.; Sun, Y.S.; Morisada, Y. Flexible control of the microstructure and mechanical properties of friction stir welded joints. *Mater. Design* **2013**, *46*, 348–354. [CrossRef]
24. Yoon, S.; Ueji, R.; Fujii, H. Effect of rotation rate on microstructure and texture evolution during friction stir welding of Ti-6Al-4V plates. *Mater. Character.* **2015**, *106*, 352–358. [CrossRef]
25. Zhou, L.; Liu, H.J.; Liu, P.; Liu, Q.W. The stir zone microstructure and its formation mechanism in Ti-6Al-4V friction stir welds. *Scripta Mater.* **2009**, *61*, 596–599. [CrossRef]
26. Liu, H.J.; Zhou, L.; Liu, Q.W. Microstructural characteristics and mechanical properties of friction stir welded joints of Ti-6Al-4V titanium alloy. *Mater. Design* **2010**, *31*, 1650–1655. [CrossRef]
27. Liu, H.J.; Zhou, L. Microstructural zones and tensile characteristics of friction stir welded joint of TC4 titanium alloy. *Trans. Nonferrous. Met. Soc. China* **2010**, *20*, 1873–1878. [CrossRef]
28. Zhou, L.; Liu, H.-J.; Wu, L.-Z. Texture of friction stir welded Ti-6Al-4V alloy. *Trans. Nonferrous Met. Soc China* **2014**, *24*, 368–372. [CrossRef]
29. Farnoush, H.; Bastami, A.A.; Sadeghi, A.; Mohandesi, J.A.; Moztarzadeh, F. Tribological and corrosion behavior of friction stir processed Ti-CaP nanocomposites in simulated body fluid solution. *J. Mech. Beh. Biomed. Mater.* **2013**, *20*, 90–97. [CrossRef] [PubMed]
30. Esmaily, M.; Mortazavi, S.N.; Todehfalah, P.; Rashidi, M. Microstructural characterization and formation of α' martensite phase in Ti-6Al-4V alloy butt joints produced by friction stir and gas tungsten arc welding processes. *Mater. Design* **2013**, *47*, 143–150. [CrossRef]
31. Pasta, S.; Reynolds, A.P. Residual stress effects on fatigue crack growth in a Ti-6Al-4V friction stir welds *Fatigue Fract. Eng. Mater. Struct.* **2008**, *31*, 569–580. [CrossRef]
32. Steuwer, A.; Hattingh, D.G.; James, M.N.; Singh, U.; Buslaps, T. Residual stress, microstructure and tensile properties in Ti-6Al-4V friction stir welds. *Sci. Technol. Weld. Join.* **2012**, *17*, 525–533. [CrossRef]
33. Muzvidziwa, M.; Okazaki, M.; Suzuki, K.; Hirano, S. Role of microstructure on the fatigue crack propagation behavior of a friction stir welded Ti-6Al-4V. *Mater. Sci. Eng. A* **2016**, *652*, 59–68. [CrossRef]
34. Yoon, S.; Ueji, R.; Fujii, H. Microstructure and texture distribution of Ti-6Al-4V alloy joints friction stir welded below β-transus temperature. *J. Mater. Process. Technol.* **2016**, *229*, 390–397. [CrossRef]
35. Ma, Z.Y.; Pilchak, A.L.; Juhas, M.C.; Williams, J.C. Microstructural refinement and property enhancement of cast light alloys via friction stir processing. *Scripta Mater.* **2008**, *58*, 361–366. [CrossRef]
36. Zhang, W.; Liu, H.; Ding, H.; Fujii, H. Superplastic deformation mechanism of the friction stir processed fully lamellar *Ti-6Al-4V alloy. Mater. Sci. Eng. A.* **2020**, *785*, 139390. [CrossRef]
37. Ramulu, M.; Edwards, P.D.; Sanders, D.G.; Reynolds, A.P.; Trapp, T. Tensile properties of friction stir welded and friction stir welded-superplastically formed Ti-6Al-4V butt joints. *Mater. Design* **2010**, *31*, 3056–3061. [CrossRef]
38. Sanders, D.G.; Ramulu, M.; Edwards, P.D. Superplastic forming of friction stir welds in Titanium alloy 6Al-4V: Preliminary results. *Mater. Sci. Eng. Technol.* **2008**, *39*, 353–357. [CrossRef]
39. Edwards, P.D.; Sanders, D.G.; Ramulu, M. Simulation of tensile behavior in friction stir welded and superplastically formed Titanium 6Al-4V alloy. *J. Mater. Eng. Perform.* **2010**, *19*, 510–514. [CrossRef]
40. Sanders, D.G.; Ramulu, M.; Klock-McCook, E.J.; Edwards, P.D.; Reynolds, A.P.; Trapp, T. Characterization of superplastically formed friction stir weld in titanium 6Al-4V: Preliminary results. *J. Mater. Eng. Perform.* **2008**, *17*, 187–192. [CrossRef]
41. Edwards, P.D.; Sanders, D.G.; Ramulu, M.; Grant, G.; Trapp, T.; Comley, P. Thinning behavior simulations in superplastic forming of friction stir processed titanium 6Al-4V. *J. Mater. Eng. Perform.* **2010**, *19*, 481–487. [CrossRef]
42. Pilchak, A.L.; Williams, J.C. Microstructure and texture evolution during friction stir processing of fully lamellar Ti-6Al-4V. *Metall. Mater. Trans. A* **2011**, *42*, 773–794. [CrossRef]
43. Humphreys, F.J. Quantitative metallography by electron backscattered diffraction. *J. Microsc.* **1999**, *195*, 170–185. [CrossRef]
44. Ji, S.; Li, Z.; Wang, Y.; Ma, L. Joint formation and mechanical properties of back heating assisted friction stir welded Ti-6Al-4V alloy. *Mater. Design* **2017**, *113*, 37–46. [CrossRef]

45. Buffa, G.; Fratini, L.; Schneider, M.; Merklein, M. Micro and macro mechanical characterization of friction stir welded Ti-6Al-4V lap joints through experiments and numerical simulation. *J. Mater. Process. Technol.* **2013**, *213*, 2312–2322. [CrossRef]
46. Yoon, S.; Ueji, R.; Fujii, H. Effect of initial microstructure on Ti-6Al-4V joint by friction stir welding. *Mater. Design* **2015**, *88*, 1269–1276. [CrossRef]
47. Fonda, R.W.; Bingert, J.F.; Colligan, K.J. Development of grain structure during friction stir welding. *Scripta Mater.* **2004**, *51*, 243–248. [CrossRef]
48. Fonda, R.W.; Knipling, K.E. Texture development in friction stir welds. *Sci. Technol. Weld. Join.* **2011**, *16*, 288–294. [CrossRef]
49. Park, S.H.C.; Sato, Y.S.; Kokawa, H. Basal plane texture and flow pattern in friction stir weld of a magnesium alloy. *Metall. Mater. Trans. A* **2003**, *34*, 987–994. [CrossRef]
50. Reynolds, A.P.; Hood, E.; Tang, W. Texture in friction stir welds of Timetal 21S. *Scripta Mater.* **2005**, *52*, 491–494. [CrossRef]
51. Li, S. Orientation stability in equal channel angular extrusion. Part. II: Hexagonal close-packed material. *Acta Mater.* **2008**, *56*, 1031–1043. [CrossRef]
52. Beausir, B.; Toth, L.S.; Neale, K.W. Ideal orientations and persistence characteristics of hexagonal close packed crystals in simple shear. *Acta Mater.* **2007**, *55*, 2696–2705. [CrossRef]
53. Mironov, S.Y.; Salischev, G.A.; Myshlayaev, M.M.; Pippan, R. Evolution of misorientation distribution during warm 'abc' forging of commercial-purity titanium. *Mater. Sci. Eng. A* **2006**, *418*, 257–267. [CrossRef]
54. Dyakonov, G.S.; Mironov, S.; Semenova, I.P.; Valiev, R.Z.; Semiatin, S.L. Microstructure evolution and strengthening mechanisms in commercial purity titanium subjected to equal-channel angular pressing. *Mater. Sci. Eng. A* **2017**, *701*, 289–301. [CrossRef]
55. Mironov, S.; Sato, Y.S.; Kokawa, H. Development of grain structure during friction stir welding of pure titanium. *Acta Mater.* **2009**, *57*, 4519–4528. [CrossRef]
56. Gey, N.; Hubert, M. Characterization of the variant selection occurring during the $\alpha \rightarrow \beta \rightarrow \alpha$ phase transformations of a cold rolled titanium sheet. *Acta Mater.* **2002**, *50*, 277–287. [CrossRef]
57. Wang, S.C.; Aindow, M.; Starink, M.J. Effect of self-accommodation on α/α boundary populations in pure titanium. *Acta Mater.* **2003**, *51*, 2485–2503. [CrossRef]
58. Dyakonov, G.S.; Mironov, S.; Semenova, I.P.; Valiev, R.Z.; Semiatin, S.L. EBSD analysis of grain-refinement mechanisms operating during equal-channel angular pressing of coppercial-purity titanium. *Acta Mater.* **2019**, *173*, 174–183. [CrossRef]
59. Prakash, D.G.L.; Ding, R.; Morat, R.J.; Jones, I.; Withers, P.J.; Quinta da Fonseca, J.; Preuss, M. Deformation twinning in Ti-6Al-4V during low strain rate deformation to moderate strains at room temperature. *Mater. Sci. Eng. A* **2010**, *527*, 5734–5744. [CrossRef]
60. Semiatin, S.L.; Levkulich, N.C.; Salem, A.A.; Pilchak, A.L. Plastic flow during hot working of Ti-7Al. *Metall. Mater. Trans A* **2020**, 1–16, in press. [CrossRef]

14

Effect of Fe Content on the As-Cast Microstructures of Ti–6Al–4V–xFe Alloys

Ling Ding [1], Rui Hu [2], Yulei Gu [1], Danying Zhou [1], Fuwen Chen [1], Lian Zhou [1] and Hui Chang [1,*]

1 Tech Institute for Advanced Materials & College of Materials Science and Engineering, Nanjing Tech University, Nanjing 210009, China; dingling2013@njtech.edu.cn (L.D.); 825721910@njtech.edu.cn (Y.G.); zhoudanying@njtech.edu.cn (D.Z.); fuwenchen@njtech.edu.cn (F.C.); zhoul@c-nin.com (L.Z.)

2 State Key Laboratory of Solidification Processing, Northwestern Polytechnical University, Xi'an 710072, China; rhu@nwpu.edu.cn

* Correspondence: ch2006@njtech.edu.cn

Abstract: In this work, the evolution of the solidification microstructures of Ti–6Al–4V–xFe (x = 0.1, 0.3, 0.5, 0.7, 0.9) alloys fabricated by levitation melting was studied by combined simulative and experimental methods. The growth of grains as well as the composition distribution mechanisms during the solidification process of the alloy are discussed. The segregation of the Fe element at the grain boundaries promotes the formation of a local composition supercooling zone, thus inhibiting the mobility of the solid–liquid interface and making it easier for the grains to grow into dendrites. With the increase in Fe content, the grain size of the alloy decreased gradually, while the overall decreasing trend was mitigated. The segregation of Fe was more obvious than that of Al and V, and the increase in Fe content had less effect on the segregation of Al and V.

Keywords: titanium alloy; simulation; boundary; segregation

---

1. Introduction

Titanium alloys have become excellent structural materials in many fields in recent years, especially in the field of aerospace applications due to their high specific strength, corrosion resistance, and other advantages [1]. Fe, as a common β-eutectoid alloy element in titanium alloys, which is even stronger than Cr, has a great influence on the solid/liquid transformation point. The increase in Fe content may cause a β-spot. Generally, the Fe content in titanium alloy is less than 5.5 wt% [2].

According to the previous research, the mechanical properties of titanium alloys can be effectively improved by adding an appropriate amount of Fe [3,4]. Kudo et al. studied the influence of microstructure on the formability of a Ti–Fe alloy [5] and found that the formability of a Ti–Fe alloy increased with the decrease in the size of the prior β phase region. Bermingham et al. found that the addition of an appropriate amount of Fe can effectively refine the grains of titanium alloys [6]. It was considered that the segregation of Fe provided the undercooling needed to inhibit grain growth and activate adjacent nuclei. It is obvious that the addition of an appropriate amount of Fe in the titanium alloys can affect the morphology of the original beta grains during solidification, thus influencing the mechanical properties. Ehtemam designed and manufactured Ti–11Nb–xFe (x = 0.5, 3.5, 6, 9 wt%) alloys by cold crucible levitation melting to study the effect of Fe addition on its phase transformation, microstructure, and mechanical properties [7]. The results showed that the Ti–11Nb–0.5Fe alloy had a typical dual phase microstructure of α + β and the volume fraction of the β phase could be increased by increasing the Fe content. However, the formation and growth of the original beta grains during the solidification process of titanium alloys are difficult to observe experimentally, so it is not easy to verify the mechanism of Fe on the grain morphology.

Through the phase field simulation, the microstructure evolution during the solidification process as well as the influence of element content on the microstructure of the alloys can be examined. The phase field model is a powerful tool to describe the complex evolution of the interface between the matrix and new phases in the non-equilibrium state based on the unified control equations in the whole system [8,9], which is suitable for describing solid–liquid phase transformation [10]. However, the simulation of microstructure evolution with the phase field method relies on the data of the temperature field parameters and thermophysical parameters of the related elements. The electromagnetic-thermal coupled simulation conducted by Kermanpur et al. [11] and the multi physical field coupling simulation conducted by Li [12] provided the data needed for the temperature field of the microstructure simulation.

For the simulation of microstructure, Kundin et al. used the phase field method to simulate the solidification of the Ti–Fe alloy [13], and Gong et al. studied the microstructure evolution of a Ti–6Al–4V alloy by the phase field method [14–16]. As for the related thermophysical parameters, Nakajima used the tracer diffusion method and Mossbauer spectrum to study the diffusion of Fe in the β-titanium alloy [17]. It is considered that the diffusion mechanism of Fe in β-titanium alloy is an extremely rapid interstitial diffusion. Chen et al. used the DICTRA software (Thermo-Calc Software Solna, Sweden) to strictly evaluate the experimental diffusion data to determine the atomic mobility of the BCC phase in the Ti–Al–Fe system [18]. Through the comprehensive comparison between the calculated and the experimental diffusion coefficients, a better consistency is obtained. The developed mobility of atoms is verified by good prediction of the mutual diffusion behavior observed in the diffusion couple experiment in the existing literatures.

In this work, the effect of Fe content on the microstructure of Ti–6Al–4V–xFe (x = 0.1, 0.3, 0.5, 0.7, 0.9) alloys produced by levitation melting was studied by the phase field method and verified by experiments. Levitation melting is often used in the laboratory research of titanium alloys due to the small size and uniform composition of the ingot. The Ti–6Al–4V alloy is the most widely used titanium alloy (α + β type) with good comprehensive mechanical properties, which is composed of a vanadium rich BCC phase (body centered cubic, β) and aluminum rich HCP phase (hexagonal close packed, α) [19].

2. Model and Experiments

2.1. *Phase Field Model*

As the temperature change calculated according to the simulation is small, the following assumptions were made:

(1) The diffusion coefficients of Al, V, and Fe in the solid phase and the liquid phase did not change in the simulation.
(2) The temperature gradient and cooling rate in the whole process remained invariant.

Dendritic growth and grain growth models were established using MICRESS 6.3 (ACCESS e.V. Aachen, Germany) software. The dendrite growth model had a mesh size of 600 × 600, a cell resolution of 0.1 μm, and a minimum time step of 1×10^{-3} s. The initial condition was considered to be 1 for the initial grain. The grain growth model had a grid size of 1000 × 1000, a cell resolution of 5 μm, and a minimum time step of 1×10^{-2} s. Figure 1 shows a schematic of the modeled domain, which was in the middle of ingot. Set 10 initial grain levels to randomly generate grains according to grain radius and distribution density.

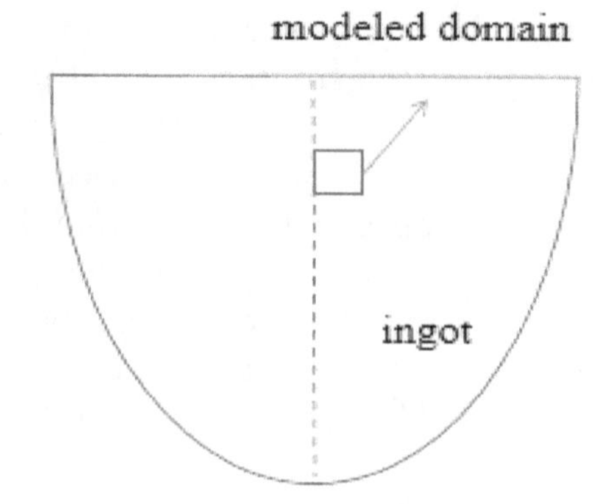

Figure 1. Schematic of the modeled domain.

Multiphase field theory is a computational method to describe the evolution of multiphase field parameters $\varphi_\alpha(\vec{x},t)$ in time and space [20]. At the solid–liquid interface, 0 and 1 represent the liquid phase and solid phase, respectively, and φ_α changes continuously between 0 and 1 with an interface thickness η. Based on the principle of minimum free energy, the multiphase field equation of MICRESS was used [21]:

$$\dot{\varphi}_\alpha = \sum_\beta M_{\alpha\beta}(\vec{n})\left(\sigma^*_{\alpha\beta}(\vec{n})K_{\alpha\beta} + \frac{\pi}{\eta}\sqrt{\varphi_\alpha\varphi_\beta}\Delta G_{\alpha\beta}(\vec{c},T)\right) \tag{1}$$

$$K_{\alpha\beta} = \varphi_\beta\nabla^2\varphi_\alpha - \varphi_\alpha\nabla^2\varphi_\beta + \frac{\pi^2}{\eta^2}(\varphi_\alpha - \varphi_\beta) \tag{2}$$

where $M_{\alpha\beta}$ is the mobility of the interface of the interface orientation, described by the normal vector $\vec{n}$. $\sigma^*_{\alpha\beta}$ is the effective anisotropic surface energy, and $K_{\alpha\beta}$ is about the local curvature of the interface. $\Delta G_{\alpha\beta}$ is the thermodynamic driving force, which is a function of the composition $\vec{c}$, and the diffusion equation can be described as:

$$\dot{\vec{c}} = \nabla\sum_{\alpha=1}^{N}\varphi_\alpha\vec{D}_\alpha\nabla\vec{c}_\alpha \tag{3}$$

where $\vec{D}_\alpha$ is the multicomponent diffusion coefficient matrix for phase α.

The boundary conditions are based on the symmetric boundary of the MICRESS software. The phase field value of the boundary element is defined to be the same as the second adjacent element in the analog domain, thereby revealing that a plane of symmetry is crossing through the center of the outermost element of the region. This condition is similar to an isolation condition that moves half a unit. The interface thickness is 5 cells.

The simulated interface energy can use common interface energy [22]. The phase diagram data required for the simulation are directly extracted from the Thermo-Calc 2015b (Thermo-Calc Software Solna, Sweden) TTTi3 database. The solid phase diffusion coefficients of the Al and V are calculated from the MOBTI1 database, and the liquid phase diffusion coefficients of Al and V are estimated. Since there are no diffusion data of Fe in the MOBTI1 database, a kinetic database containing Fe was prepared by Chen's study of β phase diffusion kinetics of a Ti–Al–Fe alloy [18], and the data obtained were imported into MICRESS to calculate the solid phase diffusion coefficient of Fe. The liquid phase diffusion coefficient of Fe was derived from the solid phase diffusion coefficient of Fe with reference to Kundin's study [13]. The partitial physical parameters are shown in Table 1.

Table 1. Partial physical parameters [13,18,22].

Physical Parameters	Ti–6Al–4V–xFe
Interface energy σ (J/cm^2)	2×10^{-5}
Al Liquid diffusion coefficient D_l (cm^2/s)	1.5×10^{-5}
Al Solid diffusion coefficient D_s (cm^2/s)	1.3×10^{-7}
V Liquid diffusion coefficient D_l (cm^2/s)	5×10^{-5}
V Solid diffusion coefficient D_s (cm^2/s)	6.9×10^{-7}
Fe Liquid diffusion coefficient D_l (cm^2/s)	1×10^{-4}
Fe Solid diffusion coefficient D_s (cm^2/s)	2×10^{-5}
Molar volume V (cm^3/mol)	11.2
Calculated temperature T (K)	1950
Anisotropic strength η	0.05

2.2. Experiments

The chemical composition of the Ti–6Al–4V–xFe samples (melted by Levitation melting to obtain a hemispherical ingot of about 800 g with a diameter of 90 mm, furnace cooling) is shown in Table 2.

Table 2. Mass fraction of each element.

Alloys	Al (wt%)	V (wt%)	Fe (wt%)	O (wt%)
Ti–6Al–4V	5.98	4.10	0.03	0.083
Ti–6Al–4V–0.1Fe	5.92	4.05	0.13	0.110
Ti–6Al–4V–0.3Fe	5.99	4.09	0.33	0.084
Ti–6Al–4V–0.5Fe	5.95	4.07	0.52	0.076
Ti–6Al–4V–0.7Fe	5.92	4.02	0.73	0.081
Ti–6Al–4V–0.9Fe	5.99	4.10	0.91	0.033

A 5 mm thick flat plate was cut by wire electrode cutting in the middle of the ingot. Six 15 × 15 mm squares were cut from the center of the ingots. The samples were electrolytic polished (using $HClO_4$:C_2H_5OH = 3:57 electrolyte) and quickly washed in alcohol and distilled water.

The metallographic photographs were obtained with an optical microscope (OM, Carl Zeiss, Jena, Germany). Line scan and surface scan images of the grain boundary of the Ti–6Al–4V–xFe alloys were obtained by electron probe micro analysis (EPMA, JEOL, Tokyo, Japan). As the primary β grain of the alloy is larger and the grain boundary is finer, the grain boundary is easily confused with the precipitated α lamellae structure, making it difficult to find the grain boundary in backscattered electron (BSE) mode. However, electropolishing (electropolishing is slightly corrosive) and secondary electron image (SEI) mode are used to find the original β grain boundary. Due to the precision limitation of EPMA, when the Fe content is low, it is difficult to measure it accurately. Therefore, Ti–6Al–4V–0.5Fe and Ti–6Al–4V–0.9Fe alloys were selected for surface scanning on the triangular crystal surface, and Ti–6Al–4V–0.5Fe, Ti–6Al–4V–0.7Fe, and Ti-6Al-4V-0.9Fe alloys were selected for line scanning through the grain boundary to obtain the corresponding element concentration distribution.

3. Results and Discussions

3.1. Effect of Fe Content on the Microstructure of Single Crystal

First, we studied the growth of the single grain. In the process of the alloy growing, the solute concentration in the liquid phase at the front of the solid–liquid interface decreased with the increase in distance from the interface, and the corresponding liquidus temperature T_L changed from low to high. When the curve of the liquidus temperature T_L was higher than the actual temperature T_Q line in the liquid phase, the composition supercooled zone will be formed in the liquid phase at the front of the solid–liquid interface.

With the solidification layer moving inward, the heat dissipation ability of the solid phase was gradually weakened. The internal temperature gradient tended to be gentle. The solute atoms in the liquid phase were enriched, so the component supercooling in front of the interface increased. As the distribution coefficient of Al and V elements is close to that of Ti and their content is relatively low, the alloy is similar to pure metal if there is no Fe element in the alloy. Therefore, the component supercooling was not obvious and the grain was nearly plane growth, as shown in Figure 2a (the color bar represents the field parameters field parameters φ, and 0 and 1 represent the liquid phase and solid phase respectively). When the Fe content increased to 0.9 wt%, the component supercooled region at the front of the interface was larger. The protruding part continued to grow into the supercooled liquid phase. At the same time, branches grew on its side, and the grain growth tended to be dendrite. With the increase in Fe content, the growth rate of the whole grain decreased. In a certain concentration range, Fe content has a great influence on the morphology of Ti–6Al–4V grains. As shown in Figure 3, in the early stage of solidification, the grain surface was relatively stable. The solid surface formed a bulge and gradually extended with time to the supercooled zone. Due to the small temperature gradient (5 K/cm) of the suspension melting, equiaxed grains were finally formed.

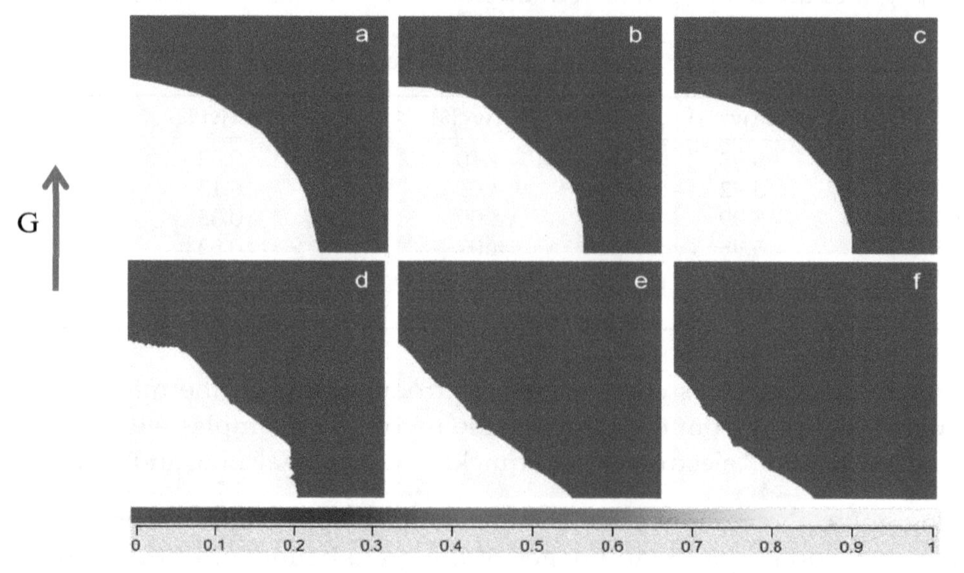

Figure 2. Effect of Fe content on the morphology of a single grain: (**a**) Ti–6Al–4V, (**b**) Ti–6Al–4V–0.1Fe, (**c**) Ti–6Al–4V–0.3Fe, (**d**) Ti–6Al–4V–0.5Fe, (**e**) Ti–6Al–4V–0.7Fe, (**f**) Ti–6Al–4V–0.9Fe.

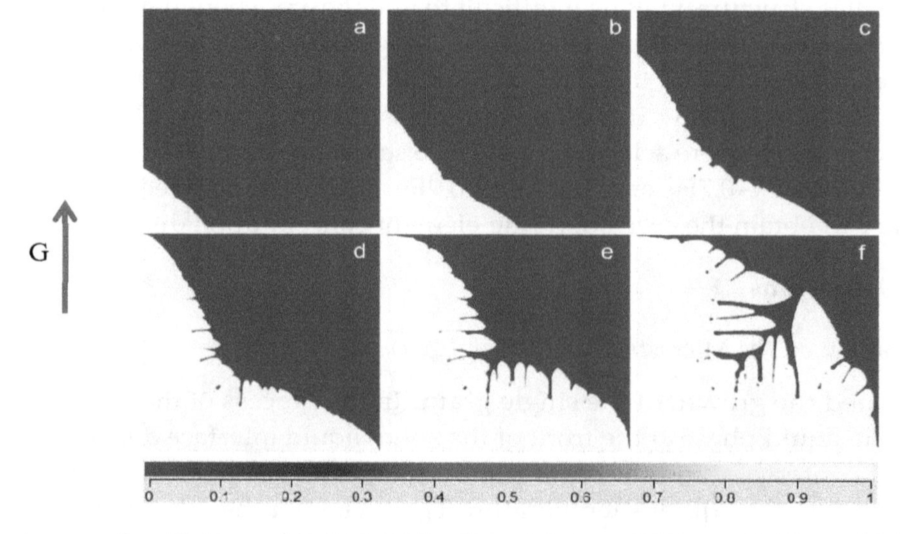

Figure 3. Grain growth with time of Ti–3Al–0.9Fe alloy: (**a**) 4 ms, (**b**) 8 ms, (**c**) 12 ms, (**d**) 16 ms, (**e**) 20 ms, (**f**) 30 ms.

The influence of the increase in Fe content on the component supercooling was discussed. The existence of supercooling zone depends on the temperature gradient at the solid–liquid interface determined by the external heat flux,

$$G = \left.\frac{dT_L}{dx'}\right|_{x'=0} \tag{4}$$

where G is the temperature gradient at the solid–liquid interface determined by the external heat flux; T_L is the actual temperature of the liquid phase at the front of the interface; and x' is the direction of the temperature gradient. In equilibrium, there is $G = -mG_c$, where G_c is the concentration gradient. When $G \geq G_c$, the liquid phase in the front interface is in the state of component supercooling. According to the study of Kurz et al. [23], by assuming that there is no convection in the liquid phase and only diffusion, the critical condition of component supercooling can be rewritten as

$$\frac{G}{v} \geq \frac{mC_0(1-k_0)}{Dk_0} \tag{5}$$

where m is the slope of liquidus; D is the diffusion coefficient of liquid phase; v is the migration rate of interface; C_0 is the initial composition; and k_0 is the distribution coefficient. For the Ti–6Al–4V–xFe alloy in this paper, if Ti is the solvent and Al, V, and Fe are the solute, then the liquid surface is a function of the concentration of Al, V, and Fe in the liquid phase, $T_L = T_L(C_{Al}, C_V, C_{Fe})$. The temperature gradient of the liquid melting point at the solid–liquid interface is:

$$\left.\frac{dT_l}{dx'}\right|_{x'=0} = m_{Al}\left(\frac{dC_{Al}}{dx'}\right)_{x'=0} + m_V\left(\frac{dC_V}{dx'}\right)_{x'=0} + m_{Fe}\left(\frac{dC_{Fe}}{dx'}\right)_{x'=0} \tag{6}$$

where m_{Al} is slope of the liquidus of C_{Al}, $m_{Al} = \frac{dT_L}{dC_{Al}}$, $m_V = \frac{dT_L}{dC_V}$, and $m_{Fe} = \frac{dT_L}{dC_{Fe}}$.

In the equilibrium state, the solute mass at the solid–liquid interface is conserved. Assuming that there is no interaction between Al, V, and Fe, there is

$$D_{Al}\left(\frac{dC_{Al}}{dx'}\right) = -v\left(\frac{C_{0Al}}{k_{Al}} - C_{0Al}\right) \tag{7}$$

$$D_V\left(\frac{dC_V}{dx'}\right) = -v\left(\frac{C_{0V}}{k_V} - C_{0V}\right) \tag{8}$$

$$D_{Fe}\left(\frac{dC_{Fe}}{dx'}\right) = -v\left(\frac{C_{0Fe}}{k_{Fe}} - C_{0Fe}\right) \tag{9}$$

where D_{Al}, D_V, and D_{Fe} are the liquid diffusion coefficients of the corresponding element; C_{0Al}, C_{0V}, and C_{0Fe} are the initial concentrations of the corresponding element; k_{Al}, k_V, and k_{Fe} are the partition coefficients of the corresponding elements. Substitute Equations (7)–(9) into Equation (6), and the actual temperature gradient G is greater than or equal to $\left.\frac{dT_l}{dx'}\right|_{x'=0}$:

$$\frac{G}{v} \geq -\frac{m_{Al}C_{0Al}(1-k_{Al})}{D_{Al}k_{Al}} - \frac{m_V C_{0V}(1-k_V)}{D_V k_V} - \frac{m_{Fe}C_{0Fe}(1-k_{Fe})}{D_{Fe}k_{Fe}} \tag{10}$$

According to Equation (10), due to $k_{Fe} < 1$, the component supercooling is easier to achieve when C_{0Fe} increases. Therefore, the increase in Fe content will promote the formation of the component supercooling zone, which will affect the morphology of the grains.

3.2. *Effect of Fe Content on the Microstructure of Multiple Grains*

The growth of several grains with different Fe content was simulated by MICRESS. Figure 4 shows the effect of Fe content on grain size (the color bar represents the mass fraction of Al). With higher Fe content, the shape of grains is more complex and the grain size is more refined. Due to the low directional temperature gradient in the levitation melting, the overall appearance of equiaxed crystal appears. The crystal interface is always composed of crystal faces with smaller interface energy The interface energy is smaller at the wide crystal face, while the energy of narrow crystal face at the edge is larger. Therefore, the crystal morphology tends to be spherical polyhedron in a stable state. Figure 5 shows as the time goes on, the liquid phase almost disappeared at 0.85 s, and an equiaxed crystal with larger grains was obtained. For the titanium alloy, the BCC phase of the cubic crystal system was first formed during solidification, and the optimal growth direction was the <001> crystal direction.

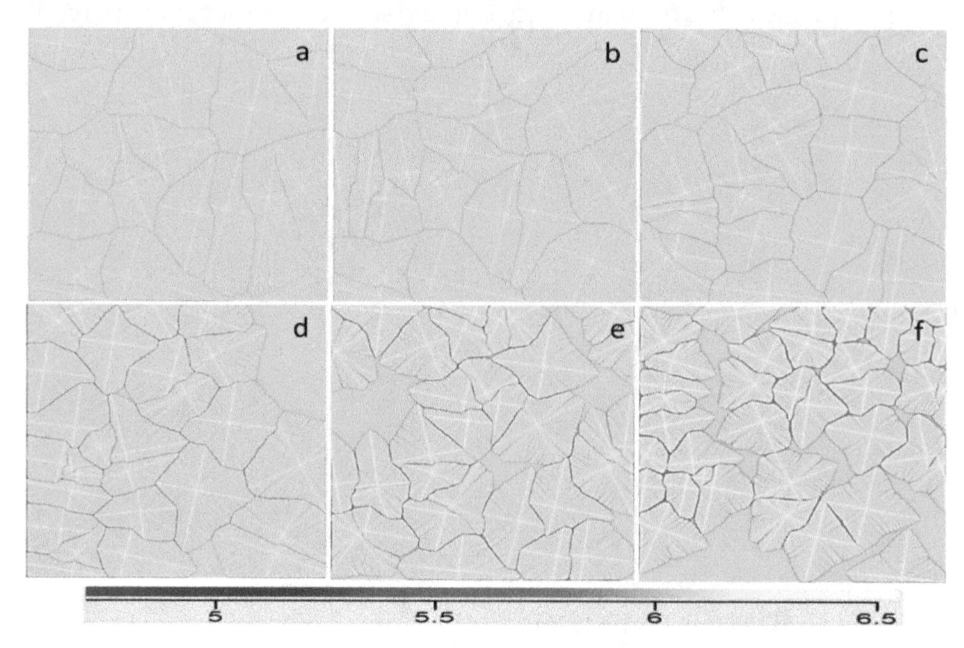

Figure 4. Effect of Fe content on grain size at 2 s: (**a**) Ti–6Al–4V–0Fe, (**b**) Ti–6Al–4V–0.1Fe, (**c**) Ti–6Al–4V–0.3Fe, (**d**) Ti–6Al–4V–0.5Fe, (**e**) Ti–6Al–4V–0.7Fe, (**f**) Ti–6Al–4V–0.9Fe.

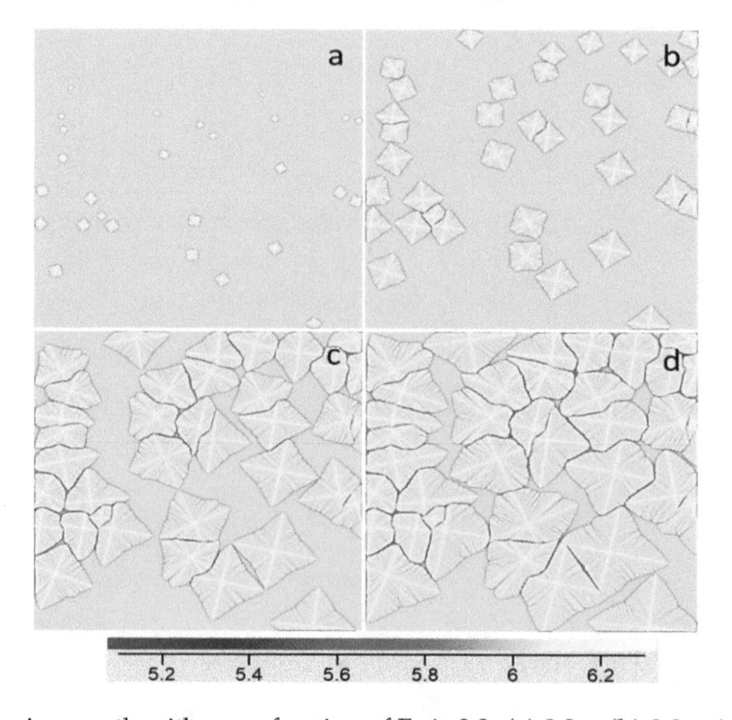

Figure 5. The grain growth with mass fraction of Fe is 0.9: (**a**) 0.3 s, (**b**) 0.8 s, (**c**) 1.5 s, (**d**) 2.0 s.

For the Ti–6Al–4V–xFe alloy, there was a large solute concentration gradient in the solid–liquid interface at the front edge of the polyhedron, and its diffusion rate was faster than that of the large plane crystal surface with a small solute concentration gradient at the front edge of the interface, resulting in the gradual change of the crystal from an octahedron to a star. This trend was more obvious at the region with a higher Fe content. Compared with Figure 4d, the segregation of Fe at the front of the solid–liquid interface in Figure 4f was stronger, and the resulting local supercooling slowed down the interface migration rate.

The microstructure of each direction was very different due to the different influence of the solute diffusion field and temperature diffusion field in four <001> directions. Figure 6 shows the effect of Fe content on the grain growth rate. When the Fe content exceeds 0.3 wt%, the growth rate of the alloy begins to decrease significantly. If the Fe content reaches 0.9 wt%, more time is needed for the liquid phase to disappear.

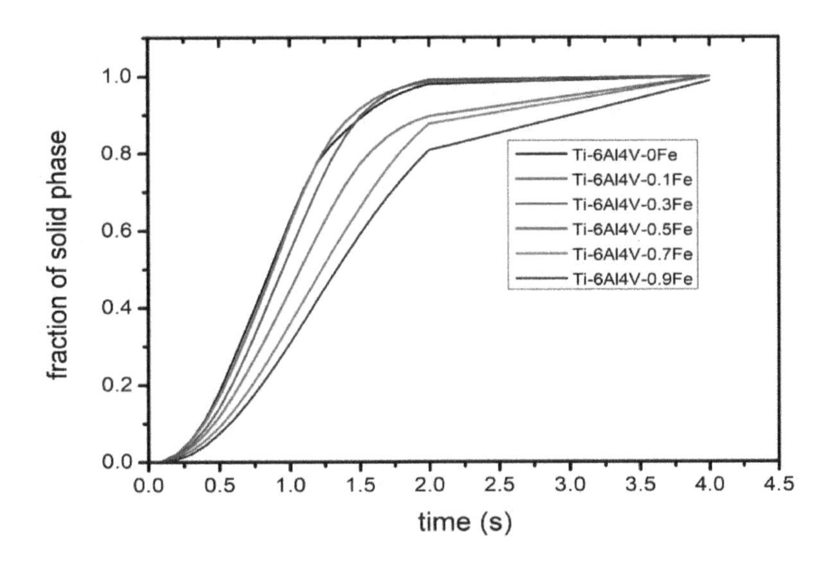

Figure 6. Effect of Fe content on grain growth rate.

In the growth process, the gap between the grains is large, and the growth speed of the grains is slow, which may provide more space for the growth of small grains and reduce the annexation of grains. Therefore, the increase of Fe in the experiment made the grains more refined.

Figure 7 shows the grain size of the alloy obtained by levitation melting. In a certain range, with the increase in Fe content, the grain size of the alloy gradually decreases. According to the number of grains and the cut-off area, the average grain radius is simply estimated, as shown in Figure 8. Compared with the simulated grain size, the experimental result was larger, which is due to the limited simulation time, while the experimental grains completed the grain growth. When there was no Fe in the alloy, as shown in Figure 7a, the grain size was the largest and the grain distribution was relatively uniform. The grain radius was about 2.29 mm, and the shape of the grain was close to circular. With the increase in Fe content, the grain size of the alloy decreased gradually, while the overall decreasing trend was mitigated. When the Fe content was 0.9 wt%, the average grain radius was the smallest (about 1.03 mm).

With the increase in Fe content, the distribution of grains was no longer uniform. Some small grains were distributed at the junction of larger grains, and the morphology of grains was close to a complex polygon. It can be considered that the addition of Fe changes the size and distribution of the grains and affects the shape of the grains, which verifies the simulation results.

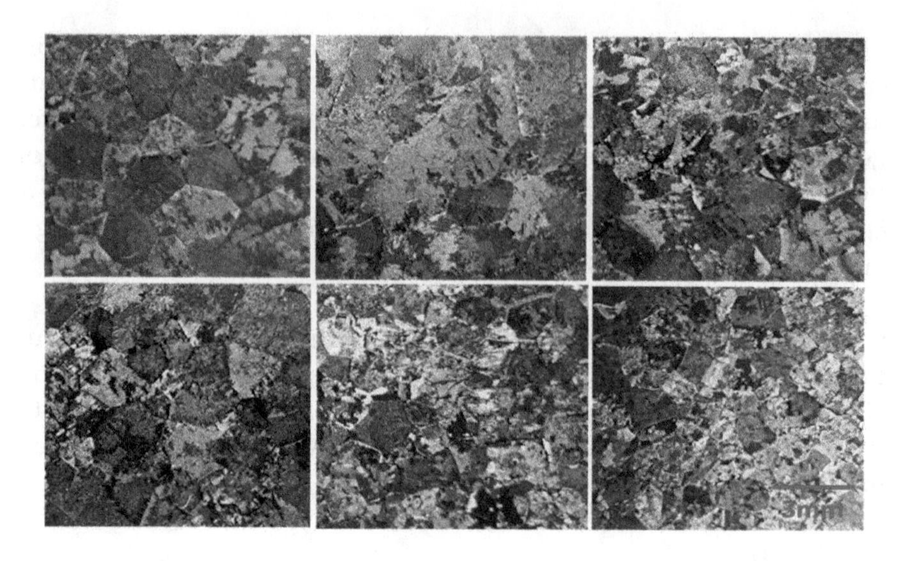

Figure 7. Effect of Fe content on grain size: (**a**) 0Fe, (**b**) 0.1Fe, (**c**) 0.3Fe, (**d**) 0.5Fe, (**e**) 0.7Fe, (**f**) 0.9Fe.

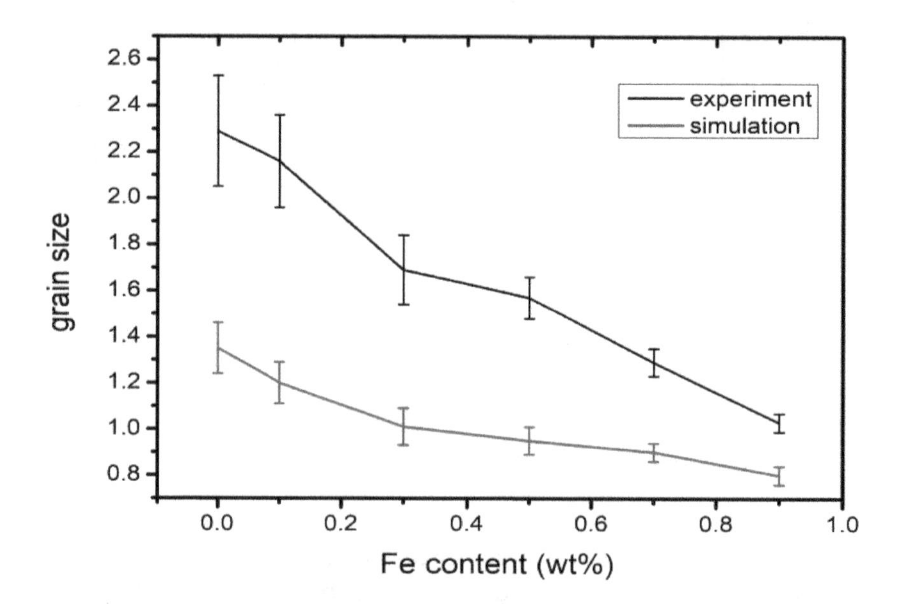

Figure 8. The change in the average grain size with Fe content.

3.3. Element Distribution in Ti–6Al–4V–xFe Alloy

As shown in Figure 9, the Fe composition distribution along the green lines was obtained and demonstrated in Figure 9e. As the Fe content increased from Figure 9a–d, the maximum solute concentration $C_L{}^*$ of Fe in the liquid phase at the solid–liquid boundary continued to rise (here represented by mass fraction), which were 0.67, 1.12, 1.48, and 1.63 wt%, respectively, corresponding to the four peaks in Figure 9e. The segregation ratio S_R was 4.01, 4.15, 4.00, and 3.54, respectively, and the overall segregation trend was reduced. Within a certain range, the diffusion distance δ_n of Fe (Figure 9f) in the liquid phase had a linear relationship with the Fe content in the alloy, and the relationship can be fitted as:

$$\delta_n = 31.1C_0 + 29.5 \tag{11}$$

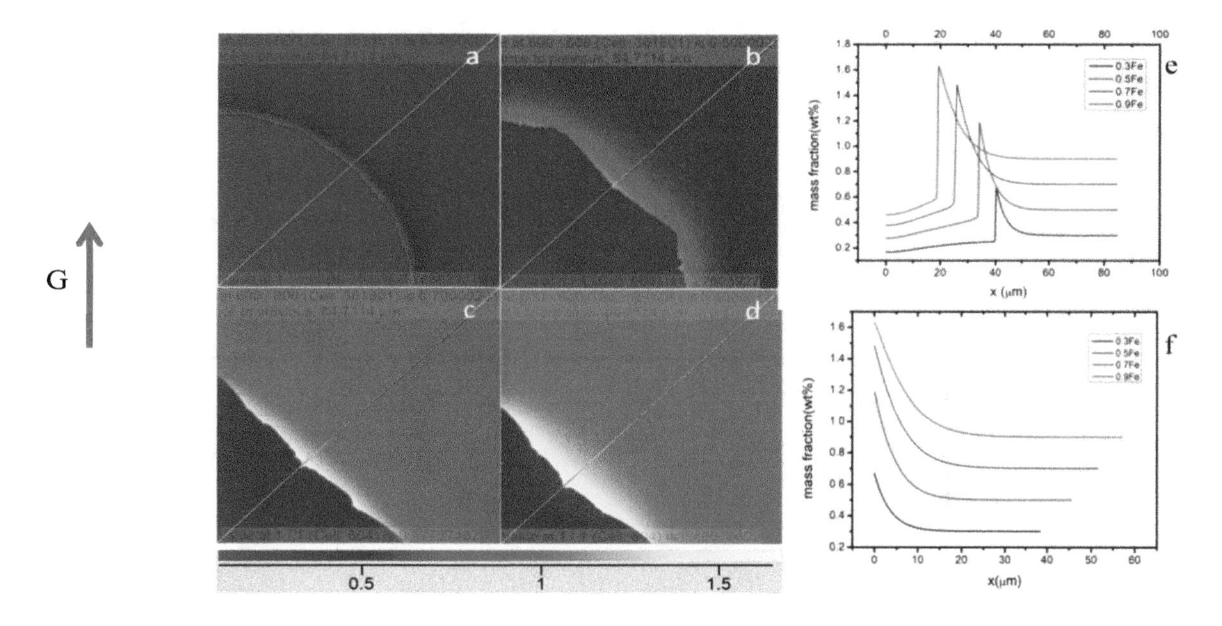

Figure 9. The change of Fe mass fraction in the direction perpendicular to the larger plane of grain under different Fe content: (**a**) 0.3Fe, (**b**) 0.5Fe, (**c**) 0.7Fe, (**d**) 0.9Fe, (**e**) liquid and solid phase, (**f**) liquid phase.

According to classical theory, for convective solute distribution, under directional solidification conditions, there is:

$$D_L \frac{d^2C_L}{dx^2} + v\frac{dC_L}{dx'} = 0 \tag{12}$$

when $x = 0$, $C_L = C_L^* < C_0/k_0$, and when $x = \delta_n$, $C_L = C_0$.

Defining $\frac{dC_L}{dx} = z$, $\frac{dz}{dx} = \frac{d^2C_L}{dx^2}$, then $\frac{dz}{z} = -\frac{v}{D_L}d$. After inserting the boundary conditions into the function, we can obtain:

$$C_L = \left(1 - \frac{1 - e^{-\frac{v}{D_L}x}}{1 - e^{-\frac{v}{D_L}\delta n}}\right)(C_L^* - C_0) + C_0 \tag{13}$$

where k_0 is the partition coefficient; x is the diffusion distance; v is the interface moving rate; and D_L is the liquid diffusion coefficient.

Three assumptions were made: (1) there is only diffusion (no convection) in the liquid phase; (2) the diffusion distance δ_n at the thin solid–liquid interface are infinity; and (3) the components of the liquid phase outside the solute enrichment layer keep the original concentration C_0 unchanged during the solidification process. Under these assumptions, the maximum solute concentration is $C_L^* = C_0/K_0$ in the stable liquid phase, and the solute distribution equation of the stable state can be simplified as:

$$C_L = C_0\left[1 + \left(\frac{1-k_0}{k_0}\right)e^{-\frac{v}{D_L}x}\right] \tag{14}$$

The composition of the liquid phase outside the solute enrichment layer is no longer C_0, but gradually increases in the case of limited liquid volume due to convection in the outer diffusion layer during the actual solute redistribution process. As the actual $C_L^* < C_0/K_0$, the solute concentration calculated by Equation (13) is higher, as shown in Figure 10. In this work, since the solidification process of levitation melting was not directional solidification, the direction of the temperature gradient has little effect on the grain morphology. As shown in Figure 9, the grain growth speed was slow in the direction perpendicular to the larger plane of the grain, which had an angle of 45° with the relative temperature gradient direction. The solute distribution value should be between Equations (13) and (14). For the levitation melting of the Ti–6Al–4V–xFe alloy with a slow growth rate, the modified distribution equation of solute in the steady state can be proposed according to the results of phase field simulation:

$$C_L = 0.79C_0\left[1 + \left(\frac{1 - k_0}{k_0}\right)e^{-\frac{v}{D_L}x}\right] + 0.11 \tag{15}$$

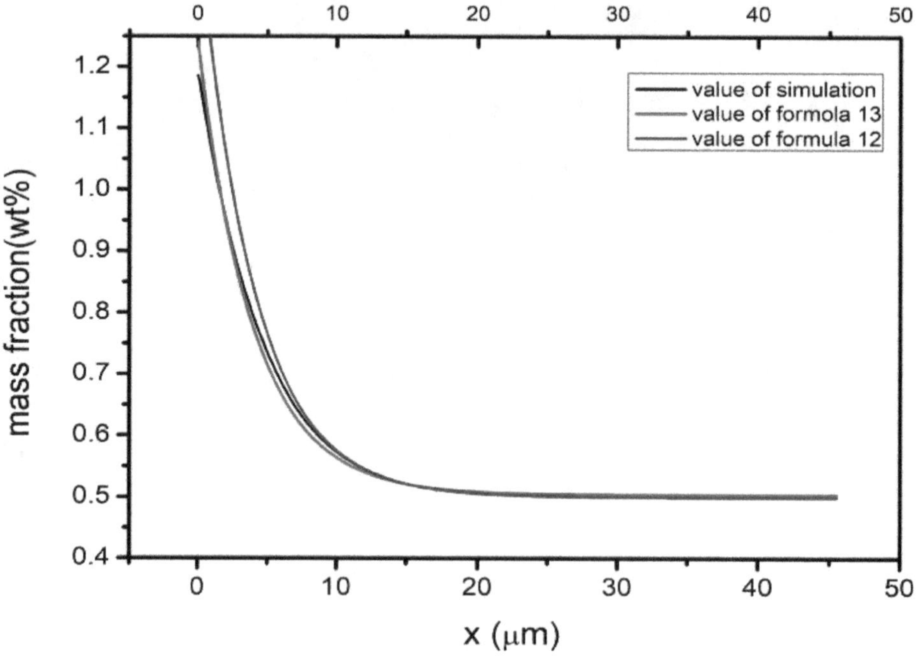

Figure 10. Solute distribution of Ti–6Al–4V–0.5Fe in the liquid phase.

The agreement of Equation (14) with the simulation results was close to 90%, while the agreement of the modified equation with the simulation results was close to 97%.

The composition at the triangular grain boundaries of Ti–6Al–4V–0.5Fe and Ti–6Al–4V–0.9Fe alloys were scanned by EPMA, and the results are shown in Figure 11. The segregation of Ti at the grain boundary was not obvious. The overall distribution presents a homogeneous contrast due to the matrix material of Ti. Compared with Figure 11a, Figure 11b shows that there was a certain segregation of the Al element in α lamellae. The most serious segregation was in the β grain boundary, while the lowest content was at the edges of the β grain boundary.

In the solidification process of the titanium alloy, the solid–liquid phase transformation first occurs, forming β original grains, and growing continuously with the decrease in temperature. The amount of liquid phase gradually decreases and concentrates at the boundary of β grains at the end of the solid–liquid phase transformation, as presented in Figure 5d. As the solidification proceeded, the liquid phase finally disappeared, forming the original β grain, as indicated in Figure 4a. With the slow decrease in temperature (i.e., non-quenching), the BCC phase in the high temperature state of the Ti alloy was gradually transformed into the HCP phase (i.e., β/α transformation, forming primary α phase). The α lamellar structure (about 0.5–2 μm) was formed in the original β grain; and the remaining β phase was distributed at the boundary of the α lamellar. The morphology of the β original grain was retained without any deformation in the end. As a result, the microstructure shown in

Figure 11a was formed. During the cooling process, a relatively wide α lamellar structure (about 2–3 μm) was formed from the β grain boundaries. The remaining β phase was distributed at the edges.

In the same way, V and Fe, as β stable elements, concentration increased from the inner area to the edges of α lamellae. Due to the wide β grain boundary, the segregation at the edge of the β grain boundary was more obvious. Since Fe is a stronger β stable element than V, the segregation of Fe was more obvious. Comparing the β grain boundaries in Figure 11a,c, Figure 11c was finer (about 1–2 μm), which may be attributed to the grain refinement of Fe. The distribution trend of Figure 11b was the same as in Figure 11d.

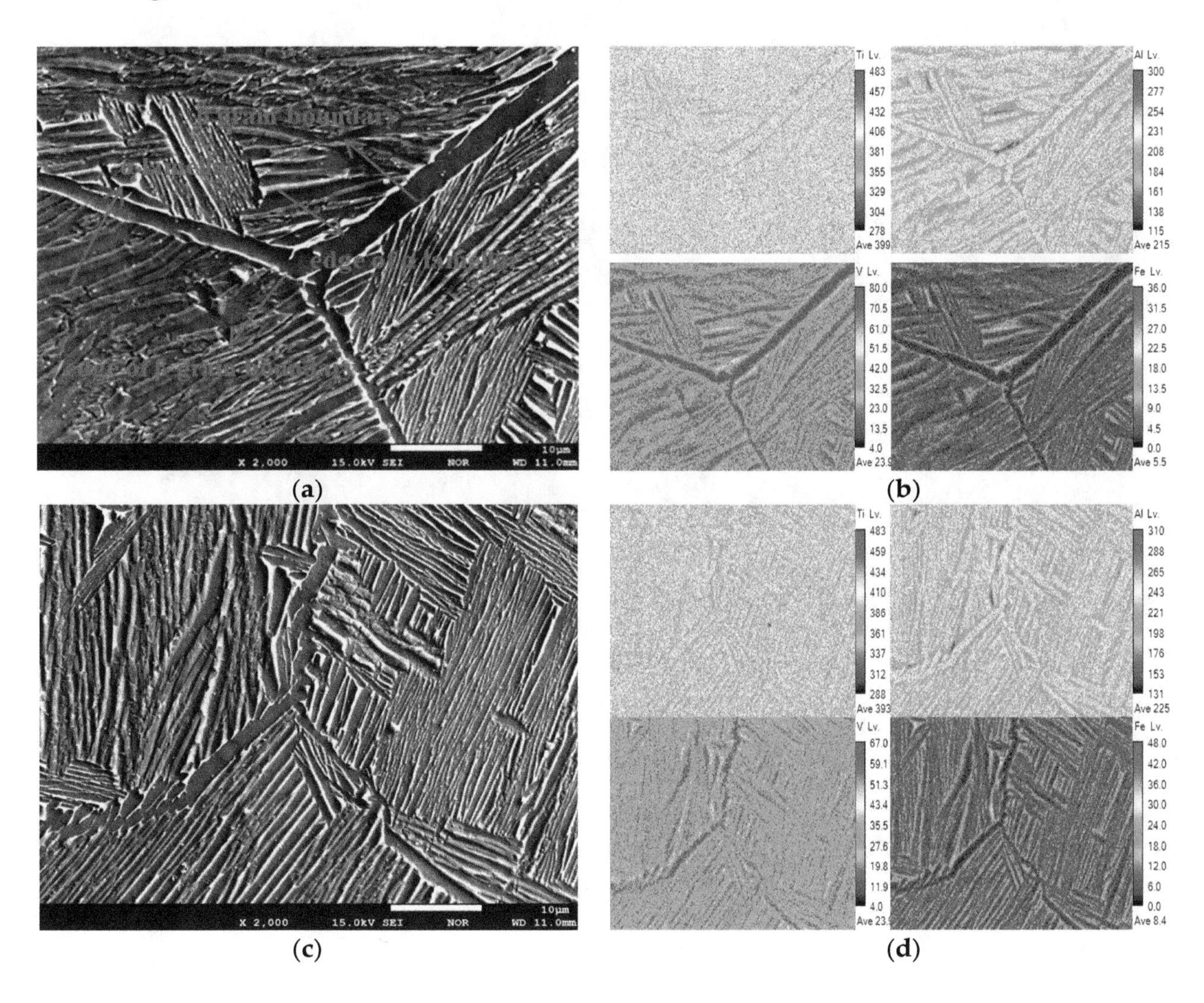

Figure 11. Results of the electron probe micro analysis surface scan. (**a**) Secondary electron image of Ti–6Al–4V–0.5Fe, (**b**) Concentration distribution of Ti–6Al–4V–0.5fe, (**c**) Secondary electron image of Ti–6Al–4V–0.9Fe, (**d**) Concentration distribution of Ti–6Al–4V–0.9Fe.

Figure 12 compares the simulated with the experimental values of the Fe composition distribution. As the simulation process does not complete the β/α transformation, the segregation of Fe is mainly concentrated in the residual liquid phase between β grains. Comparing the segregation degree of the simulated and experimental values, the segregation degree of Fe in the simulation was not more than three times that of the nominal composition, whereas the segregation degree of the experimental value was close to six times the nominal composition. This may be attributed to the decrease in β phase amount in β/α transformation and the further compression of the range of Fe segregation distribution.

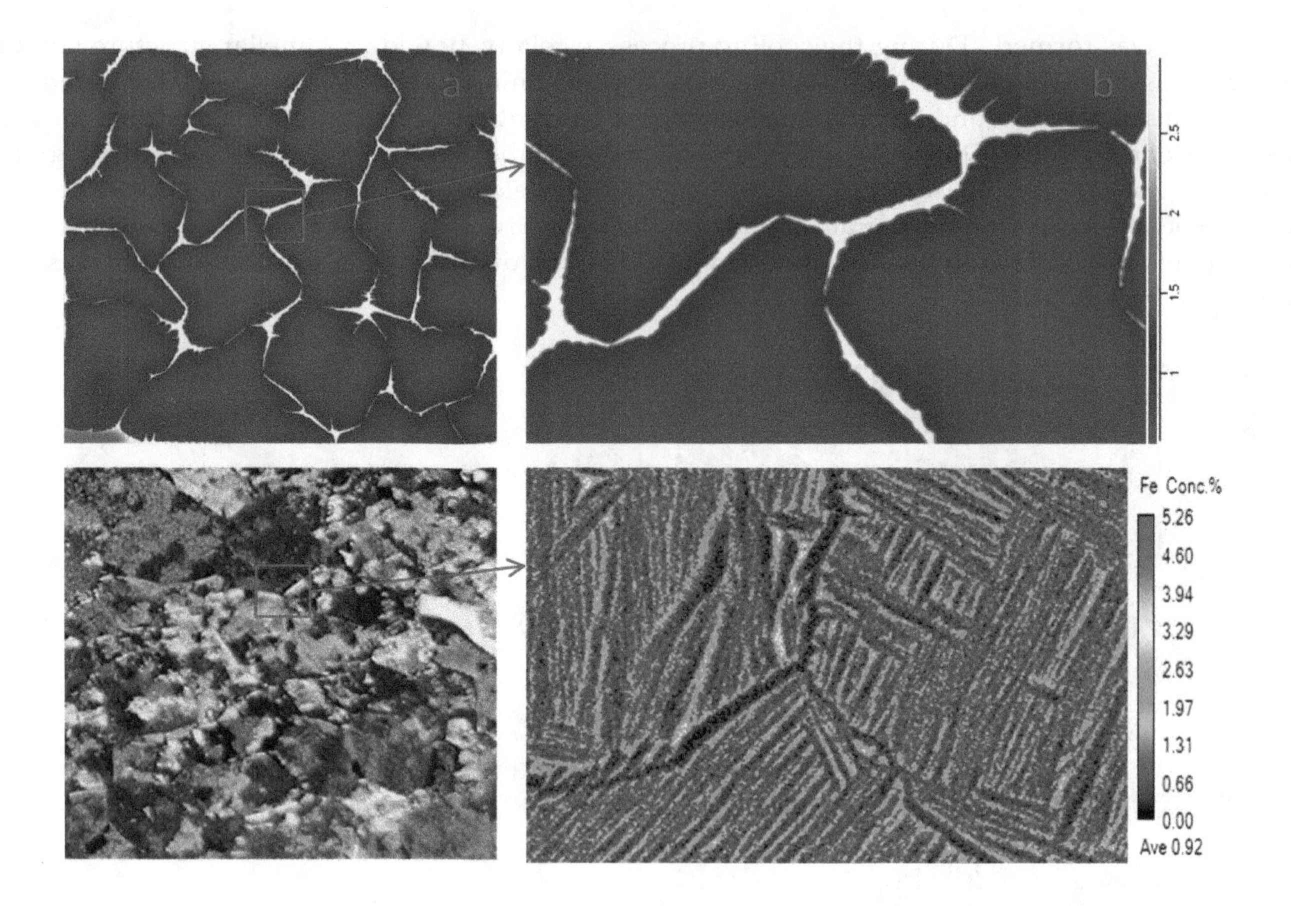

Figure 12. Comparison of simulated and experimental values of Fe concentration in Ti–6Al–4V–0.9Fe alloy. (**a**) Selection of simulation position, (**b**) Simulated value of Fe concentration, (**c**) Selection of experimental position, (**d**) Experimental value of Fe concentration.

Through line scans crossing the grain boundaries of Ti–6Al–4V–0.5Fe, Ti–6Al–4V–0.7Fe, and Ti–6Al–4V–0.9Fe alloys, it can be seen from Figure 13b that the fluctuation range of Al composition in the alloy ranged from 9.78 to 11.07 at%, V ranged from 2.19 to 7.98 at%, and Fe from 0.18 to 1.81 at%. Compared with the average composition, the fluctuation values of Al, V and Fe were 8%, 91%, and 202%, respectively. Obviously, the segregation of Fe was greater than V, while the segregation of V was greater than Al.

For three samples, the mean values of Fe composition were 0.60 at%, 0.76 at%, and 0.87 at% with the standard deviations of 0.31, 0.39, and 0.40, respectively. Considering that the grain boundary of the Ti–6Al–4V–0.9Fe sample was less and the overall element distribution was more uniform, the segregation of Fe was still slightly larger. It can be seen that in a certain range, with the increase in Fe content in the alloy, the segregation of Fe tended to increase; however, the influence of Fe content on the segregation of Al and V elements was negligible. With the increase in Fe content, the trends of Fe segregation in the simulation and experiment were the opposite. It is considered that the segregation of Fe mainly occurs in the stage of grain growth or solid-state phase transformation, which needs further study.

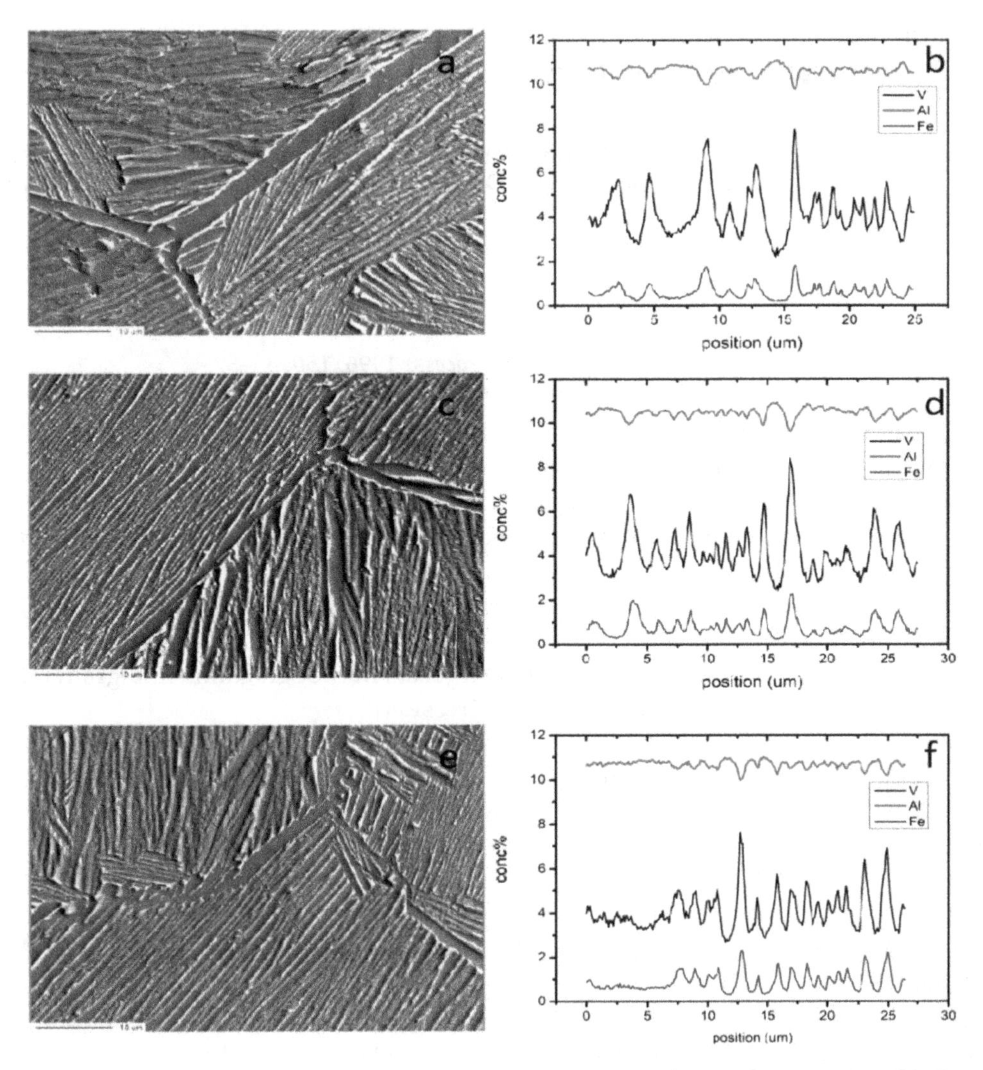

Figure 13. Results of the EPMA line scan. (**a**) Selected location of Ti–6Al–4V–0.5Fe, (**b**) Composition distribution of Ti–6Al–4V–0.5Fe, (**c**) Selected location of Ti–6Al–4V–0.7Fe, (**d**) Composition distribution of Ti–6Al–4V–0.7Fe, (**e**) Selected location of Ti–6Al–4V–0.9Fe, (**f**) Composition distribution of Ti–6Al–4V–0.9Fe.

4. Conclusions

In this work, the processes of levitation melting of five Ti–6Al–4V–xFe alloys were simulated, the effect of Fe content on the microstructure of single crystal and multi crystal was studied, and the distribution of elements in the Ti–6Al–4V–xFe alloy was discussed. Some simulation results were verified by experiments. The specific conclusions are as follows:

(1) The segregation of Fe element at the grain boundary of Ti–6Al–4V–xFe alloys can inhibit the interface mobility, thus promoting the formation of a local supercooling zone and making the grains easier to grow into dendrites.

(2) With the increase of Fe content, the grain size of the alloy decreased gradually. When there was no Fe in the alloy, the grain size was the largest (radius close to 2.29 mm), the grains were more uniform, and the shape of the grain was close to circular. The grain size decreased gradually with an increase in the Fe content and the overall decrease trend slowed down. When the Fe content was 0.9, the average grain radius was the smallest, which was about 1.03 mm.

(3) With the increase in Fe content, the distance of diffusion layer δ_n increased in the liquid phase. Within a certain range, there was a linear relationship between them.

(4) The segregation of Fe was more obvious than that of Al and V. With the increase in Fe content, the segregation of Fe increased, but there was less of an effect on Al and V.

Author Contributions: Conceptualization, H.C.; data curation, Y.G.; formal analysis, L.D.; funding acquisition, L.Z.; investigation, D.Z.; methodology, F.C.; software, L.D.; supervision, R.H. All authors have read and agreed to the published version of the manuscript.

References

1. Allen, P. Titanium alloy development. *Adv. Mater. Process.* **1996**, *150*, 35–37.
2. Lütjering, G.; Williams, J.C. *Titanium*; Springer: Berlin/Heidelberg, Germany, 2007.
3. Xin-ping, Z.; Si-rong, Y.; Zhen-ming, H.; Qiu-hua, H. Mechanical properties of new type ti-fe-mo-mn-nb-zr titanium alloy. *Chin. J. Nonferrous Met.* **2002**, *12*, 78–82.
4. Hotta, S.; Yamada, K.; Murakami, T.; Narushima, T.; Iguchi, Y.; Ouchi, C. Beta. Grain refinement due to small amounts of yttrium addition in.Alpha.+.Beta. Type titanium alloy, sp-700. *ISIJ Int.* **2006**, *46*, 129–137. [CrossRef]
5. Kudo, T.; Murakami, S.; Itsumi, Y. Influence of microstructure on formability in ti-fe alloy. *J. Mater. Process. Technol.* **2010**, *60*, 33–36.
6. Bermingham, M.J.; Mcdonald, S.D.; Stjohn, D.H.; Dargusch, M.S. Segregation and grain refinement in cast titanium alloys. *J. Mater. Res.* **2009**, *24*, 1529–1535. [CrossRef]
7. Ehtemam-Haghighi, S.; Liu, Y.; Cao, G.; Zhang, L.-C. Phase transition, microstructural evolution and mechanical properties of ti-nb-fe alloys induced by fe addition. *Mater. Des.* **2016**, *97*, 279–286. [CrossRef]
8. Boettinger, W.J.; Warren, J.A.; Beckermann, C.; Karma, A. Phase-field simulation of solidification. *Ann. Rev. Mater. Res.* **2002**, *32*, 163–194. [CrossRef]
9. Gyoon Kim, S.; Tae Kim, W.; Suzuki, T.; Ode, M. Phase-field modeling of eutectic solidification. *J. Cryst. Growth* **2004**, *261*, 135–158. [CrossRef]
10. Suzuki, T.; Ode, M.; Kim, S.G.; Kim, W.T. Phase-field model of dendritic growth. *J. Cryst. Growth* **2002**, *237*, 125–131. [CrossRef]
11. Kermanpur, A.; Jafari, M.; Vaghayenegar, M. Electromagnetic-thermal coupled simulation of levitation melting of metals. *J. Mater. Process. Technol.* **2011**, *211*, 222–229. [CrossRef]
12. Li, H.; Wang, S.; He, H.; Huangfu, Y.; Zhu, J. Electromagnetic-thermal-deformed-fluid-coupled simulation for levitation melting of titanium. *IEEE Trans. Magn.* **2016**, *52*, 1–4. [CrossRef]
13. Kundin, J.; Kumar, R.; Schlieter, A.; Choudhary, M.A.; Gemming, T.; Kühn, U.; Eckert, J.; Emmerich, H. Phase-field modeling of eutectic ti–fe alloy solidification. *Comput. Mater. Sci.* **2012**, *63*, 319–328. [CrossRef]
14. Gong, X.; Chou, K. Phase-field modeling of microstructure evolution in electron beam additive manufacturing. *JOM J. Miner. Met. Mater. Soc.* **2015**, *67*, 1176–1182. [CrossRef]
15. Sahoo, S.; Chou, K. Phase-field simulation of microstructure evolution of ti–6al–4v in electron beam additive manufacturing process. *Addit. Manuf.* **2016**, *9*, 14–24. [CrossRef]
16. Wu, L.; Zhang, J. Phase field simulation of dendritic solidification of ti-6al-4v during additive manufacturing process. *JOM* **2018**, *70*, 2392–2399. [CrossRef]
17. Nakajima, H.; Ohshida, S.; Nonaka, K.; Yoshida, Y.; Fujita, F.E. Diffusion of iron in β ti-fe alloys. *Scr. Mater.* **1996**, *34*, 949–953. [CrossRef]
18. Chen, Y.; Li, J.; Tang, B.; Kou, H.; Segurado, J.; Cui, Y. Computational study of atomic mobility for bcc phase in ti–al–fe system. *Calphad* **2014**, *46*, 205–212. [CrossRef]
19. Tiley, J.S.; Shiveley, A.R.; Pilchak, A.L.; Shade, P.A.; Groeber, M.A. 3d reconstruction of prior β grains in friction stir–processed ti–6al–4v. *J. Microsc.* **2014**, *255*, 71–77. [CrossRef]
20. Böttger, B.; Eiken, J.; Apel, M. Phase-field simulation of microstructure formation in technical castings—A self-consistent homoenthalpic approach to the micro–macro problem. *J. Comput. Phys.* **2009**, *228*, 6784–6795. [CrossRef]
21. Eiken, J.; Bottger, B.; Steinbach, I. Multiphase-field approach for multicomponent alloys with extrapolation scheme for numerical application. *Phys. Rev. EStat. NonlinearSoft Matter Phys.* **2006**, *73*, 066122. [CrossRef] [PubMed]

22. Guo, J.; Li, X.; Su, Y.; Wu, S.; Li, B.; Fu, H. Phase-field simulation of structure evolution at high growth velocities during directional solidification of ti55al45 alloy. *Intermetallics* **2005**, *13*, 275–279. [CrossRef]
23. Kurz, W.; Fisher, D.J. Dendrite growth at the limit of stability: Tip radius and spacing. *Acta Metall.* **1981**, *29*, 11–20. [CrossRef]

15

Digital Design, Analysis and 3D Printing of Prosthesis Scaffolds for Mandibular Reconstruction

Khaja Moiduddin *, Syed Hammad Mian, Hisham Alkhalefah and Usama Umer

Advanced Manufacturing Institute, King Saud University, Riyadh 11421, Saudi Arabia; syedhammad68@yahoo.co.in (S.H.M.); halkhalefah@ksu.edu.sa (H.A.); usamaumer@yahoo.com (U.U.)
* Correspondence: kmoiduddin@gmail.com

Abstract: Segmental mandibular reconstruction has been a challenge for medical practitioners, despite significant advances in medical technology. There is a recent trend in relation to customized implants, made up of porous structures. These lightweight prosthesis scaffolds present a new direction in the evolution of mandibular restoration. Indeed, the design and properties of porous implants for mandibular reconstruction should be able to recover the anatomy and contour of the missing region as well as restore the functions, including mastication, swallowing, etc. In this work, two different designs for customized prosthesis scaffold have been assessed for mandibular continuity. These designs have been evaluated for functional and aesthetic aspects along with effective osseointegration. The two designs classified as top and bottom porous plate and inner porous plate were designed and realized through the integration of imaging technology (computer tomography), processing software and additive manufacturing (Electron Beam Melting). In addition, the proposed designs for prosthesis scaffolds were analyzed for their biomechanical properties, structural integrity, fitting accuracy and heaviness. The simulation of biomechanical activity revealed that the scaffold with top and bottom porous plate design inherited lower Von Mises stress (214.77 MPa) as compared to scaffold design with inner porous plate design (360.22 MPa). Moreover, the top and bottom porous plate design resulted in a better fit with an average deviation of 0.8274 mm and its structure was more efficiently interconnected through the network of channels without any cracks or powder material. Verily, this study has demonstrated the feasibility and effectiveness of the customized porous titanium implants in mandibular reconstruction. Notice that the design and formation of the porous implant play a crucial role in restoring the desired mandibular performance.

Keywords: mandibular reconstruction; scaffolds; reconstruction plate; finite element analysis; 3D printing; titanium alloy

---

1. Introduction

Mandibular reconstruction is recognized as the most challenging and significant procedures by maxillofacial surgeons. It can be attributed to the strict requirements demanded by patients, in terms of anatomy, outer profile of the mandible and optimal restoration of oral functions [1–4]. The problem of mandibular reconstruction is further escalated owing to a rapid increase in mandibular defects due to modern human skeletal diversity and chewing behavior [5]. Generally, the mandibular continuity defect involves a complete bone loss and is caused by infection, trauma, lesion, osteonecrosis and resection of benign and malignant tumors [1]. The timely and adequate rehabilitation of mandibular defect is crucial to prevent impairment of masticatory function, loss of speech, cosmetic deformity and to essentially maintain the patient's quality of life. Certainly, the titanium plate with autogenous bone transplantation can be regarded as the primary standard and a reliable treatment for mandibular reconstruction [6]. In spite of the availability of reconstruction techniques related to autogenous bone graft, perfect mandibular reconstruction is still not possible and remains a challenge. Generally, the available standard commercial reconstruction plates (implants) are employed in mandibular reformation. These plates are manufactured using traditional methods such as casting and the powder metallurgical process, which are time consuming processes [7]. Furthermore, the standard plates are straight and they need bending in order to align them along the mandible curved bone. This not only raises the operative (or surgery) time, but also involves the tedious task of repeatedly adapting and revising the plate according to the patient's anatomy. Since, it is a trial and error procedure, the possibility of discrepancies between the bone and plate interface increases, which in turn causes implant failure as well as discomfort to the patient. Therefore, it is indispensable to utilize custom made implants, which not only reduce disproportion and mismatch, but also result in improved appearance and actualization. The personalized implant design not only enhances fitting accuracy, but also minimizes the surgical time in contrast to standard plates.

Recent developments in tissue and scaffold engineering represents a contemporary prospect and a new application in the evolution of mandibular restoration. Scaffolds can be combined with solid parts and fabricated as an implant. Ideally, the scaffolds should be highly porous, crack free and biocompatible with tissue ingrowth [8]. As reported by numerous clinical studies, the titanium scaffold (porous structure) can achieve long term bone fixation and promote full bone ingrowth when compared to the solid or bulk part [9,10]. In addition, solid titanium implants due to variation in mechanical properties as compared to bone may lead to bone resorption, which induces stress shielding effect on its surrounding bone and eventually leads to implant failure [11]. The impeccable porosity influences cell behavior and the interconnected channels of pores stimulate the vascularization [12]. The encouragement of early osseointegration is critically important for the success of implantation, otherwise longer healing time would lead to implant failure [13].

With advancements in engineering technology, including medical modeling software and three-dimensional (3D) printing or additive manufacturing, it is now possible to design and fabricate customized implants with better accuracy and in a shorter period of time. The unification of data acquisition, image processing, as well as modeling and additive manufacturing, have made it possible to comprehend tailor-made implants according to the patient's requirements. Undoubtedly, the implementation of integrated techniques can save a lot of money for medical practitioners as well as revamp the quality of life for a large number of people [14]. The agreeable effect in mandible restoration depends on many aspects of the implant, including its design, fabrication technology, biomechanical properties, accuracy, surface integrity and weight. Certainly, 3D printing techniques have emerged as a promising potential in the development of bone reconstruction, rehabilitation and in the field of surgery [15]. Among several 3D printing techniques, electron beam melting (EBM) has been regarded as the fast and successful method for the fabrication of titanium medical implants from computer-aided design (CAD) models with Food and Drug Administration (FDA) and Conformité Européene (CE) approval [16]. EBM technique, which was first commercialized in 1997 by ARCAM AB, fabricate parts by melting metal powder in a layer-by-layer fashion [17]. It has increasingly been used for the fabrication of 3D titanium alloy scaffolds for medical applications with complex architecture [18,19]. Mandibular bone is not a uniform and regular structure, but rather a curved and special structure. Therefore, very few researchers have attempted to custom design prosthesis for mandibular reconstruction [20,21] and very limited information is available on the study of mandibular scaffold. In addition, no clear evidence and investigation are available in the biomechanical, structural integrity and fitting evaluation of mandibular prosthetic scaffolds.

In this study, two different types of custom specific mandibular prosthesis scaffolds have been designed, fabricated and evaluated for their performance. These two designs were categorized as top and bottom porous plate and inner porous plate. In the top and bottom porous plate design, the mesh or porous structure was attached on the top and bottom of the plate, whereas in the inner porous plate design, the porous structure was inside the plate. An extensive integrated methodology has been utilized for the realization of the patient-specific porous implant. The part fabrication using EBM was supplemented with computer tomography (CT) for image acquisition and processing software for implant modeling. The two scaffold designs were also analyzed to determine their biomechanical effect under the mastication process using Finite Element Analysis (FEA), surface integrity using micro-CT scans as well as fitting accuracy and appearance utilizing the 3D comparison technique.

2. Methodology

The typical flowchart as shown in Figure 1, demonstrates the methodology adopted in this work. It was based on six primary steps: Data acquisition, customized implant design and modeling, virtual assembly, FEA, part fabrication and evaluation. This approach was prominent because it involved interaction between the engineering and medical fields right from the patient diagnosis until the mandibular reconstruction. The authors in this methodology have emphasized the importance of communication between the engineering and medical departments. In the current study, the medical practitioners were customers, therefore, they were engaged in each and every stage during the entire process. These communication links are evidently specified by using red circles in the Figure 1. These communications acted as a feedback loop to get the assessment or the criticism from the medical people. Of course, the engineers had to explain various aspects and engineering terms or analysis to medical professionals before every session. This communication or information exchange helped to improve the overall results by minimizing design revision and preventing implant failure.

2.1. Data Acquisition

A forty-year-old patient with deformities and a lesion in the left mandibular area attended the emergency department of the university hospital. Upon diagnosis and a series of tests by the medical doctor, the patient was subjected to a non-invasive CT scans. The non-invasive CT can be defined as a medical procedure which does not involve any deterioration of the skin, internal body as well as the destruction of healthy tissues. During the course of patient diagnosis, it was found that the patient was suffering from mandibular continuity defect with a loss of portion of the bone resulting in a gap of ~2 cm or more. It is a patient-specific defect which is larger in size. The CT images were acquired using a Promax 3D "Cone beam computer tomography machine" (Planmeca, Helsinki, Finland) [22]. The minimum resolution model (voxel size) was 0.10 mm^3. It was implemented under the following conditions: Voltage—54–90 kV, Current—1–14 mA, Focal spot 0.4 mm, detector resolution 127 μm, scan time 18–26 s. The radiologist performed the CT scan on the patient and saved the scanned images in Digital Imaging and Communications in Medicine (DICOM) format which is a universal stored format for medical images. The DICOM files containing a series of two-dimensional (2D) images, stored in a database, did not provide a perfect picture of the anatomical structure. Several medical modeling and image processing software available in the market were used to convert the 2D images into a 3D anatomical model. MIMICS 17.0® (Materialise Interactive Medical Image Control System; Materialise NV, Leuven, Belgium) was used in this study. The 2D images of DICOM files were imported into MIMICS® which stacked the 2D images over each other and developed a typical 3D model. In medical CT imaging, the Hounsfield unit (HU) represents the grayscale from black to white with a range from −1024 (minimum value) to 3071 (maximum value). A custom thresholding Hounsfield unit of 282 to 2890 HU was used for bone identification. Segmentation by thresholding technique was used to select the soft and hard tissue by defining the range of the threshold value. Figure 2 illustrates the patient mandibular tumor in a different view.

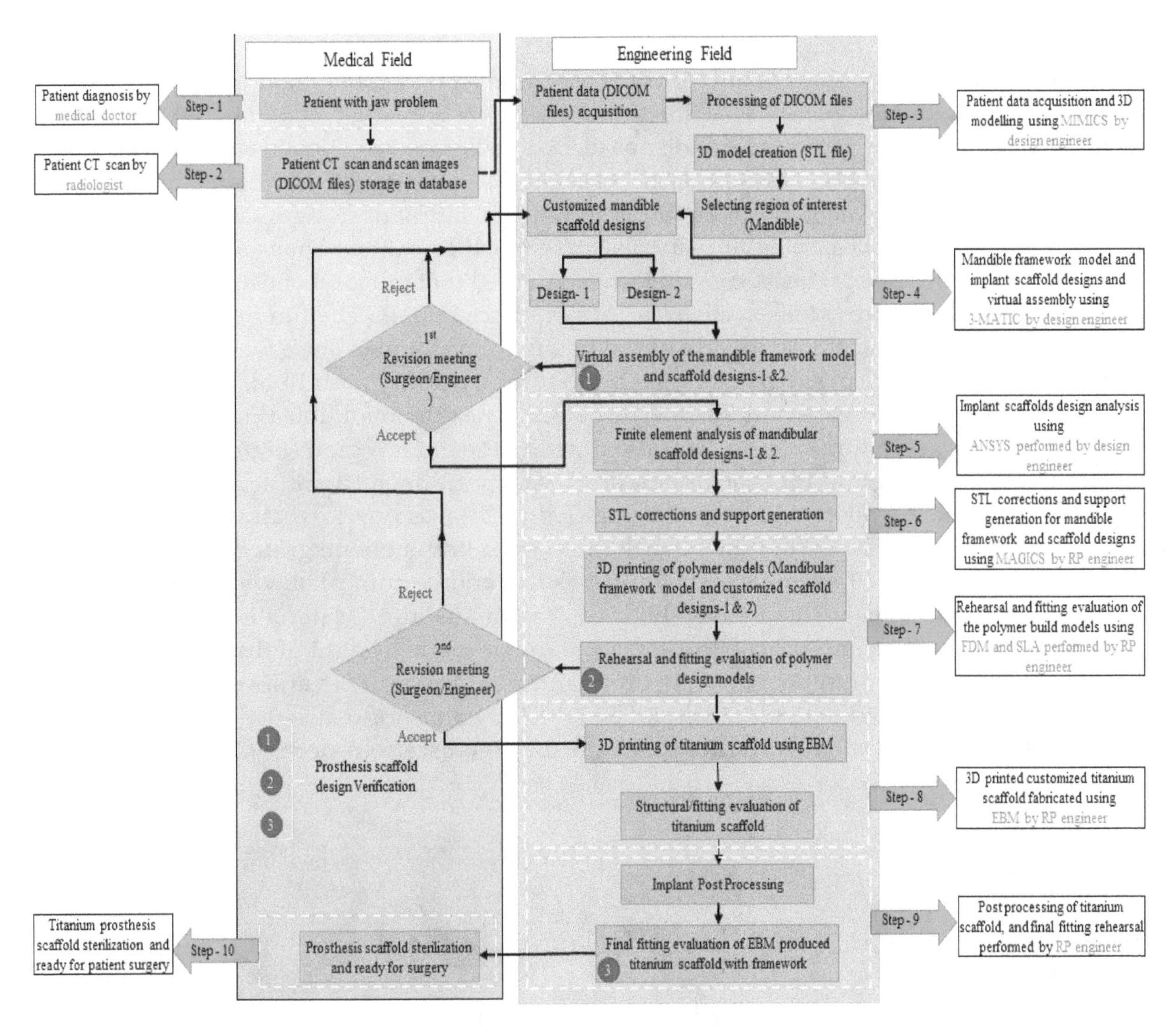

Figure 1. The proposed methodology for design, analysis and fabrication of customized mandibular prosthesis scaffolds. Note: The red circles indicate the formal meetings between the engineering and medical department for scaffold design verification and evaluation.

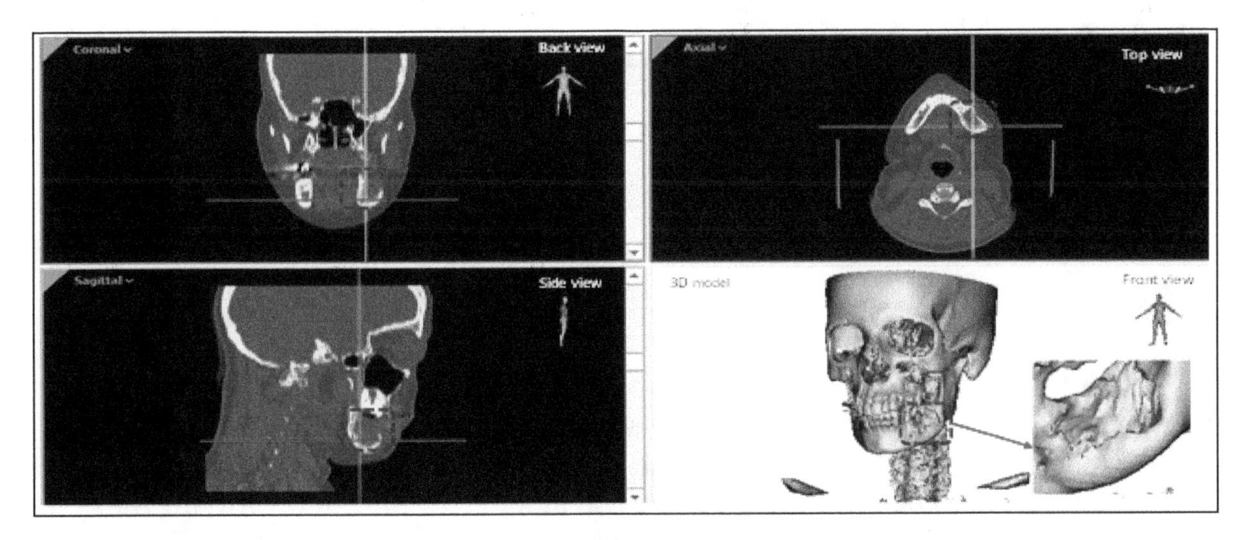

Figure 2. Patient anatomical model depicting the tumor region in different planes.

2.2. Customized Implant Design and Modeling

The region growing technique using MIMICS was used to extract the region of interest (mandible) from the surrounding tissues. Figure 3a–e illustrates the region growing techniques, where the full face mask was segregated to the region of interest in mandible Figure 3e. The obtained tumor mandible without teeth was then saved as a Standard Tessellation Language (STL) file. The STL file was imported into 3-Matic® (Materialise, Leuven, Belgium) for implant design. Mirror reconstruction design technique is the most common implant design where the healthy bone is mirrored and replaced over the defective bone. Several research studies have proved that mirror reconstruction technique has successfully restored and provided excellent facial symmetry [23,24]. The tumor on the left mandible (Figure 3f) was resected and the right side of the healthy mandibular bone was mirrored as shown in Figure 3g. The symmetrical sides were merged to form a healthy mandible. Wrapping operation was performed to nullify the gaps and voids. The obtained healthy mandible (Figure 3h) was used for the implant design by selecting (Figure 3i) and extracting the outer region (Figure 3j) for customized implant design. Smoothing and trimming operations were performed to get the implant design shape as shown in Figure 3k. An offset thickness of 2 mm (Figure 3l) was provided and two implant designs with one inner bone graft carrier and the other with top and bottom bone graft carrier were designed as shown in Figure 3m,m′. The inner plate and thick top and bottom plate were patterned into the porous structure (scaffold) using dode thick (Figure 3n) from Magics® (Materialise, Belgium) as shown in Figure 3o. The dode thick mesh structure was used to reduce the weight of the mandibular implant and to provide good adhesion between the bone and the implant. Several research articles have proved that titanium scaffold with a porosity of 500–1000 microns influence the osseointegration and faster bone healing [25,26]. Figure 3p illustrates the designed scaffold pore (900 microns) and strut (300 microns) size.

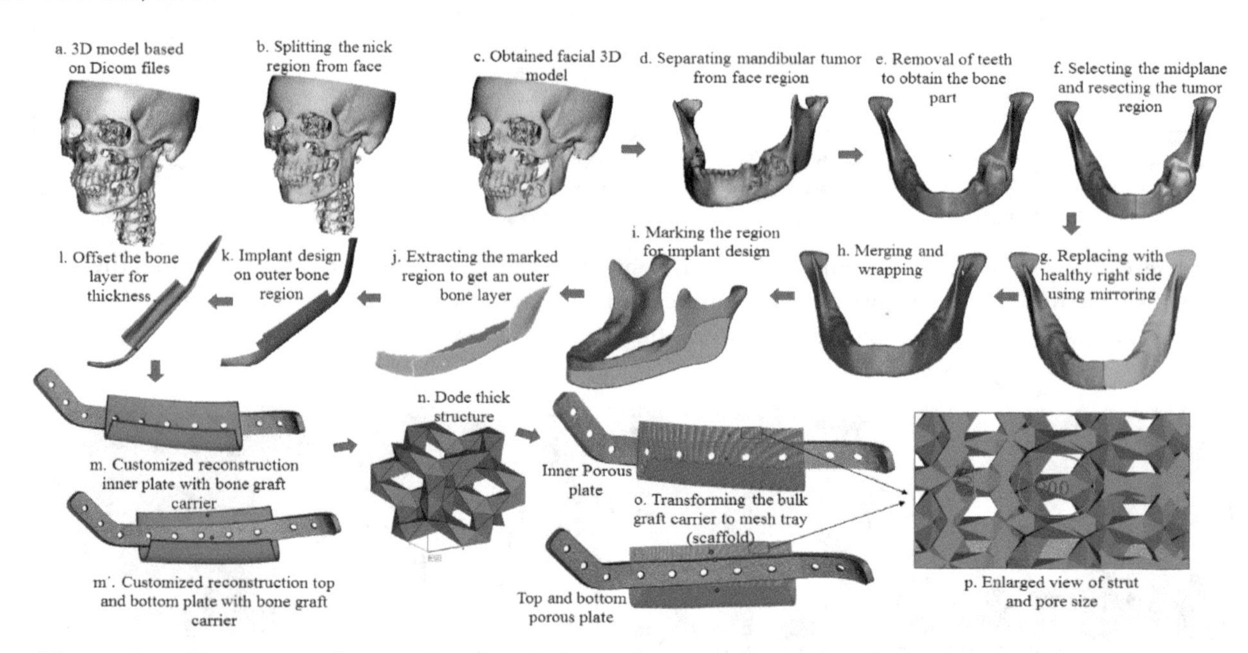

Figure 3. Sequence of steps in the design of customized prosthesis scaffold (implant) for mandibular defects.

2.3. Virtual Assembly

The two designed prosthesis scaffolds were virtually assembled and aligned with the mandibular framework model for fitting and assembly evaluation as shown in Figure 4. Formal meetings used to take place between the engineering and medical field for evaluating and verifying the design as indicated by red circles (Figure 1). Any error or void in-between the implant and the bone would result in the redesigning of the implant. The virtual assembly also helped with surgical guidance, understanding the surgical anatomy and real world preoperative surgery scenario to improve the reliability and safety of the surgical process.

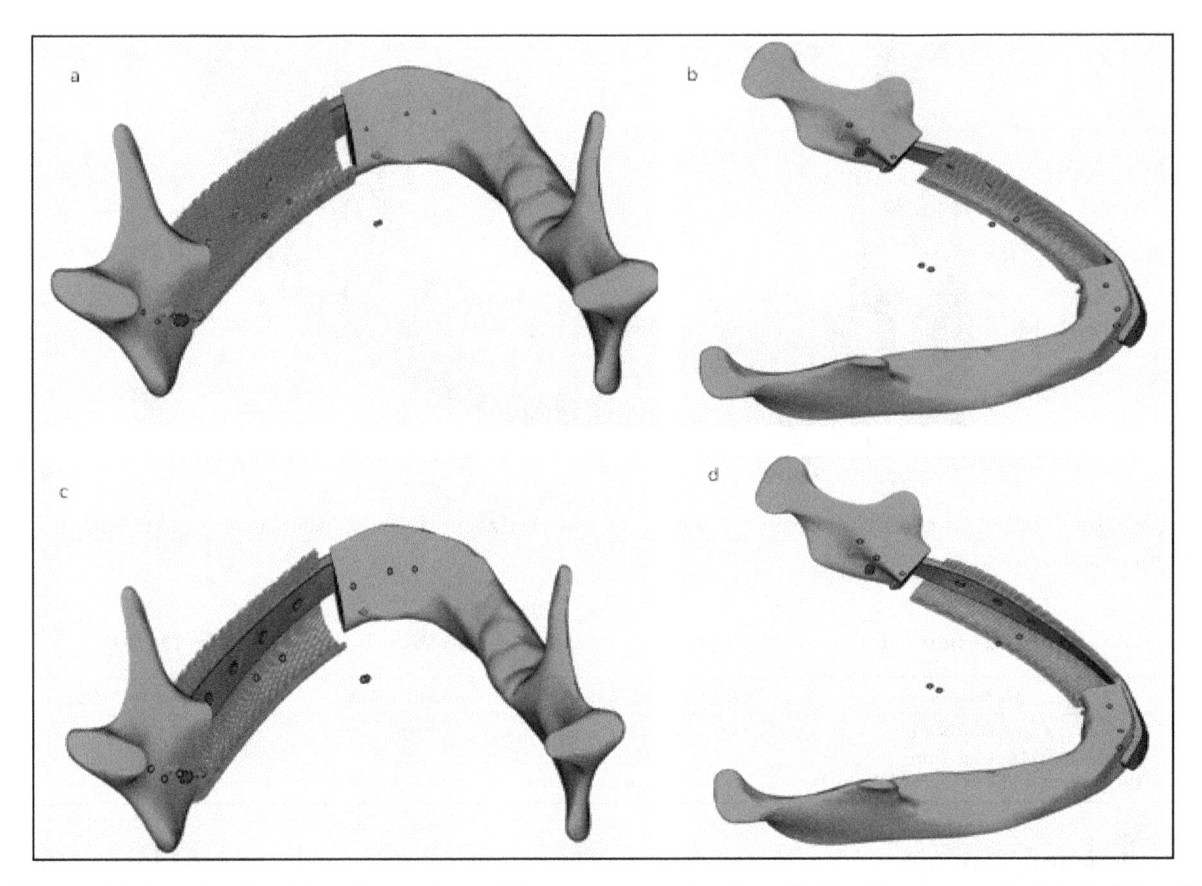

Figure 4. Posterior (back) and top view of the two customized scaffolds: Inner porous plate (**a**,**b**) and (**c**,**d**) top and bottom porous plate.

The designed reconstruction scaffolds were incorporated with countersink medical screw holes with three screws on the condyle side and three screws on the chin area. The countersink holes were designed for the complete immersion of the screw head inside the screw hole in order to provide a better aesthetic effect. Figure 5 illustrates the virtual assembly of the mandibular framework model containing the cortical and trabecular bone with scaffold fitted with six screws. The error free designed scaffold and the framework model were saved as a Standard for the Exchange of Product model data (STP) file for analysis.

2.4. *Finite Element Analysis*

Once the designed scaffolds were examined for fitting and conformance in the virtual assembly, the FEA model was created to evaluate their functionality as well as the biomechanical effect of clenching on the prosthesis scaffold. The FEA was employed because it is recognized as one of the crucial tools to emulate and predict the behavior of the CAD model in real scenarios. It was first used in the aerospace industry but quickly spread throughout a wide range of sciences including medicine and dentistry [27]. A finite element model (FEM) consisting of the temporomandibular model and two designed scaffolds was created using Ansys® software. In this study, the sustained clenching and masticatory muscle activity using three muscular forces (masseter, medial pterygoid and temporalis) were simulated. The material properties of the cortical bone, trabecular bone, screws and scaffold were adapted from the literature study and were assumed as homogeneous, isotropic and linear elastic [28,29]. The Young's modulus, Poisson's ratio and yield strength of the simulated study are presented in Table 1.

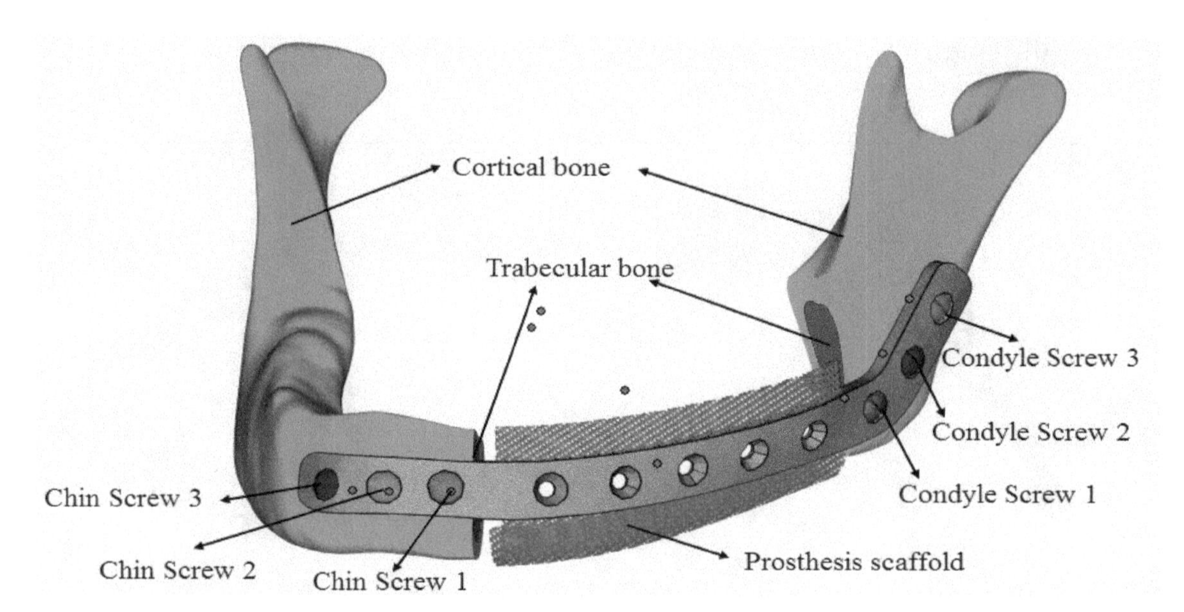

Figure 5. Global view of virtual design assembly of customized prosthesis scaffold on the mandibular framework model.

Table 1. Mechanical properties of study materials used in FE model. Data from [28,29].

Materials	Young's Modulus (MPa)	Poisson's Ratio	Yield Strength (MPa)
Compact Bone	13,700	0.3	122
Trabecular Bone	1370	0.3	2
Prosthesis scaffold, (Ti6Al4V ELI)	120,000	0.3	930

For clenching simulation, the superior part of both condyles was constrained in all directions. The displacement in the molar region as shown in Figure 6 was restrained in the upper region to simulate chewing. While the biting forces acted axially, the molar movement was kept at near zero displacement. This restraint was perpendicular to the occlusal plane (Z-direction), while allowing freedom of movement in the horizontal plane (X and Y direction). The FEM was meshed with the 10-node 3D tetrahedral element.

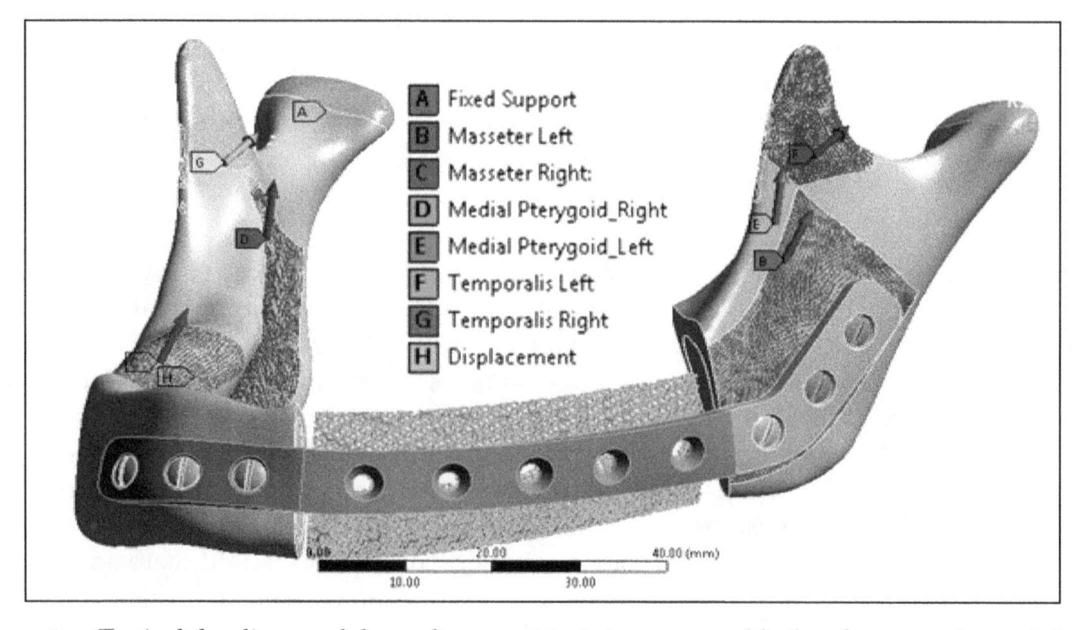

Figure 6. Typical loading and boundary constraints on mandibular framework model with prosthesis scaffold.

As shown in Figure 7, the triangle surface mesher strategy with program controlled patch conforming method was used in order to refine the mesh at the area of fixation and to obtain more accurate results. The magnitude and boundary condition of the masticatory forces were derived from the literature study [30,31]. The interface between the scaffold-bone and screw-scaffold-bone were considered as bonded. The clenching movement was simulated in the FEM with muscular forces and their vectors are presented in Table 2.

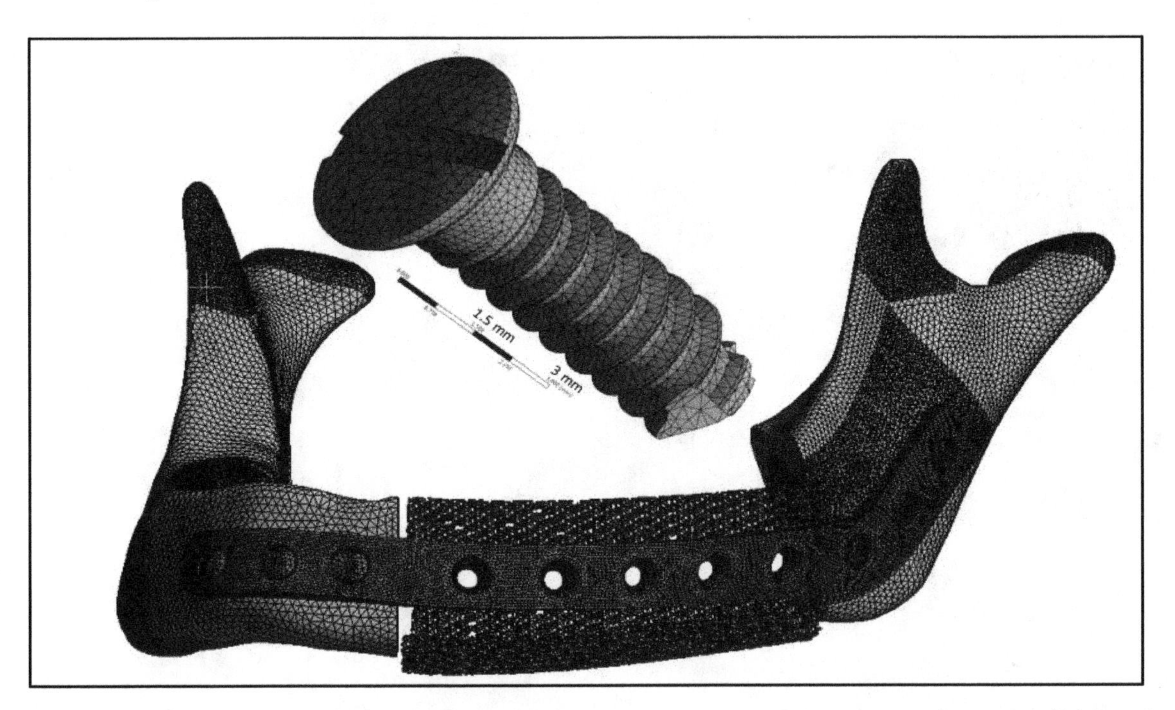

Figure 7. Meshing on the simulated mandibular framework model with prosthesis scaffold and a close-up view of screw meshing.

Table 2. Magnitude and functional direction of masticatory muscles in Newton's (N). Data from [30,31].

Masticatory Muscles	X (N)	Y (N)	Z (N)
Masseter	50	−50	200
Medial pterygoid	0	−50	100
Temporalis	0	100	200

2.5. Fabrication

In this study, 3D printing was used for the fabrication of customized prosthesis scaffolds. Two types of materials—polymer and metal—were used in the fabrication. The polymer 3D printing was used for the testing and fitting evaluation (virtual assembly), whereas metal (Ti6Al4V ELI) was used for the patient prosthesis implant. For polymer-based 3D printing, Stratasys-fused deposition modeling (FDM) machine and FORMLABS-2 a (stereolithography) SLA machine were used. ARCAM's EBM machine (EBM A2, ARCAM AB, Mölndal, Sweden) was used for printing titanium metal scaffolds.

2.5.1. Polymer Fabrication

The FDM machine as shown in Figure 8a was used to print mandibular framework models (Figure 8b) using ABS (acrylonitrile butadiene styrene) material which is a common thermoplastic resin with good functional properties [32]. FDM works on additive manufacturing process where the ABS material unwound from the coil and is heated to melting point and extruded in a layer-by-layer fashion to produce 3D objects. Formlabs-2 3D printer as shown in Figure 8c was used to fabricate the mandibular prosthesis scaffold (Figure 8d) which used the liquid resin material. Formlabs-2 form works on laser-based SLA principle where the laser solidifies the liquid resin material in a photo-polymerization process and builds the 3D model in a layer-by-layer fashion [33]. SLA produces objects with higher resolution with more accuracy when compared to FDM due to its optimal spot size laser which is very small [34]. Formlabs-2 was used in the fabrication of mandibular scaffold as it provided higher resolution and accuracy for the complicated porous structures.

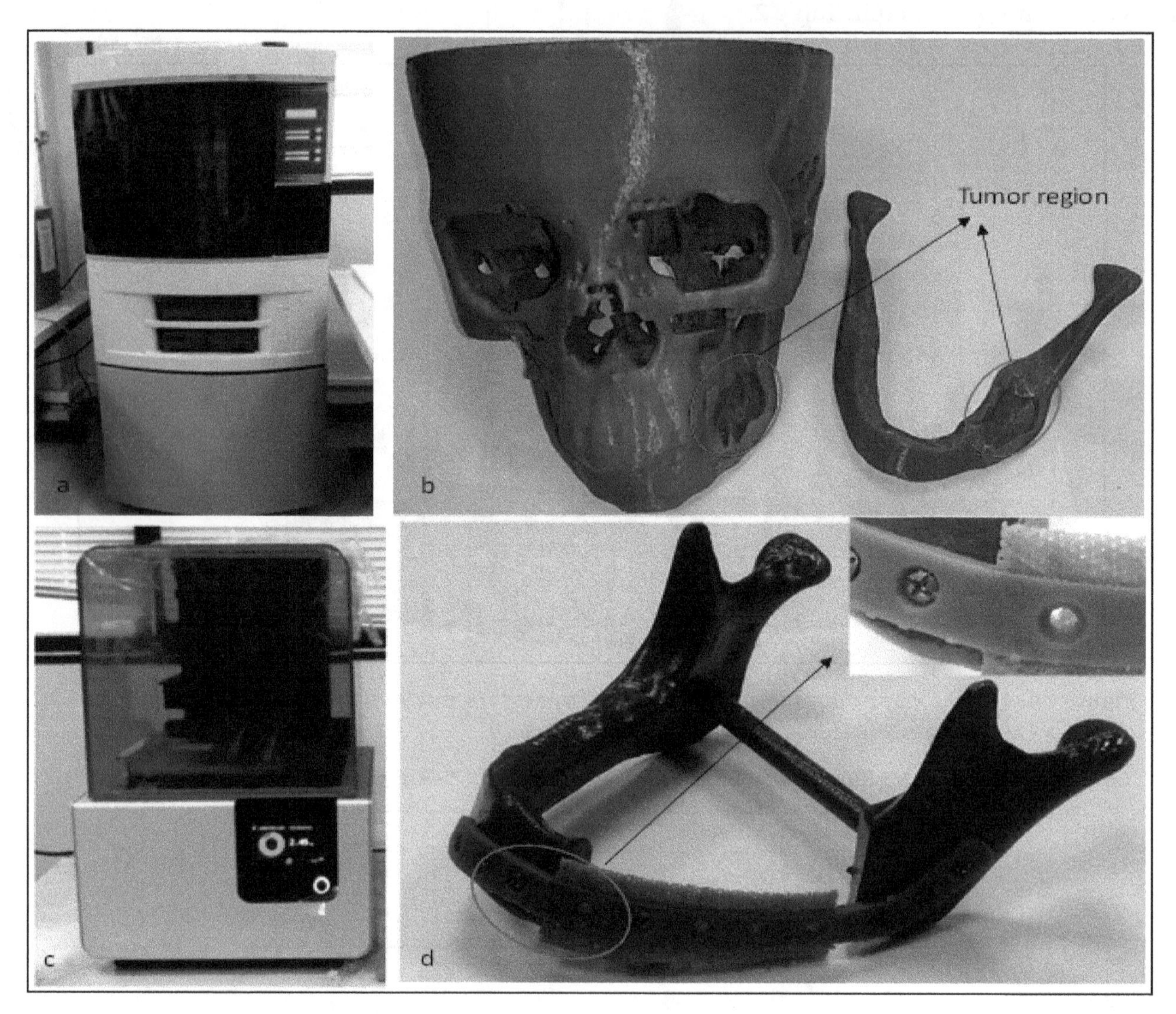

Figure 8. (**a**) Fused Deposition Modeling machine with its fabricated polymer model (**b**) indicating the tumor region and (**c**) SLA machine and its produced mandibular scaffold (**d**) with a close-up view.

2.5.2. Titanium Fabrication

It is well proven that scaffolds with elastic modulus closer to that of bone, minimizes the stress shielding effect and promotes bone-implant tissue in-growth [35,36]. Powder bed metal based 3D printing technologies such as EBM and selective laser melting (SLM) have demonstrated the capability to produce scaffolds in medical applications [37]. The EBM process in comparison requires less supporting material and minimizes post processing steps such as machining and heat treatment [36]. An EBM process is most suited for reactive metals such as titanium alloy as the complete build process takes place in a vacuum environment [38]. In addition, EBM produces parts at a much faster rate (80 cm^3/h) when compared to SLM (20–40 cm^3/h) [39]. The standard layer thickness of the printed samples using ARCAM's A2 EBM machine was 50–70 μm.

Figure 9a,b illustrates the typical working principle of the EBM process and the different components of the EBM machine respectively. The tungsten filament in the electron beam gun on reaching above 2500 °C, emits a beam of electrons which accelerates at half the speed of light and passes through a series of controlled coils (lens) and impacts the powder surface, thus melting the powder. The first (astigmatism) lens assists to keep the beam in circular and round shape regardless of its position on the build plate. Without this coil, the focus point of the beam tends to have a wider area (elliptical shape) when it is deflected towards the edge of the build region. It also eliminates electro-optical artifacts (human error). The second (focus) lens keeps the beam in focus and sharpens to a desired (0.1 mm) diameter. The third (deflection) lens scans the beam across the build area. The build process takes place inside the build chamber. Inside the build chamber, there are two hoppers which hold the metal stock powder. Metal powder is spread homogeneously over the build table using rakes. The rakes fetches the powder from either end of hoppers and spreads it evenly over the build table. The build tank lowers down in the z-direction after each melt cycle. The start plate was placed at the center of the build table which holds the build surrounded by powder. Vacuum is maintained throughout the build cycle to eliminate impurities and to prevent reactions between the reactive metals. Titanium powder (Ti6Al4V ELI) with the particle size of 50–100 mm was used in this study. The chemical composition of Ti6Al4V ELI (extra low interstitial) was made of 6.04% Al, 4.05% V, 0.013% C, 0.0107% Fe, and 0.13% O, while the rest as Titanium (in weight percent).

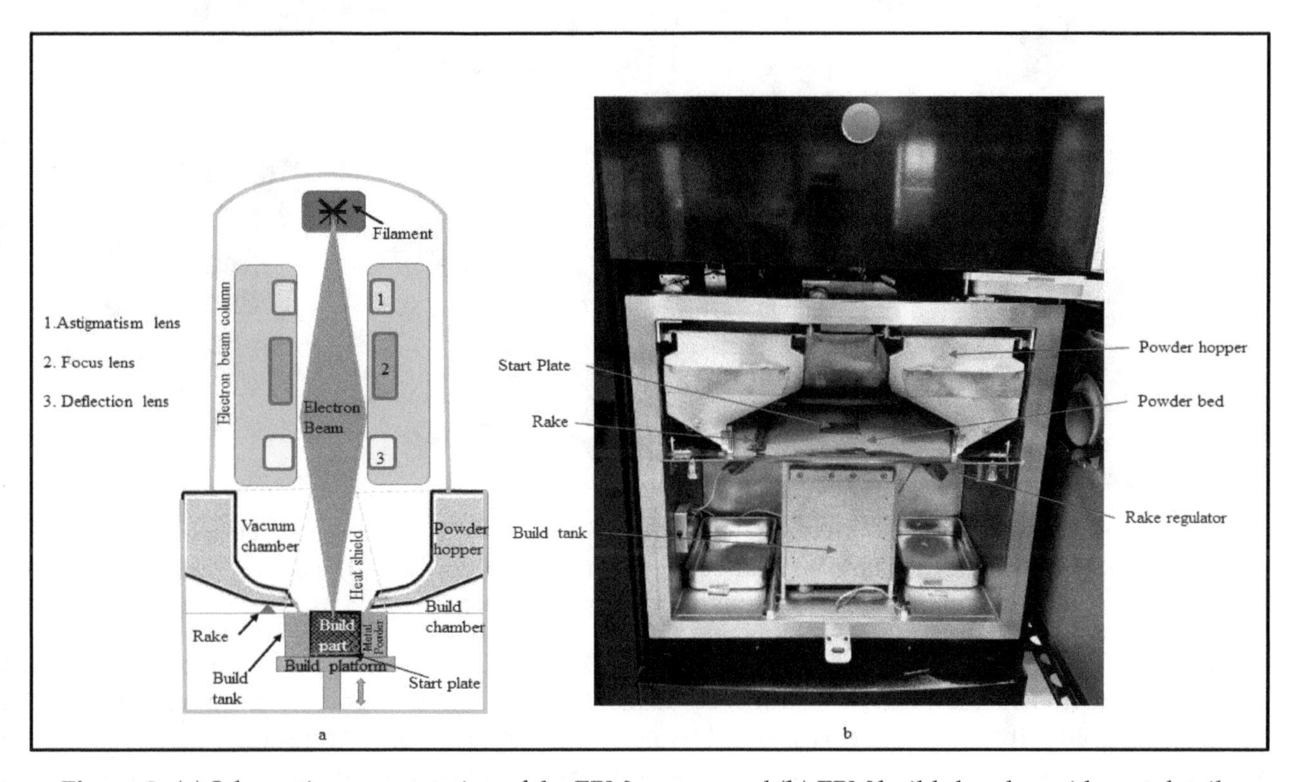

Figure 9. (**a**) Schematic representation of the EBM process and (**b**) EBM build chamber with part details.

The part fabrication in the EBM machine (ARCAM A2) as shown in Figure 10b is dependent on three phases—(1) Preheating of the metal powder. (2) Scanning and melting. (3) Lowering of build table and raking of powder.

(1). Preheating the metal powder: The Ti6Al4V ELI metal powder spread on the powder bed is preheated by multiple beams of electron at high scan speed and low beam current to reduce the internal residual stresses.
(2). Scanning and melting: The high velocity beam of electrons scans the metal powder and melts the power in line as per the defined CAD geometry. The melting process consist of two steps, melting the contours (outer and inner boundary) and infill hatching. The majority of the melting takes place in hatching where the beam current and scan speed are increased.

(3). Lowering build table and raking of powder: The build table is lowered after each melt layer cycle (50 μm) and a new layer of powder is fed from hoppers and spread evenly on the previously solidified powder layer using rakes. This process continues till the final 3D part is built.

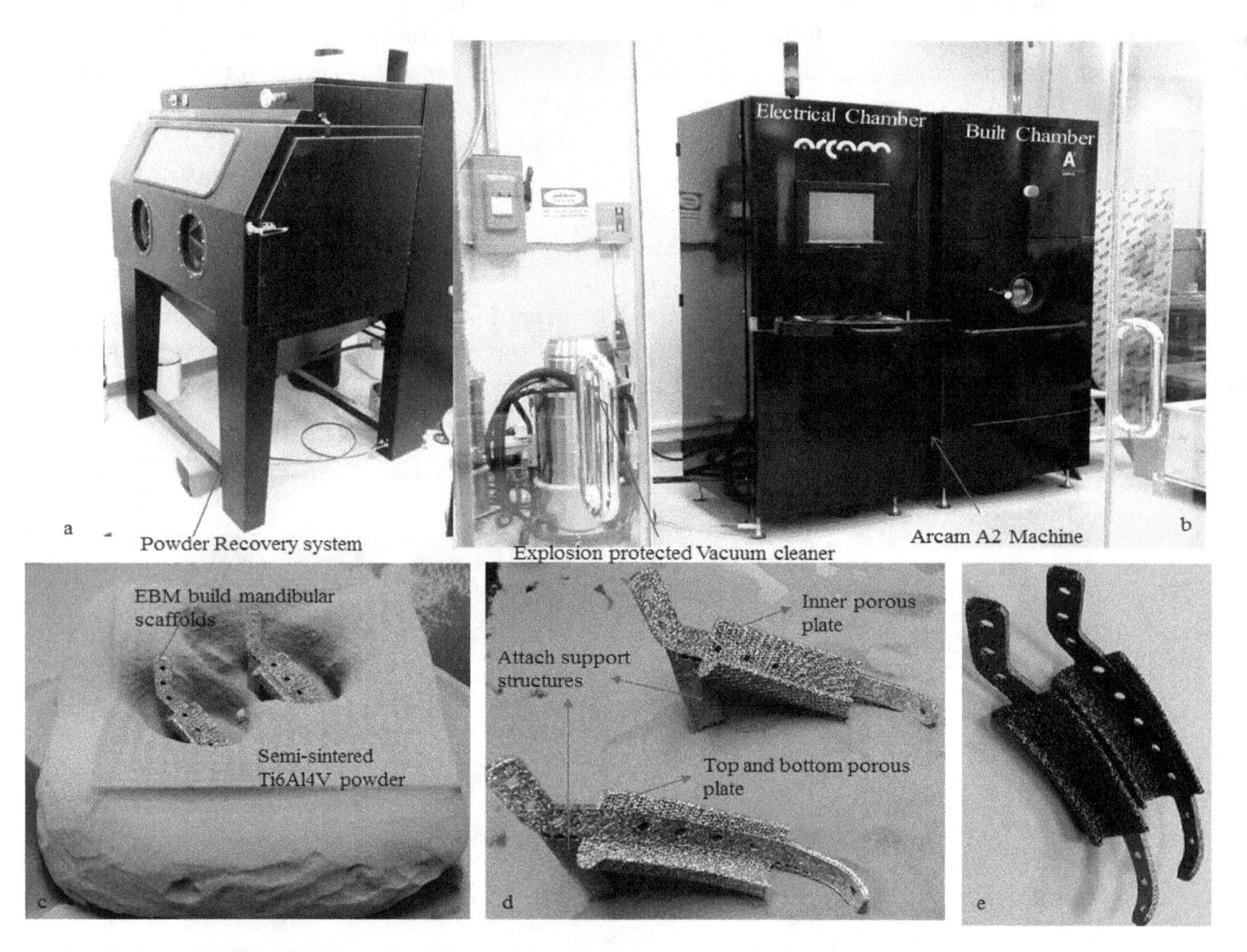

Figure 10. (**a**) PRS machine, (**b**) EBM machine with explosion protection vacuum cleaner, (**c**) EBM built mandibular prosthesis scaffold surrounded by semi-sintered powder, (**d**) titanium scaffolds with support structures and (**e**) mandibular scaffolds after support removal.

The EBM build lasted approximately 8–10 h. After build completion, the produced part (mandibular prosthesis scaffold) was allowed to cool under helium gas. Figure 10c shows the EBM build scaffold with supports surrounded by semi-sintered powder. The semi-sintered titanium powder was then blasted in powder recovery system (PRS) as shown in Figure 10a as a post processing process and to get the finished part with supports. The supports (Figure 10d) which were added to the scaffolds during the build to dissipate the heat and the overhang structures were manually removed with simple tools such as pliers. Figure 10e illustrates the final EBM built mandibular scaffolds which can be sandblasted or machined using laser ablation to achieve a smoother finish if required [40].

2.6. Evaluation and Validation

At this stage, the fabricated titanium scaffolds were investigated for structural integrity, fitting accuracy as well as the weight.

2.6.1. Micro-CT Scan on Titanium Lattice Structure

A non-destructive technique (i.e., micro-CT scan) was employed in order to examine the stochastic defects and structural integrity of the dode thick mesh structure used in scaffold design. The micro-CT scans were utilized in order to validate the quality of the dode thick structure in terms of cracks, internal trapped powder, in addition to examine the interior construction of the built struts without any physical cutting and polishing. A 15 mm solid cube (Figure 11a) was designed and transformed into a dode thick structure (Figure 11b,c) and fabricated using EBM as shown in Figure 11d. The micro-CT scanner (Bruker Skycam 1173, Kontich, Belgium) with a source voltage of 120 KV focused on the EBM fabricated cube structure with a spot size of 5 μm and with an image pixel size of 12.03 μm. Each 2D slice image of the cubic structure in the form of 512 × 512 bitmaps as output data was collected.

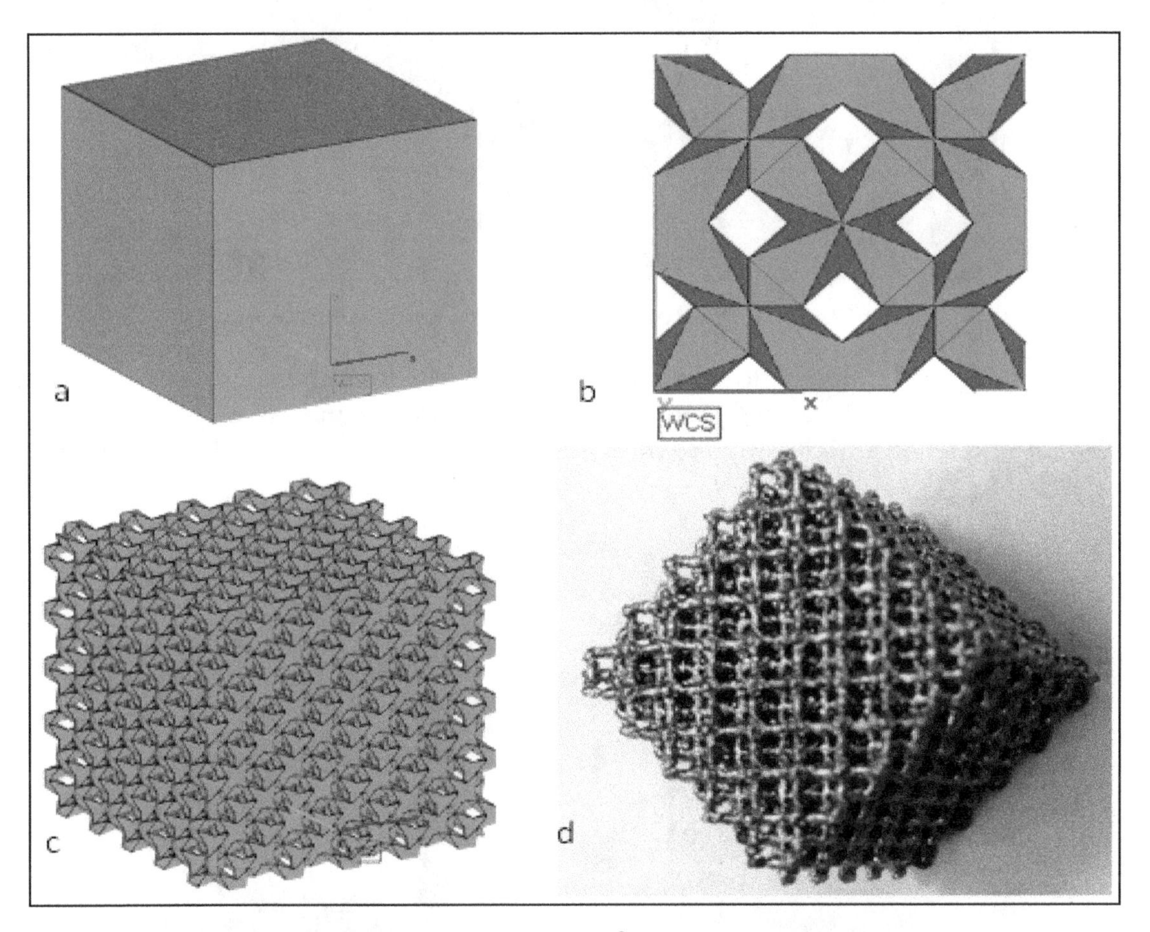

Figure 11. Cubes with unit cell structure of 15 × 15 mm^2 (**a**) solid cube, (**b**) dode thick unit cell structure, (**c**) dode thick cube structure and (**d**) EBM fabricated dode thick cube.

2.6.2. 3D Comparison

The 3D comparison technique was implemented in order to accurately compare the fitting accuracy of both the implant designs (inner porous plate and top and bottom porous plate) with respect to the mandible. The fitting accuracy of the implants was computed using Geomagics Control® [41]. The 3D comparison analysis can be considered as one of the most powerful and extensive techniques, to graphically represent the surface deviations between the reconstructed objects and the reference CAD model [42]. At the outset, the test model had to be aligned on the reference CAD model by utilizing the best fit alignment. Consequently, the analysis software automatically estimated the best fit between the test and reference object. This best fit alignment confirmed that both the test and reference objects were positioned (or fixed) in the same coordinate system. Furthermore, the statistic used in this work in order to quantify the fitting accuracy of the implants on the mandible was the average deviation. This statistic was utilized because it reported the deviation in the mandible, thereby approximating the gap between the implant (scaffold) and the mandible. In this work, the test model was acquired as a point cloud set by employing the laser scanner mounted on the Faro Platinum arm (FARO, Lake Mary, FL, USA) as shown in Figure 12.

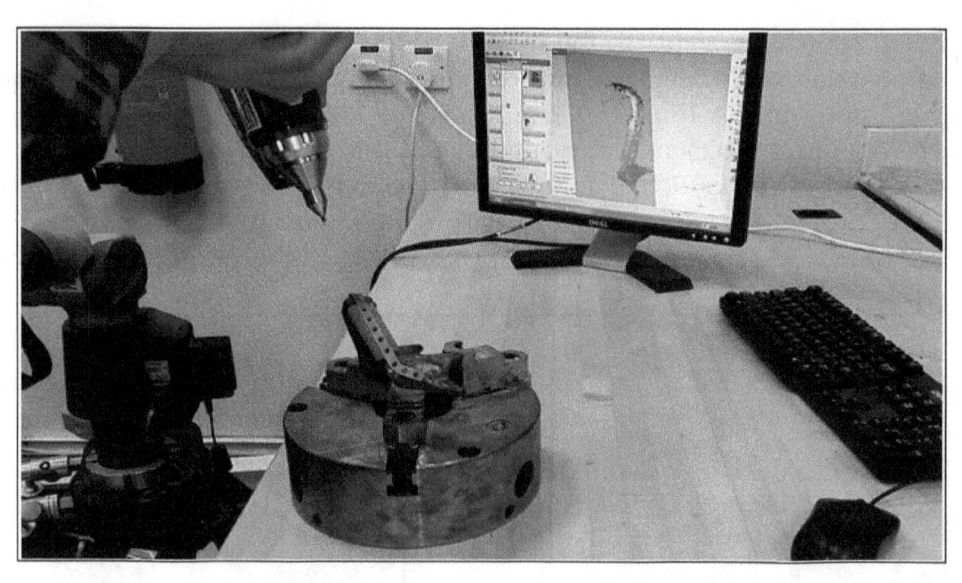

Figure 12. Acquisition of test data using a Faro Platinum arm.

As shown in Figure 13, the scaffolds were mounted on the mandible and scanned to obtain the test data. The reference model was obtained by removing the defect and imitating the healthy side on it. The reference model acquired using the mirroring technique was assumed to represent the ideal anatomical structure [23,24].

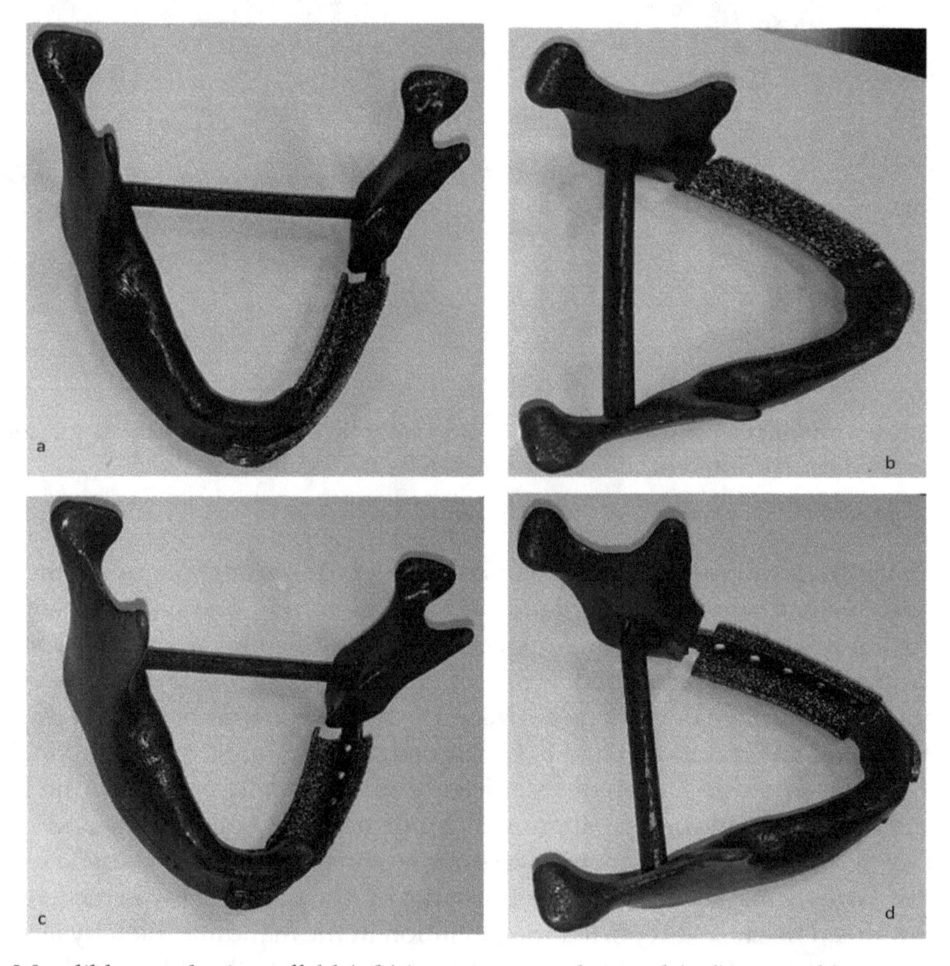

Figure 13. Mandible prosthesis scaffold (**a**,**b**) inner porous plate and (**c**,**d**) top and bottom porous plate mounted on the mandibular framework.

The outer surface of the scaffold mounted mandible were scanned and imported as STL model in Geomagics control® in order to compare it with the reference mandible. The outer surface was studied because the customized scaffolds were designed depending on the outer profile of the mandible. The 3D comparison analysis software represented the result by means of error scale through the computation of the shortest distance between the test model and the surface of the reference model.

2.6.3. Weights of the Scaffold Designs

In order to reduce the stress shielding effect between the implant and the surrounding bone, it was imperative to build lighter implants with weights closer to that of the bone being replaced [43]. The minimization of stress shielding was critical for reducing bone resorption as well as decreasing the rate of aseptic loosening. The weight of the mandibular bone to be replaced was calculated from the density formulae where volume was taken from the Magics® software (Materialise, Leuven, Belgium) and assuming density as 1600 kg/m^3 [44]. The weights of the two EBM fabricated scaffolds were measured using a digital weighing machine.

3. Results and Discussion

In this work, two customized prosthesis scaffolds were designed from the patient CT scan files. The clinical setup for both the designed scaffolds were simulated under physiological clenching conditions. The FEA analysis was essential in order to find out the continuous grabbing and chewing ability of the designed customized implants. The equivalent stresses and strains observed on both scaffolds are presented in Figure 14. The results indicated that the maximum stresses in both customized scaffolds were confined to the mesh structure and it was evident due to its lower cross sectional area.

The simulated result summary of both designed scaffolds is presented in Table 3. The analysis showed that the FEA of inner porous plate design induced higher stress concentration than the FEA of top and bottom porous plate design. In addition, the maximum stresses on both the prosthesis scaffolds were well below the yield strength (930 MPa) of the titanium alloy (Ti6Al4V ELI). On further observation, the analysis results of the screws, revealed that the condyle screws exhibited higher stresses when compared to chin screws which indicated that the stresses were transferring from the bottom chin region towards the condyle side thus satisfying the mastication process [45].

Table 3. Summary of Von Mises stress, strain and deformation of two designed scaffolds.

FEA Outcomes				Stress on Chin Screw (MPa)			Stress on Condyl Screw (MPa)		
				Screw Numbers					
Designed Implant	**Max Von Mises Stress (MPa)**	**Max Strain**	**Deformation**	**1**	**2**	**3**	**1**	**2**	**3**
Inner porous plate	360.22	0.0032	0.29852	55.85	38.26	50.52	122.9	121.74	81.5
Top & bottom porous plate	214.77	0.0068	0.31711	61.85	39.76	53.61	127.71	125.07	84.44

The most common cause for the failure of the mandibular reconstruction is either due to the reconstruction plate failure (excessive loads) or instability in the anchoring of the screws. In this study, the maximum stresses were found to be on the scaffold rather than on the screws and were well below the yield point and fatigue strength of the material. The stresses found on the screws in both the FEM were quiet less and within the failure limits, with the highest stress observed on the top and bottom screw plate. The other important parameter of the reconstruction plate design is its flexibility, to absorb the forces and chewing load conditions. The max strain on the inner porous plate was found to be 3.2 microns and the top and bottom porous plate was 6.8 microns. The maximum strain obtained on both the designed scaffolds was less and few microns. Based on the FEA results, it seems more reasonable to use prosthesis based on the top and bottom porous plate design for mandibular reconstruction, though both the plates were mechanically stable for fixation and could bear the masticatory functions.

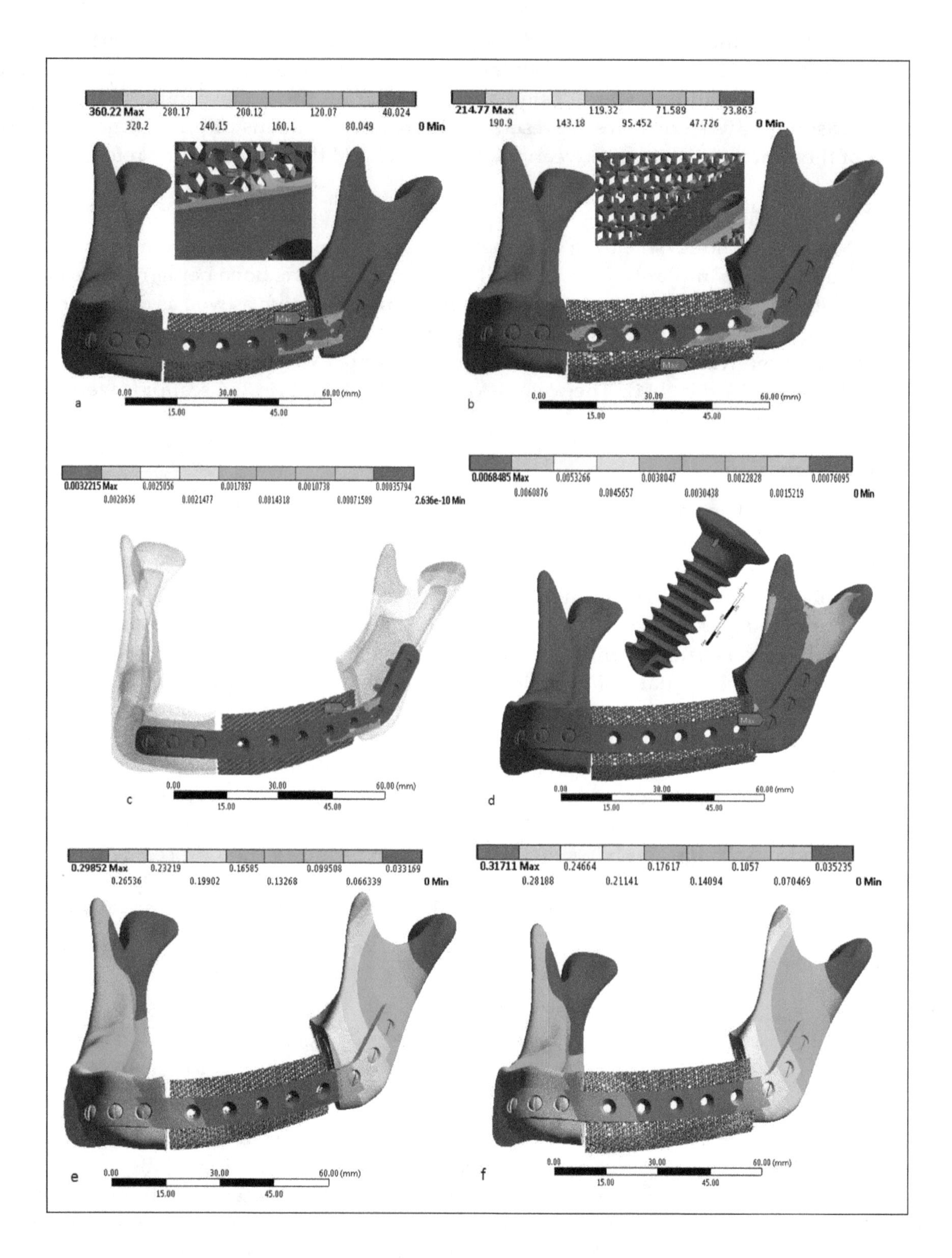

Figure 14. Von Mises stress (**top**), strain (**middle**) and deformation (**bottom**) distribution of mandibular framework model with two scaffolds (**a**,**c**,**e**) inner porous and (**b**,**d**,**f**) top and bottom porous plate.

The micro-CT scan results as shown in Figure 15 indicated that the dode thick structure was interconnected by a series of network channels and was free from any substantial internal defects such as cracks or voids. Similar results can be assumed and expected for the EBM fabricated mandibular prosthesis scaffold with dode thick structure.

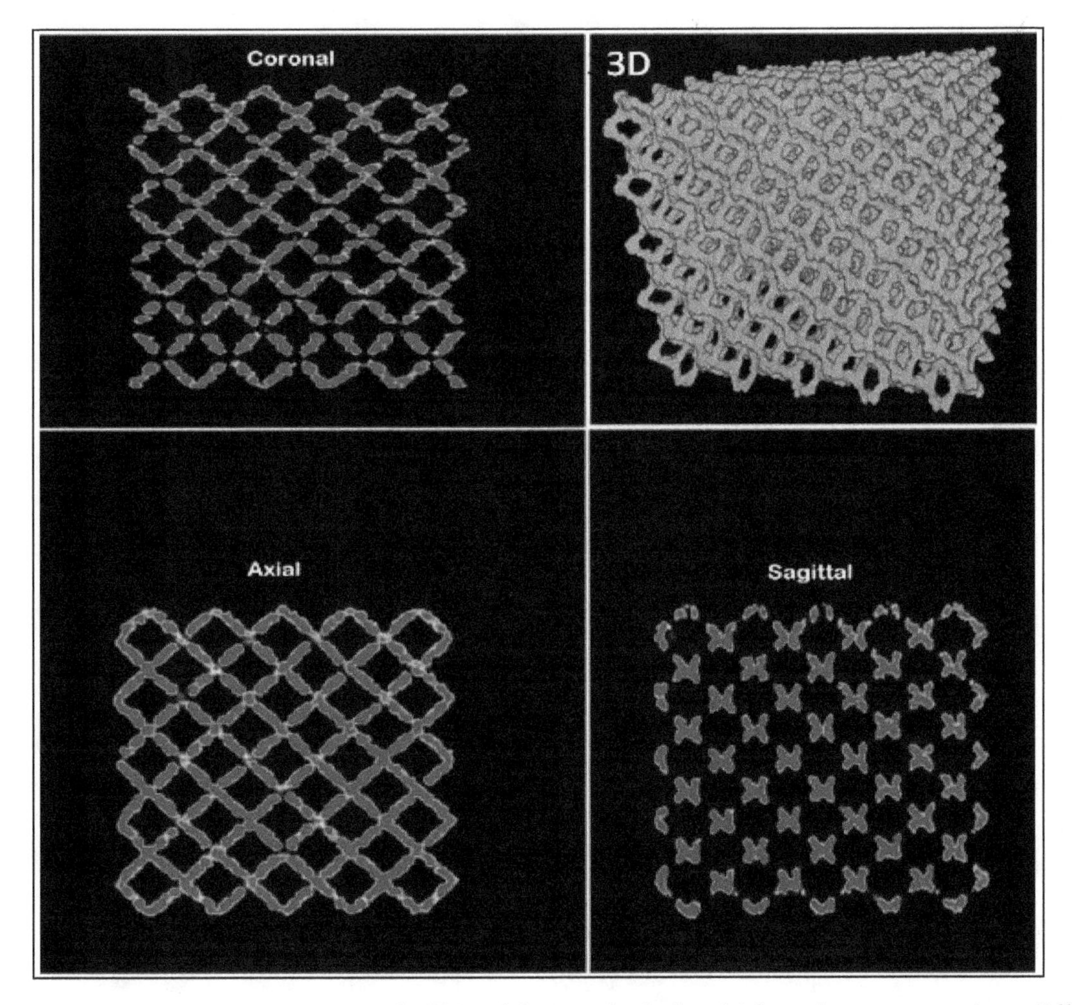

Figure 15. Micro-CT scanning of EBM fabricated dode thick cube representing different cross-sectional views.

The outcome of the 3D fitting deviation analysis has been represented graphically in Figure 16. The comprehensive investigation revealed that the scaffold with the top and bottom porous plate design provided better fitting accuracy as compared to the scaffold with inner porous plate design. An average deviation of 0.8274 mm was observed in the top and bottom porous plate design in comparison to 0.9283 mm of gap in the inner porous plate design.

The results of the weight analysis are presented in Table 4. The weight of the inner porous plate design was found to be 10.67 g and the top and bottom porous plate was 8.14 g. The weights of both reconstruction scaffolds were taken without considering the bone graft which will be placed inside the mesh carrier (tray) upon implant. Both scaffolds were low in weight and closer to that of bone properties. Certainly, this analysis confirmed that both the proposed designs possessed a lighter weight in comparison to their bone counterpart (19 g).

Table 4. Weight details of EBM fabricated scaffolds and replaced mandibular bone portion.

Parts	Replaced Bone	Inner Porous	Top and Bottom Porous
Volume (mm^3)	11879.00	2016.00	1847.00
Weight (g)	19.00	10.67	8.14

The Figure 17 illustrates the polymer and EBM fabricated titanium mandibular prosthesis scaffolds for final review before surgery.

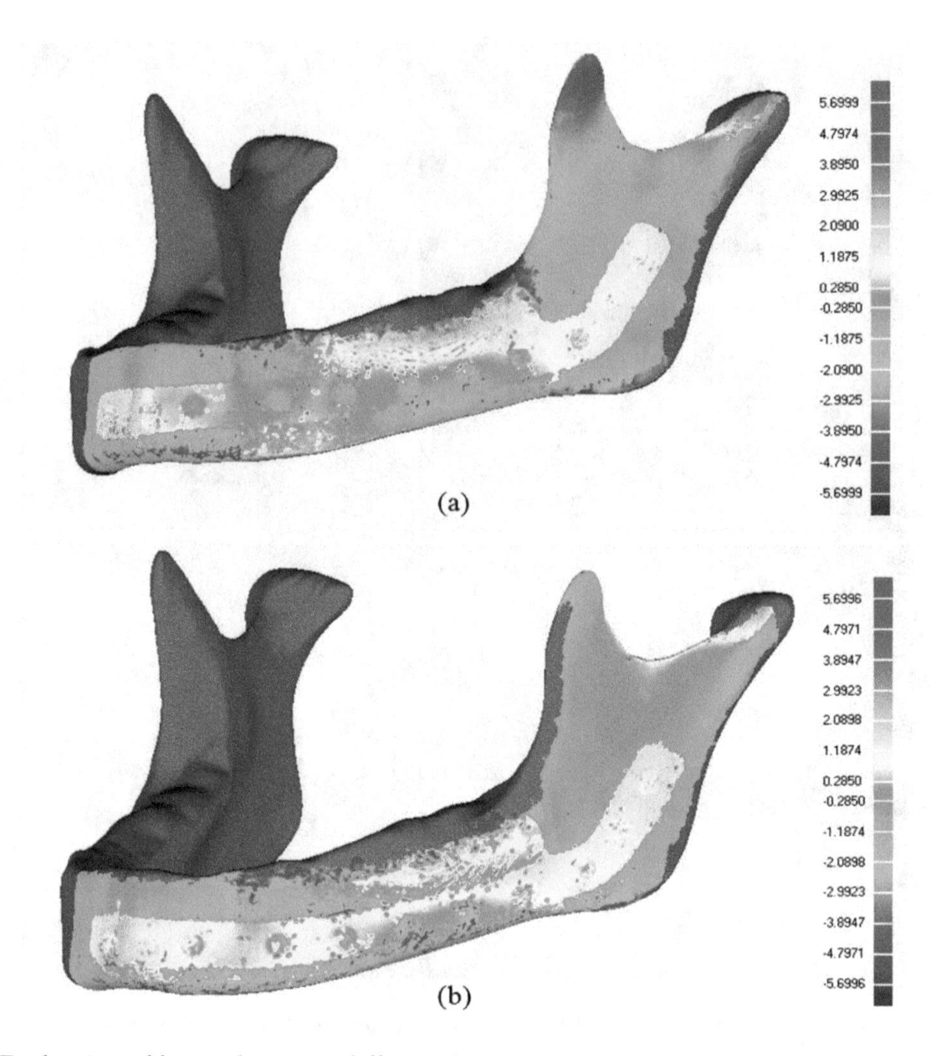

Figure 16. Evaluation of fitting deviation different designs: (**a**) Top and bottom porous; (**b**) inner porous.

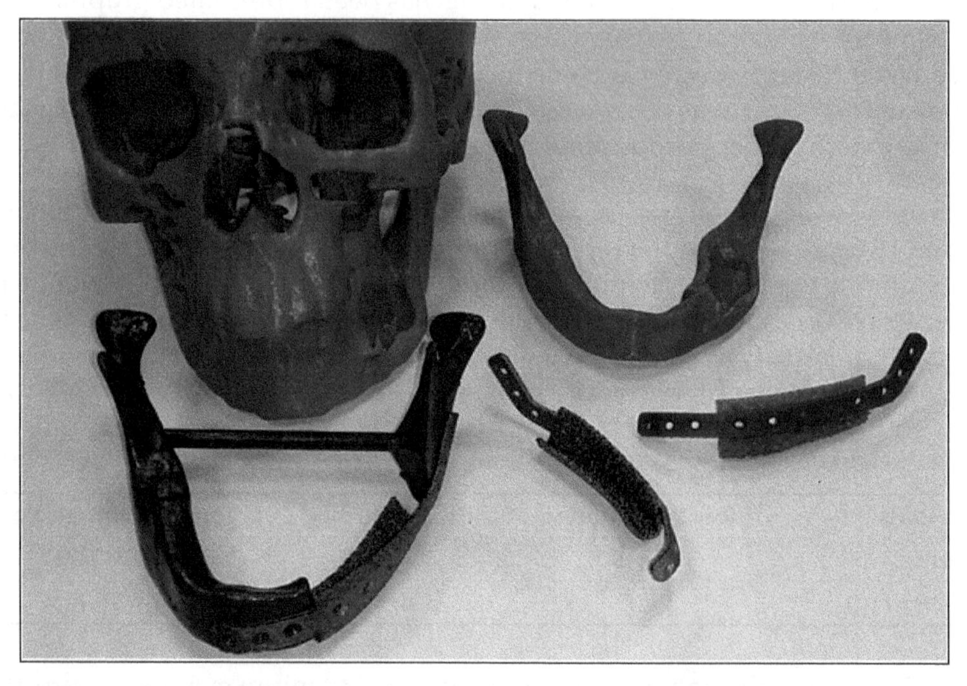

Figure 17. EBM and polymer fabricated mandibular framework models with prosthesis scaffolds.

4. Conclusions

The success of mandibular reconstruction greatly depends on its aesthetics and biomechanical properties. It emphasizes the importance of the customized implants depending on the patient's anatomy. The custom designed implants provide a better option for mandible restoration than the generic counterpart as they can fit precisely on the patient's bone. The ability to 3D print custom designed scaffolds using EBM technology, providing surface texture conducive to tissue ingrowth makes them appropriate for the personalized implants with properties closer to that of bone. In this study, two customized scaffolds based on the inner porous plate as well as the top and bottom porous plate were designed, 3D printed and evaluated for structural integrity, weight and fitting accuracy. A competent methodology has been presented to acquire the customized, pleasing and reliable mandibular implants. The methodology was exhaustive comprising of data acquisition using CT, mandible reconstruction as well as design, FEA, implant fabrication and testing.

Eventually, depending on the FEA, weight analysis and fitting accuracy evaluation, it can be inferred that the scaffold with the top and bottom porous plate is more favorable for bone reconstruction as compared to scaffold with the inner porous implant and can successfully be employed in the reconstruction of the defective mandible. Indeed, it can be asserted that the employment of prosthesis scaffolds in mandibular reconstruction satisfies the sustained need of lighter implants with accurate fitting and lesser surgical time and minimal revisions.

The customized porous implants are very effective and valuable because they provide an improved fit, enhanced osseointegration properties, lesser shielding effect and a higher implant stability. They strengthen the functional recovery of the mandibular deformities and maintain a graceful appearance on the mandible. It is mandatory that the research in this area should continue in the future for acquiring further innovative implant designs and reconstruction methods. The authors would like to expand this work by introducing new designs with different porous structures, and analyzing them for their strength and accuracy in mandible restoration. In addition, the authors would like to extend this work by including an extensive clinical (in-vivo) study in the future.

Author Contributions: K.M. conceived and designed the experiments; K.M. & S.H.M. performed the experiments; U.U. helped in the analysis; H.A. analyzed the data; K.M. & S.H.M. wrote and revised the paper.

Acknowledgments: The authors extend their appreciation to the Deanship of Scientific Research at King Saud University for funding this work through Research Group no. RG-1440-034.

References

1. Wong, R.C.W.; Tideman, H.; Kin, L.; Merkx, M.A.W. Biomechanics of mandibular reconstruction: a review. *Int. J. Oral Maxillofac. Surg.* **2010**, *39*, 313–319. [CrossRef]
2. Yan, R.; Luo, D.; Huang, H.; Li, R.; Yu, N.; Liu, C.; Hu, M.; Rong, Q. Electron beam melting in the fabrication of three-dimensional mesh titanium mandibular prosthesis scaffold. *Sci. Rep.* **2018**, *8*, 750. [CrossRef]
3. Miles, B.A.; Goldstein, D.P.; Gilbert, R.W.; Gullane, P.J. Mandible reconstruction. Curr. Opin. Otolaryngol. *Head Neck Surg.* **2010**, *18*, 317–322.
4. Hayden, R.E.; Mullin, D.P.; Patel, A.K. Reconstruction of the segmental mandibular defect: current state of the art. *Curr. Opin. Otolaryngol. Head Neck Surg.* **2012**, *20*, 231–236. [CrossRef]
5. Von Cramon-Taubadel, N. Global human mandibular variation reflects differences in agricultural and hunter-gatherer subsistence strategies. *Proc. Natl. Acad. Sci. USA* **2011**, *108*, 19546–19551. [CrossRef]
6. Yuan, J.; Cui, L.; Zhang, W.J.; Liu, W.; Cao, Y. Repair of canine mandibular bone defects with bone marrow stromal cells and porous β-tricalcium phosphate. *Biomaterials* **2007**, *28*, 1005–1013. [CrossRef] [PubMed]
7. Moiduddin, K. Implementation of Computer-Assisted Design, Analysis, and Additive Manufactured Customized Mandibular Implants. *J. Med. Biol. Eng.* **2018**, *38*, 744–756. [CrossRef]

8. Chanchareonsook, N.; Junker, R.; Jongpaiboonkit, L.; Jansen, J.A. Tissue-engineered mandibular bone reconstruction for continuity defects: A systematic approach to the literature. *Tissue Eng. Part B Rev.* **2014**, *20*, 147–162. [CrossRef] [PubMed]
9. Ryan, G.; Pandit, A.; Apatsidis, D. Fabrication methods of porous metals for use in orthopaedic applications. *Biomaterials* **2006**, *27*, 2651–2670. [CrossRef] [PubMed]
10. Wang, X.; Xu, S.; Zhou, S.; Xu, W.; Leary, M.; Choong, P.; Qian, M.; Brandt, M.; Xie, Y.M. Topological design and additive manufacturing of porous metals for bone scaffolds and orthopaedic implants: A review. *Biomaterials* **2016**, *83*, 127–141. [CrossRef] [PubMed]
11. Moiduddin, K.; Al-Ahmari, A.; Kindi, M.A.; Nasr, E.S.A.; Mohammad, A.; Ramalingam, S. Customized porous implants by additive manufacturing for zygomatic reconstruction. *Biocybern. Biomed. Eng.* **2016**, *36*, 719–730. [CrossRef]
12. Pei, X.; Zhang, B.; Fan, Y.; Zhu, X.; Sun, Y.; Wang, Q.; Zhang, X.; Zhou, C. Bionic mechanical design of titanium bone tissue implants and 3D printing manufacture. *Mater. Lett.* **2017**, *208*, 133–137. [CrossRef]
13. Raghavendra, S.; Wood, M.C.; Taylor, T.D. Early wound healing around endosseous implants: a review of the literature. *Int. J. Oral Maxillofac. Implant.* **2005**, *20*, 425–431.
14. Singare, S.; Lian, Q.; Wang, W.P.; Wang, J.; Liu, Y.; Li, D.; Lu, B. Rapid prototyping assisted surgery planning and custom implant design. *Rapid Prototyp. J.* **2009**, *15*, 19–23. [CrossRef]
15. Emadabouel, N.; Abdulrahman, A.-A.; Khaja, M.; Al Kindi, M.; Kamrani, A. A digital design methodology for surgical planning and fabrication of customized mandible implants. *Rapid Prototyp. J.* **2016**, *23*, 101–109.
16. Chua, C.K.; Wong, C.H.; Yeong, W.Y. *Standards, Quality Control, and Measurement Sciences in 3D Printing and Additive Manufacturing*; Academic Press: London, UK, 2017.
17. Electron Beam Melting—EBM Process, Additive Manufacturing. Available online: http://www.arcam.com/technology/electron-beam-melting/ (accessed on 7 July 2017).
18. Moiduddin, K.; Darwish, S.; Al-Ahmari, A.; El Watidy, S.; Mohammad, A.; Ameen, W. Structural and mechanical characterization of custom design cranial implant created using additive manufacturing. *Electron. J. Biotechnol.* **2017**, *29*, 22–31. [CrossRef]
19. Moiduddin, K.; Anwar, S.; Ahmed, N.; Ashfaq, M.; Al-Ahmari, A. Computer assisted design and analysis of customized porous plate for mandibular reconstruction. *IRBM* **2017**, *38*, 78–89. [CrossRef]
20. Narra, N.; Valášek, J.; Hannula, M.; Marcián, P.; Sándor, G.K.; Hyttinen, J.; Wolff, J. Finite element analysis of customized reconstruction plates for mandibular continuity defect therapy. *J. Biomech.* **2014**, *47*, 264–268. [CrossRef]
21. Liu, Y.; Fan, Y.; Jiang, X.; Baur, D.A. A customized fixation plate with novel structure designed by topological optimization for mandibular angle fracture based on finite element analysis. *Biomed. Eng. Online* **2017**, *16*, 131. [CrossRef]
22. Planmeca ProMax 3D Max—Dental Imaging to the Max. Available online: https://www.planmeca.com/imaging/3d-imaging/planmeca-promax-3d-max/ (accessed on 09 April 2019).
23. Arango-Ospina, M.; Cortés-Rodriguez, C.J. Engineering design and manufacturing of custom craniofacial implants. In *The 15th International Conference on Biomedical Engineering*; Goh, J., Ed.; Springer International Publishing: Basel, Switzerland, 2014; pp. 908–911.
24. Zhou, L.; Shang, H.; He, L.; Bo, B.; Liu, G.; Liu, Y.; Zhao, J. Accurate Reconstruction of Discontinuous Mandible Using a Reverse Engineering/Computer-Aided Design/Rapid Prototyping Technique: A Preliminary Clinical Study. *J. Oral Maxillofac. Surg.* **2010**, *68*, 2115–2121. [CrossRef]
25. Van Bael, S.; Chai, Y.C.; Truscello, S.; Moesen, M.; Kerckhofs, G.; Van Oosterwyck, H.; Kruth, J.-P.; Schrooten, J. The effect of pore geometry on the in vitro biological behavior of human periosteum-derived cells seeded on selective laser-melted Ti6Al4V bone scaffolds. *Acta Biomater.* **2012**, *8*, 2824–2834. [CrossRef] [PubMed]
26. Ran, Q.; Yang, W.; Hu, Y.; Shen, X.; Yu, Y.; Xiang, Y.; Cai, K. Osteogenesis of 3D printed porous Ti6Al4V implants with different pore sizes. *J. Mech. Behav. Biomed. Mater.* **2018**, *84*, 1–11. [CrossRef] [PubMed]
27. Schaller, A.; Voigt, C.; Huempfner-Hierl, H.; Hemprich, A.; Hierl, T. Transient finite element analysis of a traumatic fracture of the zygomatic bone caused by a head collision. *Int. J. Oral Maxillofac. Surg.* **2012**, *41*, 66–73. [CrossRef]
28. El-Anwar, M.I.; Mohammed, M.S. Comparison between two low profile attachments for implant mandibular overdentures. *J. Genet. Eng. Biotechnol.* **2014**, *12*, 45–53. [CrossRef]

29. Ti6Al4V ELI Titanium Alloy. 2014. Available online: http://www.arcam.com/wp-content/uploads/Arcam-Ti6Al4V-ELI-Titanium-Alloy.pdf (accessed on 27 January 2019).
30. Szucs, A.; Bujtár, P.; Sándor, G.K.B.; Barabás, J. Finite element analysis of the human mandible to assess the effect of removing an impacted third molar. *J. Can. Dent. Assoc.* **2010**, *76*, a72.
31. Simonovics, J.; Bujtár, P.; Váradi, K. Effect of preloading on lower jaw implant. *Biomech. Hungarica* **2013**, *6*, 21–28. [CrossRef]
32. What is FDM?: Fused Deposition Modeling Technology for 3D Printing | Stratasys n.d. Available online: https://www.stratasys.com/fdm-technology (accessed on 6 January 2019).
33. High Resolution SLA and SLS 3D Printers for Professionals. Formlabs n.d. Available online: https://formlabs.com/ (accessed on 7 February 2019).
34. FDM vs SLA: How does 3D Printing Technology Work? |. Pinshape 3D Printing Blog | Tutorials, Contests & Downloads 2017. Available online: https://pinshape.com/blog/fdm-vs-sla-how-does-3d-printer-tech-work/ (accessed on 7 February 2019).
35. Kumar, A.; Nune, K.C.; Murr, L.E.; Misra, R.D.K. Biocompatibility and mechanical behaviour of three-dimensional scaffolds for biomedical devices: Process–structure–property paradigm. *Int. Mater. Rev.* **2016**, *61*, 20–45. [CrossRef]
36. Horn, T.J.; Harrysson, O.L.A.; Marcellin-Little, D.J.; West, H.A.; Lascelles, B.D.X.; Aman, R. Flexural properties of Ti6Al4V rhombic dodecahedron open cellular structures fabricated with electron beam melting. *Addit. Manuf.* **2014**, *1–4*, 2–11. [CrossRef]
37. Murr, L.E.; Gaytan, S.M.; Medina, F.; Lopez, H.; Martinez, E.; Machado, B.I.; Hernandez, D.H.; Martinez, L.; Lopez, M.I.; Wicker, R.B.; Bracke, J. Next-generation biomedical implants using additive manufacturing of complex cellular and functional mesh arrays. *Philos. Trans. R. Soc. A: Math. Phys. Eng. Sci.* **2010**, *368*, 1999–2032. [CrossRef]
38. Tang, H.P.; Wang, J.; Song, C.N.; Liu, N.; Jia, L.; Elambasseril, J.; Qian, M. Microstructure, mechanical properties, and flatness of sebm Ti-6Al-4V sheet in as-built and hot isostatically pressed conditions. *JOM* **2017**, *69*, 466–471. [CrossRef]
39. Wang, M.; Li, H.Q.; Lou, D.J.; Qin, C.X.; Jiang, J.; Fang, X.Y.; Guo, Y.B. Microstructure anisotropy and its implication in mechanical properties of biomedical titanium alloy processed by electron beam melting. *Mater. Sci. Eng. A* **2019**, *743*, 123–137. [CrossRef]
40. Balza, J.C.; Zujur, D.; Gil, L.; Subero, R.; Dominguez, E.; Delvasto, P.; Alvarez, J. Sandblasting as a surface modification technique on titanium alloys for biomedical applications: abrasive particle behavior. *IOP Conf. Ser. Mater. Sci. Eng.* **2013**, *45*, 012004. [CrossRef]
41. Geomagic Control X. 3D Systems n.d. Available online: https://www.3dsystems.com/software/geomagic-control-x (accessed on 10 February 2019).
42. Hammad Mian, S.; Abdul Mannan, M.; M. Al-Ahmari, A. The influence of surface topology on the quality of the point cloud data acquired with laser line scanning probe. *Sens. Rev.* **2014**, *34*, 255–265. [CrossRef]
43. Ridtzwan, M.I.; Solehuddin, S.; Hassan, A.Y.; Shokri, A.A.; Mohamad Ibrahim, M.N. Problem of Stress Shielding and Improvement to the Hip Implant Designs: A Review. *J. Med. Sci.* **2007**, *7*, 460–467.
44. Soh, C.-K.; Yang, Y.; Bhalla, S. (Eds.) *Smart Materials in Structural Health Monitoring, Control and Biomechanics*; Springer-Verlag: Berlin, Germany, 2012.
45. Basciftci, F.A.; Korkmaz, H.H.; Üşümez, S.; Eraslan, O. Biomechanical evaluation of chincup treatment with various force vectors. *Am. J. Orthod. Dentofac. Orthop.* **2008**, *134*, 773–781. [CrossRef] [PubMed]

Mechanical Properties and In Vitro Behavior of Additively Manufactured and Functionally Graded Ti6Al4V Porous Scaffolds

Ezgi Onal [1,*], Jessica E. Frith [1], Marten Jurg [1], Xinhua Wu [1,2] and Andrey Molotnikov [1,*]

1 Department of Materials Science and Engineering, Monash University, Clayton, VIC 3800, Australia; jessica.frith@monash.edu (J.E.F.); marten.jurg@monash.edu (M.J.); xinhua.wu@monash.edu (X.W.)
2 Monash Centre for Additive Manufacturing, Monash University, Clayton, VIC 3800, Australia
* Correspondence: ezgi.onal@monash.edu (E.O.); andrey.molotnikov@monash.edu (A.M.)

Abstract: Functionally graded lattice structures produced by additive manufacturing are promising for bone tissue engineering. Spatial variations in their porosity are reported to vary the stiffness and make it comparable to cortical or trabecular bone. However, the interplay between the mechanical properties and biological response of functionally graded lattices is less clear. Here we show that by designing continuous gradient structures and studying their mechanical and biological properties simultaneously, orthopedic implant design can be improved and guidelines can be established. Our continuous gradient structures were generated by gradually changing the strut diameter of a body centered cubic (BCC) unit cell. This approach enables a smooth transition between unit cell layers and minimizes the effect of stress discontinuity within the scaffold. Scaffolds were fabricated using selective laser melting (SLM) and underwent mechanical and in vitro biological testing. Our results indicate that optimal gradient structures should possess small pores in their core (~900 μm) to increase their mechanical strength whilst large pores (~1100 μm) should be utilized in their outer surface to enhance cell penetration and proliferation. We suggest this approach could be widely used in the design of orthopedic implants to maximize both the mechanical and biological properties of the implant.

Keywords: selective laser melting; gradient structure; porous biomaterial; Ti6Al4V; mechanical properties; osteoblast

---

1. Introduction

Recent advances in additive manufacturing have revealed new possibilities for the design of the next generation of metallic biomedical implants based on lattice structures. Generally, a bone scaffold should possess four essential characteristics [1,2]: (i) biocompatibility; (ii) mechanical properties matching those of the host tissue; (iii) an interconnected porous structure for cell migration and proliferation and nutrient–waste transportation; (iv) suitable surface characteristics for cell attachment. Traditionally, large bone defects are treated with metallic implants. Several metals have been shown to fulfil the requirement of biocompatibility including cobalt-based alloys, stainless steels as well as titanium alloys [3]. However, metallic implants possess higher elastic moduli than bone, e.g., Ti6Al4V and 316 L stainless steel have a modulus of around 110 GPa and 210 GPa [4] respectively, whereas the modulus of cortical bone is in the range of 3–20 GPa [5]. The mismatch between the modulus of the bone and the implant results in the failure of the implant in the long-term due to the stress-shielding problem [6]. Therefore, matching the mechanical properties of the implant to the host bone and simultaneously providing the implant with biological performance remains a challenge. One potential strategy is to create porous metallic scaffolds where the porosity, the pore size and shape are optimized collectively to reduce the modulus while maintaining the strength of the scaffold [7].

Porous Ti6Al4V scaffolds are ideal candidates as bone scaffolds since they comply with the aforementioned requirements: they are biocompatible [8,9] and their mechanical properties can be adjusted by porosity [10]. Porous structures with an interconnected pore network are of particular interest for promoting cell migration and colonization [11] as well as tissue in-growth [12,13]. Interconnected pore network enables the flow of nutrients and oxygen to cells and tissue and promotes the formation of blood capillaries [14]. When the vascular network is not sufficient, nutrient and oxygen deficiencies cause hypoxia or necrosis [15,16]. Therefore, open-porous structures with adequate pore sizes are critical for vascularization and tissue growth. However, the optimum porous structure for orthopedic scaffolds is yet to be established and there is conflicting information regarding an optimum pore size for both enhanced bone ingrowth and mechanical strength of the scaffold. Recent reviews [17–19] summarize that optimum pore size should be 300 μm or larger for bone ingrowth and vascularization. However, whilst high porosity and/or large pore size (>800 μm) promotes flow of nutrients, vascularization and tissue growth [14,15,20], highly porous structures lack the required mechanical strength and integrity and decrease the cell seeding efficiency [15,21]. At the same time, in vitro studies have shown that scaffolds with small pore sizes (<500 μm) or low porosities are prone to pore occlusions [22].

Gradient structures present a potential solution to the opposing requirements of an optimal pore size for biological response and mechanical properties. Given an optimal pore size distribution, there is potential to develop structures that exhibit both adequate mechanical strength and tissue in-growth rate. In addition, gradient structures can mimic the natural bone in terms of its structure and mechanical properties [23]. The structure of the bone changes with the amount and direction of the applied stress [24] resulting in differences in the internal structure (porosity and composition) and mechanical properties of the bone along its dimensions. For example, the elastic modulus of trabecular bone at the ends of long bones or within the interior of vertebrae is around 0.5 GPa [25]. This variation in elastic modulus depending on the location in the bone indicates the need for development of gradient structures in bone scaffolds.

The gradient and uniform porous scaffolds can be designed using traditional CAD (Computer Aided Design) and include the use of open cellular foams [26,27] and periodic uniform unit cells based on platonic solids [28–30]. Other techniques, such as implicit surface modelling [31–33] and topology optimized scaffolds [18,34], are also gaining in popularity. The fabrication of such complex structures has recently become feasible with the advances in additive manufacturing [35]. Traditional manufacturing of porous metals such as solid or liquid state processing has limited control over the shape and size of the pores achievable through adaptation of the processing parameters. These shortcomings can be overcome through additive manufacturing which builds a three-dimensional object in layer-by-layer fashion. Selective laser melting (SLM) and electron beam melting (EBM) have both been utilized to successfully fabricate porous scaffolds [36]. Both methods rely on a computer-controlled high power energy source to selectively melt a metallic powder on each layer.

A number of studies have investigated the mechanical or biological response of metallic-based gradient porous designs produced by additive manufacturing [37–45]. The majority of these studies focused on gradient structures generated by abrupt changes between layers based on change in strut diameter or unit cell volume. For instance, Li et al. [37] studied the deformation behavior of radial dual-density rhombic dodecahedron Ti6Al4V scaffolds fabricated by EBM. They achieved radial dual-density by altering the rhombic dodecahedron unit volume between two layers which resulted in discontinuity between layers. Their finite volume method simulations revealed that the inherent discontinuity between layers resulted in stress concentration and maximum stresses at the interfaces. They concluded that continuous variation between layers are ideal to minimize the stress concentration at the interface. Nune et al. [38,39] investigated the osteoblastic functions of the scaffolds designed using Li et al. [37] work, based on gradient rhombic dodecahedron created by changing the unit volume between layers. Although their work showed promising results on cellular activity when cells were seeded from large pore side (1000 μm), there was no complementary study on the mechanical properties and therefore the adverse effect of discontinuity between layers on the mechanical properties were not covered.

Another study [40] of multiple-layer gradient structures based on changing unit cell volume also showed the mismatch between two layers. They designed gradient BCC and diamond cylinders where two different unit cell volumes were used in the outer and inner parts of the cylinder. This design approach resulted in free nodes of outer layers that are not connected to inner layers, which caused a negative effect on the mechanical properties. Another approach frequently used to generate gradient structures is based on a sharp change in strut diameter between layers [41,42]. This design principle also results in a mismatch between layers negatively effecting the mechanical properties as well as tissue ingrowth and mineralization [42].

Recently, there have been a few investigations aiming to overcome the problem of a mismatch between layers by designing continuous gradient structures which consists of gradually changing strut diameters between layers. For example, Han et al. [43] reported the mechanical properties of SLM-fabricated pure titanium Schwartz diamond unit cell and demonstrated the layer-by-layer sequential failure of these gradient scaffolds. Maskery et al. [44] also showed the layer-by-layer gradual collapse of gradient scaffolds using SLM-fabricated AlSi10Mg gradient BCC structures. These two studies highlighted that the deformation and energy absorption of gradient lattices is more predictable than the uniform lattices due to the lack of diagonal shear band formation during deformation. In another study Ti6Al4V cubic and honeycomb lattice structures [45] were combined to a gradient structure with a continuous density change and it was shown that this design had a superior energy absorption properties compared to their uniform counterparts. Although the aforementioned studies on the mechanical properties of continuous gradient structures indicate promising results; the selection of unit cells, materials as well as their in vitro and in vivo response need to be better understood.

In this work, we introduce a concept of generating continuous gradient structures by changing the strut diameter linearly across cell layers which enables a smooth transition between layers. To demonstrate the benefit of this design principle, we apply it to the BCC unit cell and create gradient structures with rising or decreasing pores sizes. These gradient structures were mirrored from the central horizontal axis to obtain a symmetrical sample. Their mechanical properties were obtained by uniaxial compression tests and cell attachment and proliferation were assessed with murine pre-osteoblast cells. It will be shown that gradient scaffold can be tailored to fulfil the mechanical properties required and simultaneously improve biological response.

2. Materials and Methods

2.1. Design and Fabrication of Ti6Al4V Gradient Cellular Structures

The lattice structures were designed with Rhinoceros v5 in the RhinoPython environment. A custom-made parametric script was developed to create the lattice models with continuous gradient structures. The scaffold structure is defined by its type (BCC) (Figure 1a) and size in each of the cardinal directions, and has a changing strut diameter based upon a polynomial equation, see Equation (1). The diameter of the strut is calculated at a minimum or three locations, evenly spaced along its length based on the location of the strut, and lofted to create a smooth transition between radii.

$$Diameter = c + \sum_{n=1}^{n,(x,y,z)} A_{n_{(x,y,z)}} \left(\left| P_{(x,y,z)} \right| - P_{(x,y,z)_0} \right)^n \tag{1}$$

Here c is constant, $A_{n_{(x,y,z)}}$ is the gradient value in the current cardinal direction, $\left|P_{(x,y,z)}\right|$ is the absolute current coordinate position in the relevant axis (x, y, or z), $P_{(x,y,z)_0}$ is the gradient origin in each of the relevant axes, defined over n polynomial terms.

In the case of linear gradient in a single axis, only a single term (A_x) is required, simplifying the Equation (1) to following:

$$Diameter = c + A_x(P_x - P_{x_0}) \quad (2)$$

It should be noted that by adding y and z terms in Equation (2) will modify the nature of the gradient and allow to produce 3D gradient scaffolds.

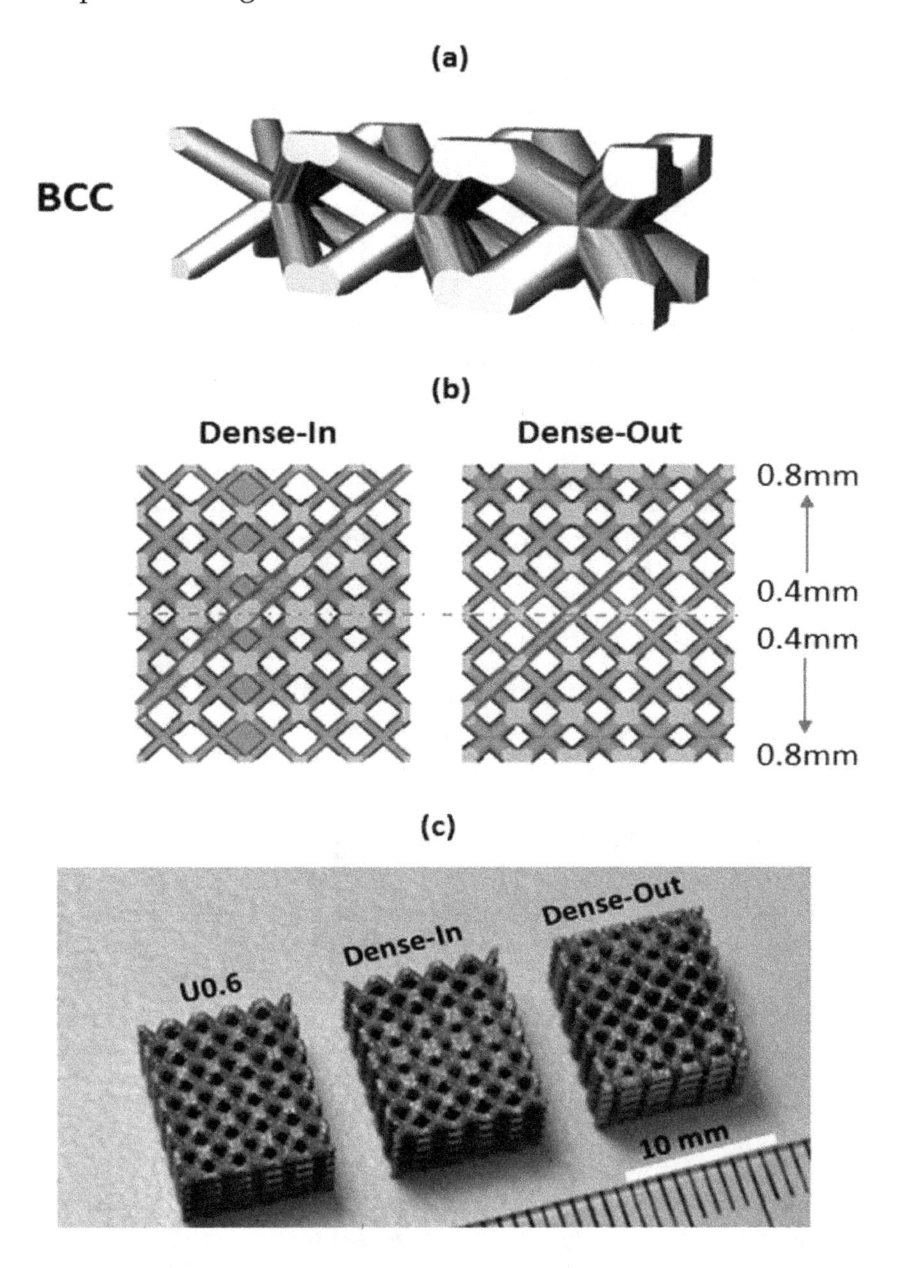

Figure 1. (**a**) Gradient BCC unit cell showing continuous transition at the unit cell junctions, (**b**) CAD view of the gradient BCC lattice scaffold; the arrows show the gradual increase in strut diameter and the highlighted areas show the gradually changing pore size and strut diameter along the scaffolds' height; Dense-In scaffold has increasing strut diameter towards the center, whereas Dense-Out follows the opposite trend; (**c**) SLM-fabricated specimens with dimensions of 10 × 10 × 12 mm^3, from left to right: uniform BCC lattice scaffold with 0.6 mm strut diameter, Dense-In and Dense-Out scaffolds.

To demonstrate the benefit of this concept, we designed two gradient structures with dimensions of 6 × 10 × 10 mm^3, mirrored in the x-axis to create lattices of 12 × 10 × 10 mm^3. In this case a gradient value A_x was set to 0.07 (mm/mm) and a constant c equal to 0.4 mm. The choice of the parameters is motivated by manufacturability of the struts and leads to a minimum diameter of 0.4 mm and a maximum of 0.82 mm over the 6-mm lattice. When gradient design was mirrored from the thinner strut plane, it is named as Dense-Out and when mirrored from the thicker strut plane, it is named as Dense-In. For comparison of the mechanical and biological response of gradient structures, we also created three uniform BCC structures which utilize the same strut diameters as present in the gradient structure, namely 0.4, 0.6, 0.8 mm. These uniform BCC structures are denoted U0.4, U0.6 and U0.8, respectively.

It should be noted that the developed script is not limited to the BCC unit cell and a large library of common unit cells was programmed allowing us to generate a smooth gradient between layers and tailor their size and local and total porosity.

The uniform and gradient BCC structures were fabricated by selective laser melting (SLM) process, described in [46], using a MLab Cusing machine (Concept Laser, Lichtenfels, Germany). Ti6Al4V-ELI powder supplied by Falcon Tech Co., Ltd. (Wuxi, China) that satisfies ASTM F136 with diameter range of 15–53 µm was used for laser melting process. The Ti6Al4V samples were fabricated using a laser power of 95 W, scan speed of 600 mm/s with a hatch distance of 0.08 mm, the beam spot size of 50 (−5, +25) µm, with oxygen content less than 0.2% in an argon atmosphere and the layer thickness of 25 µm. The samples were built on top of a solid titanium plate and were removed by using Electrical Discharge Machining (EDM, AgieCharmilles CUT 30P, GF, Losone, Switzerland). All samples were 12 mm in height with a cross-section of 10 × 10 mm^2. After the samples were removed from the build plate, they were washed in an ultrasonic bath containing ethanol for 4 h to aid in removing the unmelted particles from the surface of the samples. No further surface modifications or heat treatments were applied to the scaffolds.

2.2. Morphological Analysis

Morphological properties were characterized by three methods: scanning electron microscope (SEM), digital densitometry and gas pycnometry. Pore size and strut diameter were measured using SEM (FEI Nova NanoSEM, Thermo Fisher Scientific, Hillsboro, OR, USA) images. The average of four pore sizes and strut diameters was calculated for each specimen. The volume fraction of the samples was measured by both gas pycnometry and digital densitometry. A digital densitometry (SD-200L, AlfaMirage, Osaka, Japan) adopts the Archimedean principle with liquid displacement. Three specimens of each sample were used to measure the volume fraction. In addition to densitometry measurements, the volume fraction of each sample were measured by a gas pycnometry (AccuPyc 1330, Micromeritics, Norcross, GA, USA) with helium gas in 3 repeats. The pycnometry measures the density of solid materials by employing gas displacement and Boyle's Law of gas expansion. The volume fraction values measured by gas pycnometry were compared with the values obtained from the CAD designs.

2.3. Measurement of Mechanical Properties

A minimum of five specimens of each uniform and gradient structure were mechanically tested under compression using an Instron 5982 universal testing machine with a 100 kN load cell. Following the standard for compression test for porous and cellular metals (ISO 13314:2011), a constant cross-head velocity of 0.72 mm/min was utilized corresponding to a compression strain rate of 10^{-3} s^{-1}. The measurements were recorded after a preload of 50–70 N to avoid the initial settling of the samples between plates. A series of images were captured every 1 s during compression testing to record the deformation response of the samples.

The engineering compressive stress was calculated by normalizing the applied compression load with the initial cross-section area of each sample (10 × 10 mm^2) and the engineering strain was calculated by the displacement of the cross-heads. The stress-strain curves of each sample were analyzed and the following mechanical properties were calculated based on the guidelines of the ISO Standard 13314:2011: first maximum compressive strength (σ_{max}) (the first local maximum in the stress-strain curve), 0.2% offset yield stress (σ_y), the elastic gradient ($E_{(\sigma 20-\sigma 50)}$) (the gradient of the elastic straight line between stresses of 20 and 50 MPa). The ISO Standard recommends determining the elastic gradient between stresses of 20 and 70 MPa; however, since 70 MPa was higher than the 0.2% yield stress point for some of the samples, 50 MPa was adopted for all samples.

2.4. Cell Viability, Proliferation and Morphology

MC3T3-E1, Subclone 4, preosteoblast cells (ATCC® CRL-2593™, Manassas, VA, USA) were cultured in α-MEM medium (Gibco) supplemented with 10% fetal bovine serum (FBS, Gibco) and 1% antibiotic/antimycotic solution (Gibco) in a humidified incubator at 37 °C, 5% CO_2. As-fabricated Ti6Al4V uniform and gradient porous structures were cleaned by successive washes with ethanol and phosphate buffer saline (PBS). Subsequently, specimens with dimensions of $10 \times 10 \times 12$ mm^3 were placed in a 24-multiwell plate with 3 repeats and seeded at a density of 1×10^5 cells per specimen. In parallel, positive control wells containing SLM-fabricated solid samples were set up. Before any evaluation, all scaffolds were transferred into a new 24-multiwell plate.

Cell viability was assessed at 4 h, 4 days and 7 days after cell culture using the MTS assay (Cell Titre 96 Aqueous One Solution Cell Proliferation Assay, Promega, Madison, WI, USA). After 2.5 h incubation with MTS reagent, 100 μL of solution from each well were transferred into a 96-multiwell plate and the optical absorbance (OD) was measured at 490 nm with a microplate reader (Thermo Scientific™ Multiskan Spectrophotometer, Vantaa, Finland). The degree of cell attachment and spreading was studied by immunofluorescence staining. Samples were fixed with 4% paraformaldehyde (PFA) for 20 min and stained with ActinRed™ 555 ReadyProbes® Reagent (Molecular Probes™, Gaithersburg, MD, USA) following the manufacturer's instructions and Hoechst 33342 (10 μg/mL) (Thermo Scientific™, Rockford, IL, USA) for 30 min. After immunostaining, the top and bottom of the specimens were examined by fluorescence microscopy (Nikon Eclipse Ti, Tokyo, Japan).

Cell morphology was characterized by scanning electron microscope (SEM) at 4 h, 4 days and 7 days of cell culture. Cells on the specimens were fixed with 2.5% glutaraldehyde in 0.1 M of sodium cacodylate buffer for 2 h and postfixed in 1% osmium tetroxide for 1.5 h at room temperature. Specimens were dehydrated with a gradient series of ethanol (50%, 70%, 80%, 90%, 95% and 100%) followed by a hexamethyldisilazane (hmds) drying procedure. The specimens were sputter coated with gold and inspected using a FEI Nova NanoSEM.

2.5. Statistical Analysis

Statistical analyses were performed using SPSS Statistics 20.0 (SPSS, Inc., Chicago, IL, USA). Data were presented as mean ± standard deviation (SD). One-way analysis of variation (ANOVA) together with Tukey–Kramer post-hoc analysis were used to identify significant differences (significance threshold: $p < 0.05$).

3. Results

3.1. Morphology of Porous Scaffolds

The key morphological parameters of the scaffolds, such as pore size and strut diameter, were measured by SEM and presented in Table 1. The measured strut diameters were larger than the original designs for all samples due to the adhesion of the semimolten powder on the surface (Figure 2b). We measured average surface roughness, R_a, using a Mitutoyo Surftest SJ400 and found it to be around 10 μm. This is in agreement with the reported values of surface roughness of SLM printed lattice structures [47]. In addition, volume fraction was quantified using the gas pycnometry and densitometry (Table 1). The volume fraction values were derived directly from the pycnometry and calculated from the densitometry, based on the following equations:

$$V = (W_a - W_w)/0.9971 \quad (3)$$

$$V_f(\%) = V/V_t \quad (4)$$

Table 1. The morphometric parameters of uniform and gradient BCC structures based on CAD designs, gas pycnometry, digital densitometry and SEM.

Scaffold Name	Porosity (%)				Pore Size (mm)		Strut Diameter (mm)	
	Design	Gas Pycnometry	Densitometry	Difference (%)	Design	SEM	Design	SEM
U0.4	82	71.87 ± 0.01	71.45 ± 0.02	13	1.51	1.14 ± 0.03	0.40	0.57 ± 0.01
U0.6	64	53.06 ± 0.01	51.11 ± 0.01	17	1.26	0.94 ± 0.05	0.60	0.77 ± 0.01
U0.8	44	33.78 ± 0.01	31.86 ± 0.01	23	1.02	0.73 ± 0.03	0.80	1.06 ± 0.02
Dense-In	62	50.73 ± 0.01	49.38 ± 0.01	18	1.33 1.13 0.94	1.04 ± 0.02 0.83 ± 0.01 0.74 ± 0.07	0.40 0.61 0.82	0.59 ± 0.01 0.72 ± 0.01 0.92 ± 0.03
Dense-Out	62	51.90 ± 0.02	50.01 ± 0.01	16	0.94 1.13 1.33	0.62 ± 0.02 0.82 ± 0.02 0.98 ± 0.03	0.82 0.61 0.40	0.91 ± 0.02 0.74 ± 0.01 0.59 ± 0.01

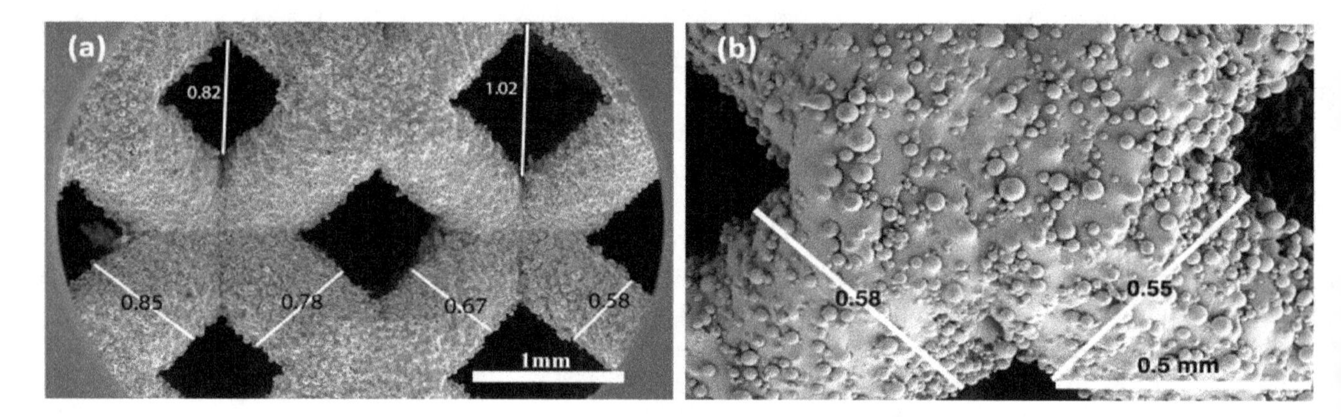

Figure 2. SEM image of (**a**) a gradient Dense-Out structure, demonstrating the change in strut diameter and pore size along the scaffold, (**b**) higher magnification view of the struts showing attached semi-molten powders.

Here, V is the volume of the scaffolds, W_a is the weight measured in air, W_w is the weight measured in water, 0.9971 is the density (g cm^{-3}) of distilled water at 25 °C and 1 atm, V_f is the volume fraction in percent, V_t is the total volume obtained from the outer dimension of the scaffolds, which is 1.2 cm^3 for all. The porosity of the scaffolds is defined as 100-Volume fraction (%).

The measured values of volume fraction were smaller than the designed volume fraction values for all samples. The difference between the original designs and pyconometry results was around 13–23%. These deviations are as expected since the as-fabricated strut diameters were larger than the original designs, resulting in smaller pore sizes than the intended geometry. This difference is characteristic to SLM process and is caused by effects such as staircase stepping due to layered manufacturing and melt pool variation due to residual stresses [48].

Strut diameter change along the gradient BCC scaffolds was noticeable in SEM images, (Figure 2a), confirming that the desired graded porosity was successfully achieved by the SLM process. The pore size of gradient Dense-In scaffold was varied from 1.14 mm to 0.74 mm, whereas it was between

0.62 mm to 0.98 mm for gradient Dense-Out scaffold. Measured porosity of the gradient Dense-In and Dense-Out scaffolds were almost identical due to symmetric design along the horizontal center plane.

3.2. Mechanical Properties of Porous Scaffolds

The compressive nominal stress–strain plots of uniform and gradient BCC structures are presented in Figure 3. The stress–strain curves exhibit characteristic stages of deformation for cellular solids [10,49], including linear elastic region, followed by plateau region with fluctuating stresses. The uniform structures showed similar behavior under compression; however, they reached different levels of maximum stress and possess different elastic moduli. The stress–strain curves for gradient structures also showed initially similar behavior to the uniform scaffolds. After the onset of plasticity an abrupt structural collapse was observed in uniform scaffolds, but not in the gradient scaffolds. Furthermore, the fluctuating degree of plateau region was more distinguished for the uniform scaffolds than the gradient scaffolds.

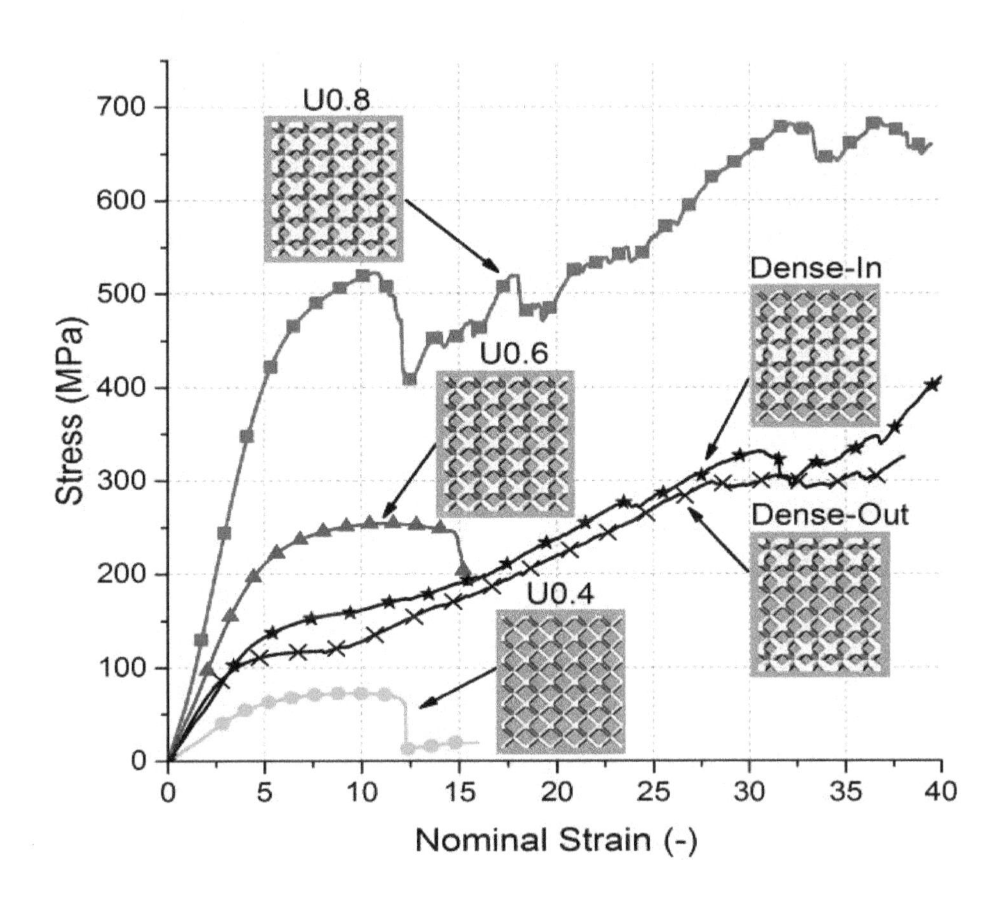

Figure 3. Nominal stress–strain curves for uniform and gradient structures.

The elastic gradient between stresses of 20 MPa and 50 MPa ($E_{(\sigma20-\sigma50)}$), the 0.2% offset yield stress (σ_y) and the first maximum compressive strength (σ_{max}) of the scaffolds are summarized (Table 2). Elastic modulus, yield stress and compressive strength increases with extension in strut diameter of the uniform structures or decrease in porosity. Aligned with the expectations of composite rule of mixtures [50], the values of elastic modulus, compressive strength and yield stress of gradient structures lie between those values of the uniform structures. No significant difference between elastic modulus of gradient structures and Uniform 0.6 sample was observed (Supplementary document Figure S1).

Table 2. The summary of the mechanical properties of uniform and gradient BCC structures measured by compression tests. (Mean ± SD).

Scaffold Name	$E_{\sigma 20-\sigma 50}$ (GPa)	σ_y (MPa)	σ_{max} (MPa)
U0.4	1.6 ± 0.2	53 ± 4	74 ± 2
U0.6	4.6 ± 0.4	192 ± 14	256 ± 4
U0.8	9.0 ± 0.6	392 ± 14	532 ± 11
Dense-In	3.9 ± 0.8	114 ± 8	150 ± 17
Dense-Out	3.5 ± 0.5	86 ± 11	128 ± 8

Relative modulus (E/E_0) against the measured volume fraction (%) is plotted in Figure 4. Elastic modulus was normalized relative to the values of solid Ti6Al4V (110 GPa). The observed average trend shows a positive power relation with the volume fraction and this trend corresponds to theoretically expected behavior of bending-dominated structures [10,51]. Results of a similar study from the literature were used to support our data [52]. The gradient structures were not considered for the power law curve since their volume fraction is similar to U0.6 specimen.

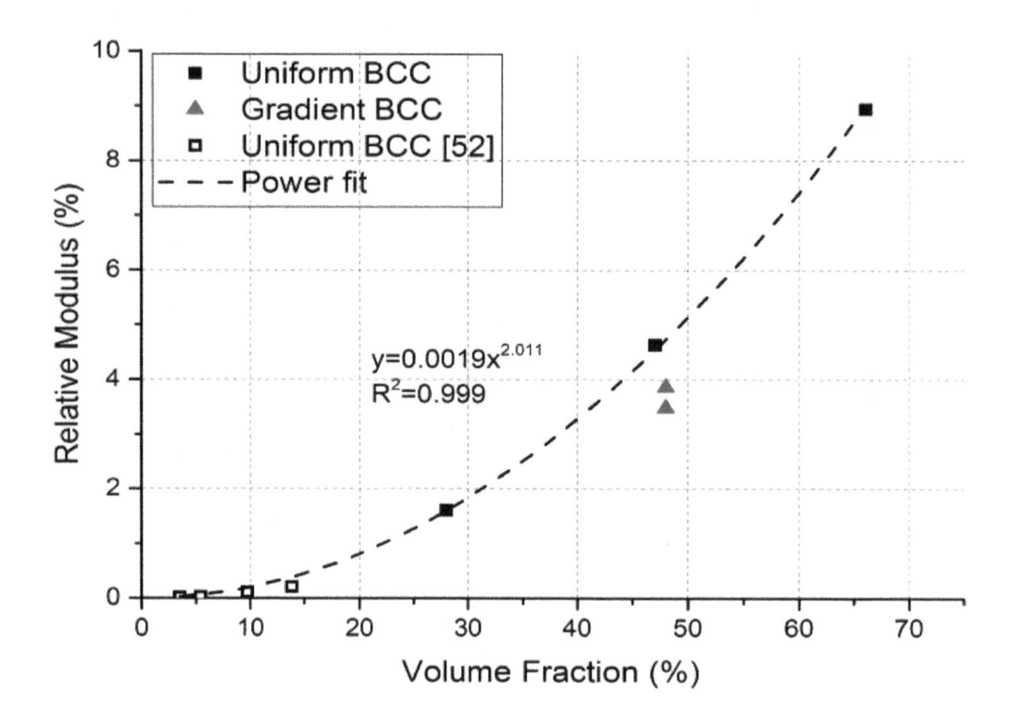

Figure 4. Relative elastic modulus vs volume fraction (%) of uniform and gradient BCC structures. Power law curve and equation was fitted on the uniform BCC structure data and demonstrates bending-dominated behavior.

Images of the initial stage and the progressive failure of uniform and gradient structures recorded during the compression tests (Figure 5) show that the major failure bands were formed at a 45° angle from the loading direction for all uniform BCC structures. For the gradient structures, the fracture initiated from the thinnest struts, that is at the top and bottom plane for Dense-In and in the middle for Dense-Out. This diagonal shear collapse of uniform structures is typical behavior of BCC structures [52–54] and other structures with different cell geometries [55,56] owing to strut bending at lattice joints [51].

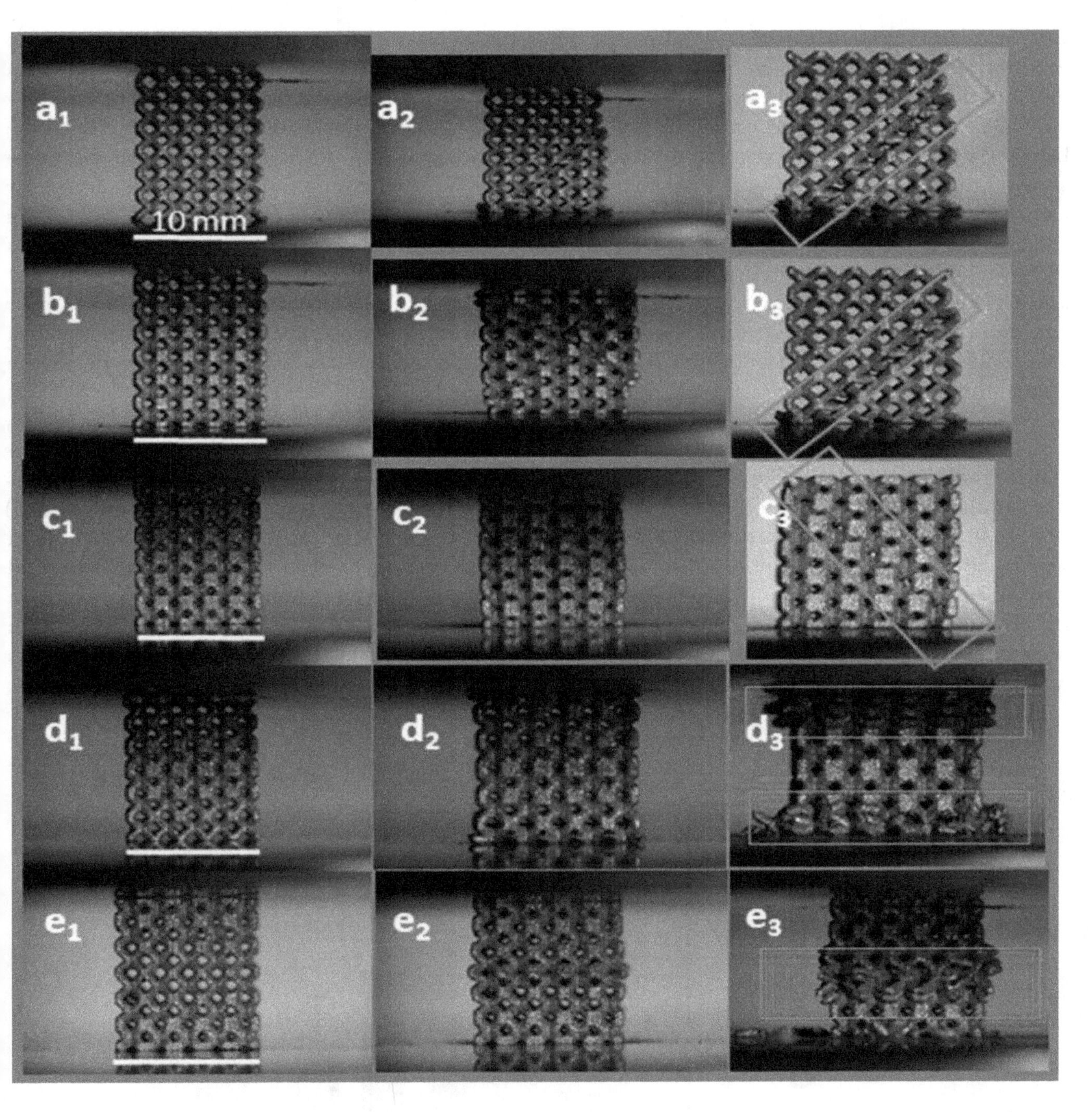

Figure 5. Failure modes of (**a**) U0.4, (**b**) U0.6, (**c**) U0.8, (**d**) Dense-In, (**e**) Dense-Out structures. Left images (subscript 1) represent the initial state and middle (subscript 2) and last right images (subscript 3) present the progressive failure. Highlights represent the observed regions of deformation and failure. (Scale bars = 10 mm).

3.3. Cellular Response to Porous Scaffolds

In order to determine the ability of the scaffolds to interact with cells, the adhesion of MC3T3-E1 preosteoblast cells on the gradient, uniform and solid scaffolds were determined by MTS assay after 4 h of incubation and showed no significant difference in cell seeding between the scaffolds (Figure 6).

In order to determine the extent of cell proliferation on the scaffolds, an MTS assay was performed after 4 and 7 days of culture. The uniform scaffolds showed a trend of decreasing cell number with increasing strut diameter at both days 4 and 7 (Figure 7). For the U0.4 and U0.6 scaffolds the number of cells approximately doubled across this time period whilst there was only a 70% increase on the

U0.8 scaffold (Figure 8), further extending the difference in cell number between the samples. For the gradient scaffolds, there were no significant differences in cell number on day 4. However, from day 4 to day 7, there was almost a 400% increase in cell numbers on the Dense-In scaffold but only 20% increase in cell numbers on the Dense-Out scaffold, resulting in significantly fewer cells on the Dense-Out scaffold as compared to the Dense-In scaffold on day 7. Although solid control sample showed the highest cell number at day 7, the percentage increase from day 4 to day 7 was largest for Dense-In scaffold. The final cell number on the Dense-Out scaffold was comparable to that of the U0.8, with both scaffolds having similar diameter of the outermost struts of the design. These results suggest that the scaffolds having thinner struts or larger pores on their outside surface (such as U0.4 and Dense-In) were more favorable for cell proliferation than the scaffolds having thicker struts or smaller pores on their outer surface. Given that the surface area of Dense-In and Dense-Out is identical as a result of their symmetrical design, it can be said that the cell viability was independent of surface area in this study.

Further to cell proliferation, cell distribution on the uniform and gradient scaffolds was studied by staining and imaging the cell nuclei and actin cytoskeleton. After 4 h of incubation, all of the scaffolds had similar cell distribution on their top surface (i.e., the surface onto which the cells were seeded) (Figure 9). The lack of cells at the bottom of the scaffolds suggests that most of the initial attachment was on the top surface. After 4 days, substantially more cells were observed both on the top and bottom surfaces of the U0.4 and Dense-In scaffolds, whereas there were no noticeable differences on the other scaffolds between 4 h and day 4 time points (Figure 10). At day 7, all the scaffolds had high density of cells on their top surface; whilst, the bottom surface of U0.6, U0.8 and Dense-Out scaffolds had almost no cells. In contrast, the bottom surface of U0.4 and Dense-In scaffolds had visibly higher cell densities.

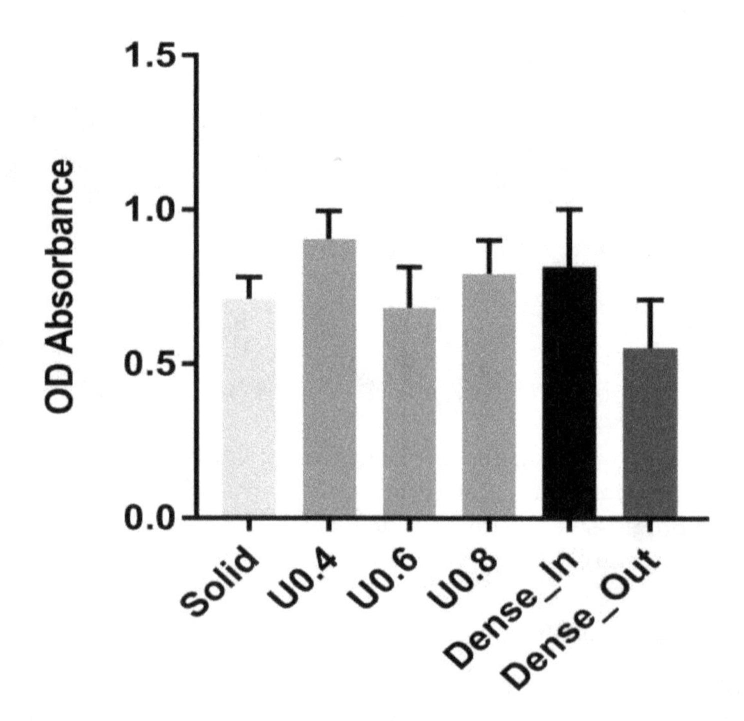

Figure 6. Adhesion of cells to uniform and gradient porous structures and to solid control sample as measured by MTS assay after 4 h. The optical absorbance (OD) was measured at 490 nm. Data are presented as mean ± SD ($n = 3$). No statistically significant differences were observed between scaffolds.

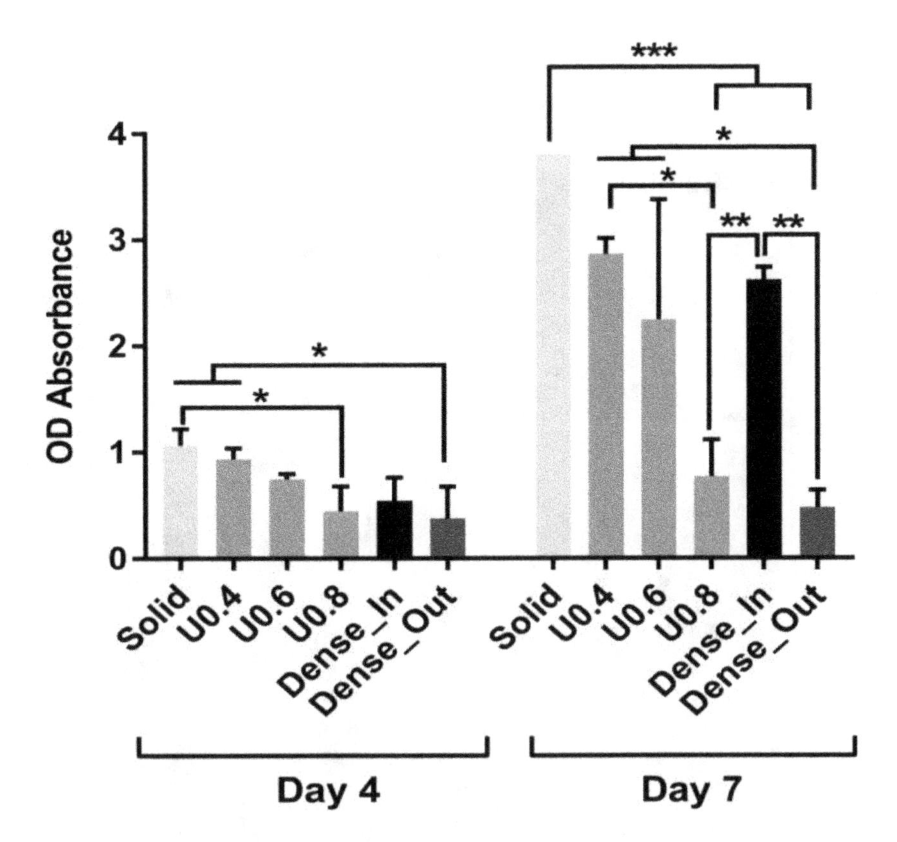

Figure 7. Cell proliferation measured by MTS assay after culturing 4 and 7 days on the uniform and gradient porous structures. The optical absorbance (OD) was measured at 490 nm. Data were presented as mean ± SD ($n = 3$). (* $p < 0.05$, ** $p < 0.01$, *** $p < 0.001$ when compared using ANOVA Tukey–Kramer post-hoc test).

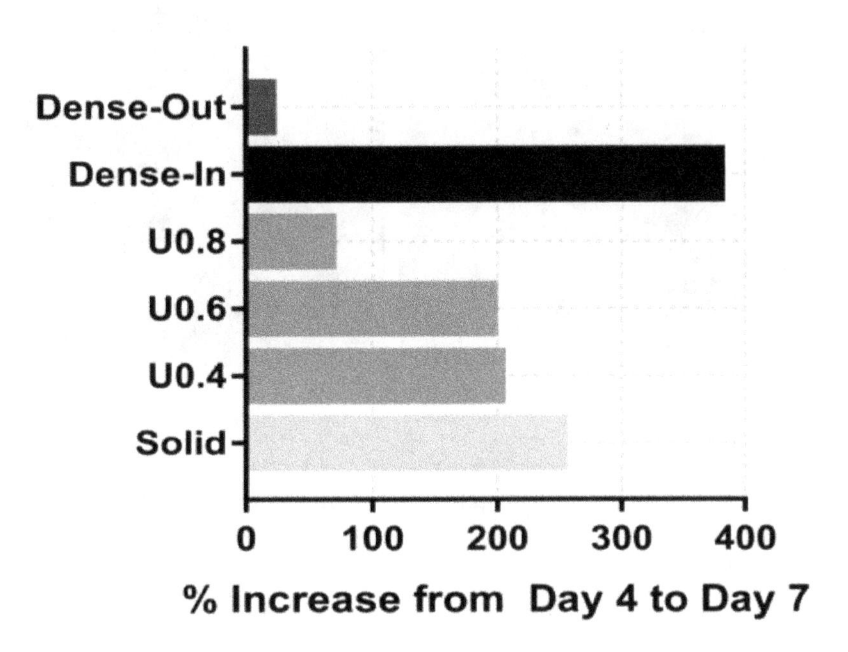

Figure 8. Percentage increase in od absorbance of the scaffolds from day 4 to day 7.

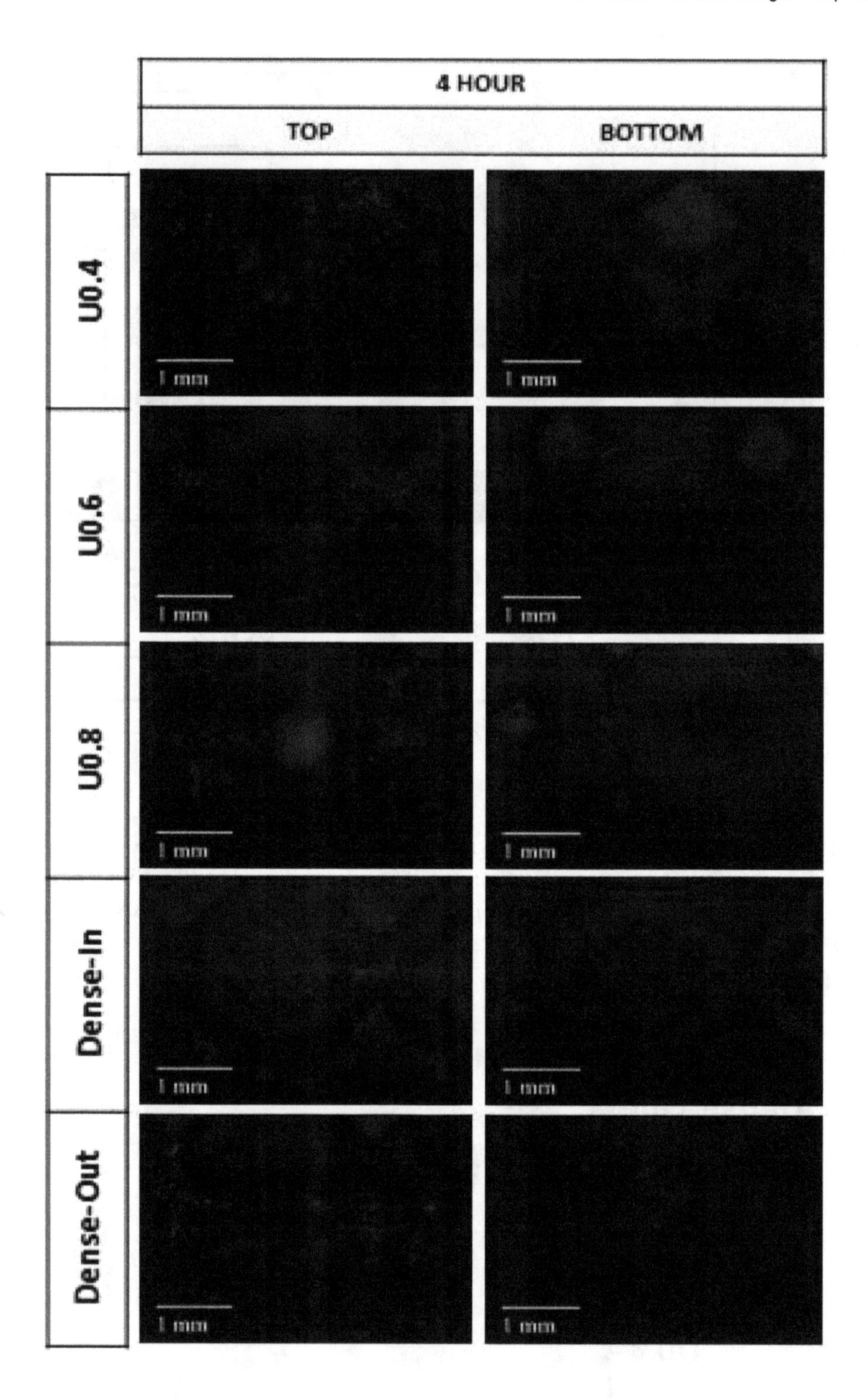

Figure 9. Fluorescence micrographs representing merged Hoechst stained nucleus (blue) and actin cytoskeleton (red) of MC3T3-E1 preosteoblast cells on the uniform and gradient BCC structures after culturing for 4 h. Top represents the side where cells were seeded onto the samples.

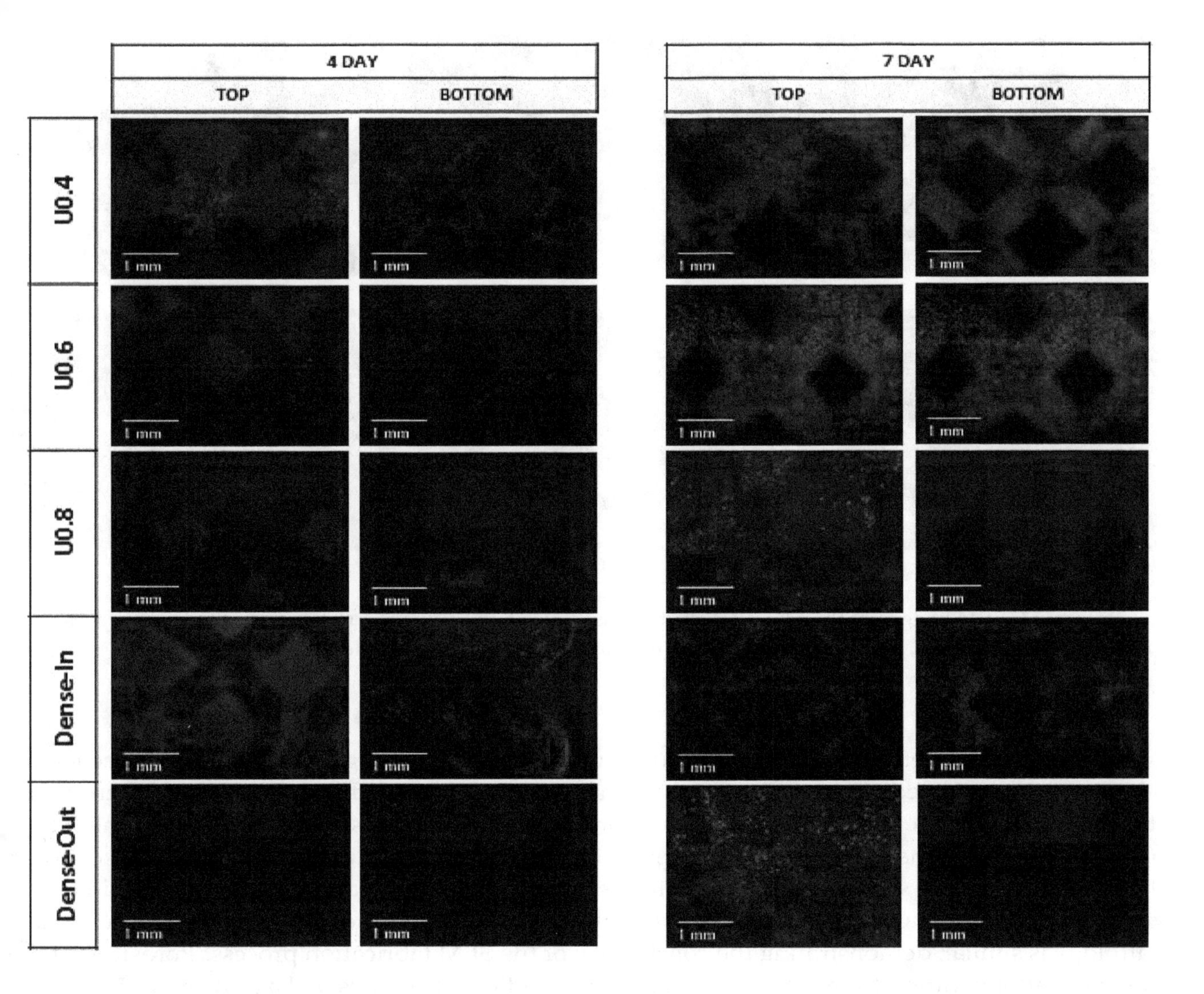

Figure 10. Fluorescence micrographs representing merged Hoechst stained nucleus (blue) and actin cytoskeleton (red) of MC3T3-E1 preosteoblast cells on the uniform and gradient BCC structures after culturing for 4 days and 7 days. Top represents the side where cells were seeded onto the samples.

Despite the difference in cell proliferation and migration on the different scaffolds, cell morphology was similar for the scaffolds when the images were taken from the top surface (Figure 9). SEM images taken from the middle and bottom part of the scaffolds supported the findings of fluorescent images showing differing cell penetration depth profiles for the varying scaffold structures (Supplementary document Figures S2–S6). The number of cells decreased in the middle and bottom parts of the U0.8 and Dense-Out scaffolds, as compared to U0.4 and Dense-In scaffolds. Cells were noticed to form a sheet-like elongated matrix (dashed line) (Figure 9). Moreover, the morphology of cells presented a high density of filopodia-like projections (red arrows) extending from the leading edges of cells and interacting with the substrate. The interaction between cells and substrate observed by SEM (Figure 11) shows that cells attach both on and between the unmelted powders, indicating that SLM process are beneficial to cell attachment and colonization.

Figure 11. SEM images of the MC3T3 preosteoblast cells after culturing for 7 days on the top surfaces of (**a**) U0.4, (**b**) U0.6, (**c**) U0.8, (**d**) Dense-In, (**e**) Dense-Out, (**f**) solid scaffolds.

4. Discussion

In this work, the effect of gradient porous structures on both biological response and mechanical behavior is investigated. The generated gradient structures, denoted as Dense-In and Dense-Out, utilize gradual change in diameter and therefore minimize the stress concentration at the lattice junctions. The designed pore sizes changed from 940 μm to 1330 μm for the gradient scaffolds. The deviation of pore size and strut diameter between CAD design and fabricated scaffolds were within an expected range [57] and was attributed to surface irregularities [58]. The deviation for each scaffold was similar, demonstrating the consistency of the SLM fabrication process. Porosities of the uniform scaffolds varied from 32% to 72%, and both gradient structures had a porosity of 50%. SEM images revealed that all scaffolds had unmelted powder attached to the surface of the struts due to layered manufacturing and melt pool variation during the SLM process.

The stiffness of the tested scaffolds varied in the range of 1.6 to 9.0 GPa, which aligns with the stiffness range of the trabecular (0.4 GPa [59,60]) and cortical bones (3–20 GPa [5,24]). The yield stress values of the scaffolds were in the range of 53 to 392 MPa, which lies in the range of cortical bones (33–193 MPa [24,61]), but it is not suitable for a replacement of trabecular bone, 2–17 MPa [60]. Considering the scaffolds presented in this study aim to be used as load-bearing implants for replacement of cortical bones, the yield stress and elastic modulus values satisfy the mechanical property requirements.

Table 2 shows that the elastic modulus, yield stress and maximum compressive strength values increase as the strut diameter increases or porosity decreases. The mechanical properties of gradient structures lie in the range of representative values of uniform structures and can be predicted based on an assumption that gradient structures are composites of uniform layers of the same diameter struts. Based on this assumption, the elastic modulus of gradient structure in uniaxial compression can be calculated through the general rule of mixtures [42,50]:

$$\frac{1}{E_{Gradient}} = \frac{1}{3E_{U0.4}} + \frac{1}{3E_{U0.6}} + \frac{1}{3E_{U0.8}} \tag{5}$$

Using Equation (1), elastic modulus of gradient structure was calculated to be 3.2 GPa, which is comparable to the measured values of 3.9 and 3.5 GPa for Dense-In and Dense-Out scaffolds, respectively.

The deformation response of uniform BCC structures follows the bending-dominated behavior with diagonal shear collapse. Interestingly, the failure mechanism of gradient structures was different due to sequential layer collapse and various deformation stages occurring simultaneously. Thinner struts reached the densification stage (when two opposite cell walls come together as the pore size decreases) while the thicker struts were still in the plateau region during the compression test. The predominant fracture band of gradient structures was initiated at the thinnest struts due to high stress concentrations on the thin strut junctions. This failure mechanism has also been noted in other studies [44,45].

Further to mechanical behavior studies, we analyzed the in vitro response of the scaffolds with preosteoblast cells. The degree of cell attachment was similar for all scaffolds, but cell proliferation and colonization were significantly different. Scaffolds with a thin strut diameter on the periphery (U0.4 and Dense-In) allowed cells to populate throughout the scaffold whereas those with a thicker outer strut (U0.6, U0.8 and Dense-Out) did not allow cells to migrate to the bottom surface, suggesting that cells were entrapped at the smaller pore size region (top surface). Consequently, the proliferation rate of cells on these scaffolds was markedly less. Although this immobilization behavior of the preosteoblast cells when seeded from small pore size region was observed in the previous studies of Nune et al. [38,39], the underlying reason for this behavior is still unknown. The smallest pore size in our scaffolds were 940 μm which is larger than the suggested pore size (100–300 μm) for cell colonization and migration [62,63]. In addition, the smallest pore size is much larger than an average size of a MC3T3-E1 pre-osteoblast cells (20–40 μm, Figure 9). It is not yet clear why thicker struts on the scaffold periphery inhibit cell activity whilst the thick struts on the interior of the Dense-In scaffold did not deter colonization of cells through the whole structure.

Our results suggest that the surface area does not affect the cell attachment and proliferation. The large surface area of the U0.8 scaffold, due to its large strut diameters, was expected to promote cell attachment and growth; however, it showed the lowest cell number at day 7. Similar behavior was observed for the gradient structures, which possessed equal surface area but showed a large difference in cell number. It is therefore likely that, parameters other than the surface area of the scaffold affected the cell colonization. Identification of the specific factors would require further clarification but could include the flow conditions and cellular aggregation [39]. In addition, it would be interesting to assess whether vascularization happens more quickly or easily with larger pores on the periphery than the smaller pores on the periphery.

In recent years, additively manufactured gradient structures for tissue engineering have been studied [37–39,41,44,64,65]; however these studies either focused on the mechanical properties or biological response independently. In this work, the biological and mechanical responses were assessed simultaneously allowing us to study the overall impact of the designed geometry of the scaffolds. Our results suggest that when designing a porous gradient structure, both biological and mechanical requirements must be considered concurrently, since their requirements are opposing. The compression test results demonstrate the benefit of utilizing smaller pore size in increasing the stiffness and strength of the porous scaffolds; whereas, the cell proliferation data suggests that scaffolds with larger pore size in their outer surface favors cell proliferation. Therefore, it can be concluded that gradient scaffolds provide a possible solution for overcoming the conflicting requirements of bone tissue implants. Gradient structures with decreasing pore size towards their center can provide the required strength and stiffness, while simultaneously promoting cell colonization throughout the whole scaffold. The Dense-In scaffold fabricated in this study has an elastic modulus of 3.9 GPa which is in the range for those of cortical bone [5]. Furthermore, this scaffold has a varying pore size ranging from 1330 μm on the outside to 940 μm at the core. These values have been previously reported to be favorable for cell colonization as well as bone ingrowth and vascularization [66–68].

In summary, an ideal scaffold for bone regeneration should facilitate cell attachment, infiltration and matrix deposition to guide bone formation [69] as well as providing initial mechanical support to the surrounding bone [70]. Porous titanium scaffolds can meet the mechanical strength and bone formation requirements without osseoinductive biomolecules [68]; however, the pore size of the scaffolds needs to be high for bone-ingrowth whereas, as the porosity increases, the mechanical strength and integrity of the structure decreases [71]. Gradient structures represent an ideal candidate to overcome these opposing requirements of high porosity and mechanical strength.

Our study demonstrated the benefit of the gradient scaffold with larger pores in its outer surface in terms of gaining optimum mechanical strength and promoting cell attachment and colonization. In addition, this framework demonstrates that mechanical properties can be tailored through gradient structure design and simultaneously improve the biological response. This approach therefore holds significant promise in the development of orthopedic implants, where the location of the implant and the corresponding loading condition can dictate the implant topology.

5. Conclusions

This work combined and assessed the in vitro behavior and mechanical response of gradient and uniform porous scaffolds for bone tissue engineering. For this purpose, five different BCC structures were fabricated using selective laser melting technology. Static mechanical properties of the gradient structures followed the rule of mixtures and the obtained values are in the range of those corresponding values for uniform structures. The mechanical properties of all studied scaffolds are comparable to the reported mechanical properties of the cortical bone. Quantitative analysis of cell viability showed higher cell colonization and proliferation rates for scaffolds with large pores (1000–1100 μm) in their outer surface after 4 and 7 days of culturing. However, when comparing the mechanical properties of structures with this comparable biological activity, the uniform U0.4 scaffold showed less than half of the respective mechanical properties for the Dense-In scaffold. The combined results of compression tests and in vitro biological analyses indicate that the Dense-In scaffold is an ideal porous structure to balance mechanical and biological performances to meet the requirements of load-bearing implants. Based on the results presented in this work, optimal gradient structures should possess small pores in their core in order to increase their mechanical integrity and strength while large pores should be utilized in their outer surface to avoid pore occlusion. We suggest that this approach could be widely used in the design of orthopedic implants to maximize both the mechanical and biological properties of the implant.

Acknowledgments: Ezgi Onal would like to acknowledge The Clive and Vera Ramaciotti Centre for Structural Cryo-Electron Microscopy. This project is funded by the ARC Research Hub for Transforming Australia's Manufacturing Industry through High Value Additive Manufacturing (IH130100008) and Jessica E. Frith is supported by an ARC DECRA (DE130100986).

Author Contributions: E.O. and A.M. conceived and designed the experiments; E.O. performed the experiments; E.O., J.E.F., X.W. and A.M. analyzed the data; M.J. contributed generating Python scripts to design structures; all authors contributed to the writing of the paper.

References

1. Hutmacher, D.W. Scaffolds in tissue engineering bone and cartilage. *Biomaterials* **2000**, *21*, 2529–2543. [CrossRef]
2. Hutmacher, D.W.; Schantz, J.T.; Lam, C.X.F.; Tan, K.C.; Lim, T.C. State of the art and future directions of scaffold-based bone engineering from a biomaterials perspective. *J. Tissue Eng. Regener. Med.* **2007**, *1*, 245–260. [CrossRef] [PubMed]
3. Niinomi, M.; Nakai, M.; Hieda, J. Development of new metallic alloys for biomedical applications. *Acta Biomater.* **2012**, *8*, 3888–3903. [CrossRef] [PubMed]

4. Andani, M.T.; Shayesteh Moghaddam, N.; Haberland, C.; Dean, D.; Miller, M.J.; Elahinia, M. Metals for bone implants. Part 1. Powder metallurgy and implant rendering. *Acta Biomater.* **2014**, *10*, 4058–4070. [CrossRef] [PubMed]
5. Bayraktar, H.H.; Morgan, E.F.; Niebur, G.L.; Morris, G.E.; Wong, E.K.; Keaveny, T.M. Comparison of the elastic and yield properties of human femoral trabecular and cortical bone tissue. *J. Biomech.* **2004**, *37*, 27–35. [CrossRef]
6. Moghaddam, N.S.; Andani, M.T.; Amerinatanzi, A.; Haberland, C.; Huff, S.; Miller, M.; Elahinia, M.; Dean, D. Metals for bone implants: Safety, design, and efficacy. *Biomanuf. Rev.* **2016**, *1*, 1. [CrossRef]
7. Al-Tamimi, A.A.; Fernandes, P.R.A.; Peach, C.; Cooper, G.; Diver, C.; Bartolo, P.J. Metallic bone fixation implants: A novel design approach for reducing the stress shielding phenomenon. *Virtual Phys. Prototyp.* **2017**, *12*, 141–151. [CrossRef]
8. Long, M.; Rack, H.J. Titanium alloys in total joint replacement—A materials science perspective. *Biomaterials* **1998**, *19*, 1621–1639. [CrossRef]
9. Abdel-Hady Gepreel, M.; Niinomi, M. Biocompatibility of Ti-alloys for long-term implantation. *J. Mech. Behav. Biomed. Mater.* **2013**, *20*, 407–415. [CrossRef] [PubMed]
10. Ashby, M.F. The properties of foams and lattices. *Philos. Trans. R. Soc. A Math. Phys. Eng. Sci.* **2006**, *364*, 15–30. [CrossRef] [PubMed]
11. Alvarez, K.; Nakajima, H. Metallic scaffolds for bone regeneration. *Materials* **2009**, *2*, 790. [CrossRef]
12. Heinl, P.; Müller, L.; Körner, C.; Singer, R.F.; Müller, F.A. Cellular Ti–6Al–4V structures with interconnected macro porosity for bone implants fabricated by selective electron beam melting. *Acta Biomater.* **2008**, *4*, 1536–1544. [CrossRef] [PubMed]
13. Marin, E.; Fusi, S.; Pressacco, M.; Paussa, L.; Fedrizzi, L. Characterization of cellular solids in Ti6Al4V for orthopedic implant applications: Trabecular titanium. *J. Mech. Behav. Biomed. Mater.* **2010**, *3*, 373–381. [CrossRef] [PubMed]
14. Rouwkema, J.; Rivron, N.C.; van Blitterswijk, C.A. Vascularization in tissue engineering. *Trends Biotechnol.* **2008**, *26*, 434–441. [CrossRef] [PubMed]
15. Kumar, A.; Nune, K.C.; Murr, L.E.; Misra, R.D.K. Biocompatibility and mechanical behaviour of three-dimensional scaffolds for biomedical devices: Process–structure–property paradigm. *Int. Mater. Rev.* **2016**, *61*, 20–45. [CrossRef]
16. Bramfeld, H.; Sabra, G.; Centis, V.; Vermette, P. Scaffold vascularization: A challenge for three-dimensional tissue engineering. *Curr. Med. Chem.* **2010**, *17*, 3944–3967. [CrossRef]
17. Perez, R.A.; Mestres, G. Role of pore size and morphology in musculo-skeletal tissue regeneration. *Mater. Sci. Eng. C* **2016**, *61*, 922–939. [CrossRef] [PubMed]
18. Wang, X.; Xu, S.; Zhou, S.; Xu, W.; Leary, M.; Choong, P.; Qian, M.; Brandt, M.; Xie, Y.M. Topological design and additive manufacturing of porous metals for bone scaffolds and orthopedic implants: A review. *Biomaterials* **2016**, *83*, 127–141. [CrossRef] [PubMed]
19. Sing, S.L.; An, J.; Yeong, W.Y.; Wiria, F.E. Laser and electron-beam powder-bed additive manufacturing of metallic implants: A review on processes, materials and designs. *J. Orthop. Res.* **2016**, *34*, 369–385. [CrossRef] [PubMed]
20. Li, J.P.; Habibovic, P.; van den Doel, M.; Wilson, C.E.; de Wijn, J.R.; van Blitterswijk, C.A.; de Groot, K. Bone ingrowth in porous titanium implants produced by 3D fiber deposition. *Biomaterials* **2007**, *28*, 2810–2820. [CrossRef] [PubMed]
21. Warnke, P.H.; Douglas, T.; Wollny, P.; Sherry, E.; Steiner, M.; Galonska, S.; Becker, S.T.; Springer, I.N.; Wiltfang, J.; Sivananthan, S. Rapid prototyping: Porous titanium alloy scaffolds produced by selective laser melting for bone tissue engineering. *Tissue Eng. Part C Methods* **2008**, *15*, 115–124. [CrossRef] [PubMed]
22. Van Bael, S.; Chai, Y.C.; Truscello, S.; Moesen, M.; Kerckhofs, G.; Van Oosterwyck, H.; Kruth, J.P.; Schrooten, J. The effect of pore geometry on the in vitro biological behavior of human periosteum-derived cells seeded on selective laser-melted Ti6Al4V bone scaffolds. *Acta Biomater.* **2012**, *8*, 2824–2834. [CrossRef] [PubMed]
23. Leong, K.F.; Chua, C.K.; Sudarmadji, N.; Yeong, W.Y. Engineering functionally graded tissue engineering scaffolds. *J. Mech. Behav. Biomed. Mater.* **2008**, *1*, 140–152. [CrossRef] [PubMed]
24. Karageorgiou, V.; Kaplan, D. Porosity of 3D biomaterial scaffolds and osteogenesis. *Biomaterials* **2005**, *26*, 5474–5491. [CrossRef] [PubMed]
25. Li, G.; Wang, L.; Pan, W.; Yang, F.; Jiang, W.; Wu, X.; Kong, X.; Dai, K.; Hao, Y. In vitro and in vivo study

of additive manufactured porous Ti6Al4V scaffolds for repairing bone defects. *Sci. Rep.* **2016**, *6*, 34072. [CrossRef] [PubMed]

26. Murr, L.E.; Gaytan, S.M.; Medina, F.; Martinez, E.; Martinez, J.L.; Hernandez, D.H.; Machado, B.I.; Ramirez, D.A.; Wicker, R.B. Characterization of Ti–6Al–4V open cellular foams fabricated by additive manufacturing using electron beam melting. *Mater. Sci. Eng. A* **2010**, *527*, 1861–1868. [CrossRef]
27. Murr, L.E.; Gaytan, S.M.; Medina, F.; Lopez, H.; Martinez, E.; Machado, B.I.; Hernandez, D.H.; Martinez, L.; Lopez, M.I.; Wicker, R.B.; et al. Next-generation biomedical implants using additive manufacturing of complex, cellular and functional mesh arrays. *Philos. Trans. R. Soc. Lond. A Math. Phys. Eng. Sci.* **2010**, *368*, 1999–2032. [CrossRef] [PubMed]
28. Arabnejad, S.; Burnett Johnston, R.; Pura, J.A.; Singh, B.; Tanzer, M.; Pasini, D. High-strength porous biomaterials for bone replacement: A strategy to assess the interplay between cell morphology, mechanical properties, bone ingrowth and manufacturing constraints. *Acta Biomater.* **2016**, *30*, 345–356. [CrossRef] [PubMed]
29. Wettergreen, M.A.; Bucklen, B.S.; Starly, B.; Yuksel, E.; Sun, W.; Liebschner, M.A.K. Creation of a unit block library of architectures for use in assembled scaffold engineering. *Comput.-Aided Des.* **2005**, *37*, 1141–1149. [CrossRef]
30. Parthasarathy, J.; Starly, B.; Raman, S. A design for the additive manufacture of functionally graded porous structures with tailored mechanical properties for biomedical applications. *J. Manuf. Process.* **2011**, *13*, 160–170. [CrossRef]
31. Bobbert, F.S.L.; Lietaert, K.; Eftekhari, A.A.; Pouran, B.; Ahmadi, S.M.; Weinans, H.; Zadpoor, A.A. Additively manufactured metallic porous biomaterials based on minimal surfaces: A unique combination of topological, mechanical, and mass transport properties. *Acta Biomater.* **2017**, *53*, 572–584. [CrossRef] [PubMed]
32. Giannitelli, S.M.; Accoto, D.; Trombetta, M.; Rainer, A. Current trends in the design of scaffolds for computer-aided tissue engineering. *Acta Biomater.* **2014**, *10*, 580–594. [CrossRef] [PubMed]
33. Kapfer, S.C.; Hyde, S.T.; Mecke, K.; Arns, C.H.; Schröder-Turk, G.E. Minimal surface scaffold designs for tissue engineering. *Biomaterials* **2011**, *32*, 6875–6882. [CrossRef] [PubMed]
34. Zhang, X.-Y.; Fang, G.; Zhou, J. Additively manufactured scaffolds for bone tissue engineering and the prediction of their mechanical behavior: A review. *Materials* **2017**, *10*, 50. [CrossRef] [PubMed]
35. Horn, T.J.; Harrysson, O.L.A. Overview of current additive manufacturing technologies and selected applications. *Sci. Prog.* **2012**, *95*, 255–282. [CrossRef] [PubMed]
36. Sidambe, A. Biocompatibility of advanced manufactured titanium implants—A review. *Materials* **2014**, *7*, 8168–8188. [CrossRef] [PubMed]
37. Li, S.; Zhao, S.; Hou, W.; Teng, C.; Hao, Y.; Li, Y.; Yang, R.; Misra, R.D.K. Functionally graded Ti-6Al-4V meshes with high strength and energy absorption. *Adv. Eng. Mater.* **2016**, *18*, 34–38. [CrossRef]
38. Nune, K.; Kumar, A.; Misra, R.; Li, S.; Hao, Y.; Yang, R. Osteoblast functions in functionally graded Ti-6Al-4V mesh structures. *J. Biomater. Appl.* **2016**, *30*, 1182–1204. [CrossRef] [PubMed]
39. Nune, K.C.; Kumar, A.; Misra, R.D.K.; Li, S.J.; Hao, Y.L.; Yang, R. Functional response of osteoblasts in functionally gradient titanium alloy mesh arrays processed by 3D additive manufacturing. *Colloids Surf. B Biointerfaces* **2017**, *150*, 78–88. [CrossRef] [PubMed]
40. Surmeneva, M.A.; Surmenev, R.A.; Chudinova, E.A.; Koptioug, A.; Tkachev, M.S.; Gorodzha, S.N.; Rännar, L.-E. Fabrication of multiple-layered gradient cellular metal scaffold via electron beam melting for segmental bone reconstruction. *Mater. Des.* **2017**, *133*, 195–204. [CrossRef]
41. Limmahakhun, S.; Oloyede, A.; Sitthiseripratip, K.; Xiao, Y.; Yan, C. Stiffness and strength tailoring of cobalt chromium graded cellular structures for stress-shielding reduction. *Mater. Des.* **2017**, *114*, 633–641. [CrossRef]
42. Van Grunsven, W.; Hernandez-Nava, E.; Reilly, G.; Goodall, R. Fabrication and mechanical characterisation of titanium lattices with graded porosity. *Metals* **2014**, *4*, 401–409. [CrossRef]
43. Han, C.; Li, Y.; Wang, Q.; Wen, S.; Wei, Q.; Yan, C.; Hao, L.; Liu, J.; Shi, Y. Continuous functionally graded porous titanium scaffolds manufactured by selective laser melting for bone implants. *J. Mech. Behav. Biomed. Mater.* **2018**, *80*, 119–127. [CrossRef] [PubMed]
44. Maskery, I.; Aboulkhair, N.T.; Aremu, A.O.; Tuck, C.J.; Ashcroft, I.A.; Wildman, R.D.; Hague, R.J.M.

A mechanical property evaluation of graded density Al-Si10-Mg lattice structures manufactured by selective laser melting. *Mater. Sci. Eng. A* **2016**, *670*, 264–274. [CrossRef]

45. Choy, S.Y.; Sun, C.-N.; Leong, K.F.; Wei, J. Compressive properties of functionally graded lattice structures manufactured by selective laser melting. *Mater. Des.* **2017**, *131*, 112–120. [CrossRef]
46. Yap, C.Y.; Chua, C.K.; Dong, Z.L.; Liu, Z.H.; Zhang, D.Q.; Loh, L.E.; Sing, S.L. Review of selective laser melting: Materials and applications. *Appl. Phys. Rev.* **2015**, *2*, 041101. [CrossRef]
47. Pyka, G.; Kerckhofs, G.; Papantoniou, I.; Speirs, M.; Schrooten, J.; Wevers, M. Surface Roughness and Morphology Customization of Additive Manufactured Open Porous Ti6Al4V Structures. *Materials* **2013**, *6*, 4737–4757. [CrossRef] [PubMed]
48. Wang, D.; Yang, Y.; Liu, R.; Xiao, D.; Sun, J. Study on the designing rules and processability of porous structure based on selective laser melting (SLM). *J. Mater. Process. Technol.* **2013**, *213*, 1734–1742. [CrossRef]
49. Rashed, M.G.; Ashraf, M.; Mines, R.A.W.; Hazell, P.J. Metallic microlattice materials: A current state of the art on manufacturing, mechanical properties and applications. *Mater. Des.* **2016**, *95*, 518–533. [CrossRef]
50. Nemat-Nasser, S.; Hori, M. *Micromechanics: Overall Properties of Heterogeneous Materials*, 2nd ed.; Elsevier: Amsterdam, The Netherlands, 1998.
51. Mazur, M.; Leary, M.; Sun, S.; Vcelka, M.; Shidid, D.; Brandt, M. Deformation and failure behaviour of Ti-6Al-4V lattice structures manufactured by selective laser melting (SLM). *Int. J. Adv. Manuf. Technol.* **2016**, *84*, 1391–1411. [CrossRef]
52. Smith, M..; Guan, Z.; Cantwell, W.J. Finite element modelling of the compressive response of lattice structures manufactured using the selective laser melting technique. *Int. J. Mech. Sci.* **2013**, *67*, 28–41. [CrossRef]
53. Gorny, B.; Niendorf, T.; Lackmann, J.; Thoene, M.; Troester, T.; Maier, H.J. In situ characterization of the deformation and failure behavior of non-stochastic porous structures processed by selective laser melting. *Mater. Sci. Eng. A* **2011**, *528*, 7962–7967. [CrossRef]
54. Cansizoglu, O.; Harrysson, O.; Cormier, D.; West, H.; Mahale, T. Properties of Ti–6Al–4V non-stochastic lattice structures fabricated via electron beam melting. *Mater. Sci. Eng. A* **2008**, *492*, 468–474. [CrossRef]
55. Zhao, S.; Li, S.J.; Hou, W.T.; Hao, Y.L.; Yang, R.; Misra, R.D.K. The influence of cell morphology on the compressive fatigue behavior of Ti-6Al-4V meshes fabricated by electron beam melting. *J. Mech. Behav. Biomed. Mater.* **2016**, *59*, 251–264. [CrossRef] [PubMed]
56. Li, S.J.; Xu, Q.S.; Wang, Z.; Hou, W.T.; Hao, Y.L.; Yang, R.; Murr, L.E. Influence of cell shape on mechanical properties of Ti–6Al–4V meshes fabricated by electron beam melting method. *Acta Biomater.* **2014**, *10*, 4537–4547. [CrossRef] [PubMed]
57. Van Bael, S.; Kerckhofs, G.; Moesen, M.; Pyka, G.; Schrooten, J.; Kruth, J.P. Micro-CT-based improvement of geometrical and mechanical controllability of selective laser melted Ti6Al4V porous structures. *Mater. Sci. Eng. A* **2011**, *528*, 7423–7431. [CrossRef]
58. Parthasarathy, J.; Starly, B.; Raman, S.; Christensen, A. Mechanical evaluation of porous titanium (Ti6Al4V) structures with electron beam melting (EBM). *J. Mech. Behav. Biomed. Mater.* **2010**, *3*, 249–259. [CrossRef] [PubMed]
59. Linde, F.; Hvid, I. The effect of constraint on the mechanical behaviour of trabecular bone specimens. *J. Biomech.* **1989**, *22*, 485–490. [CrossRef]
60. Morgan, E.F.; Keaveny, T.M. Dependence of yield strain of human trabecular bone on anatomic site. *J. Biomech.* **2001**, *34*, 569–577. [CrossRef]
61. Cullinane, D.M.; Einhorn, T.A. Biomechanics of bone. In *Principles of Bone Biology*, 2nd ed.; Raisz, L.G., Rodan, G.A., Eds.; Academic Press: San Diego, CA, USA, 2002.
62. Murphy, C.M.; Haugh, M.G.; O'Brien, F.J. The effect of mean pore size on cell attachment, proliferation and migration in collagen–glycosaminoglycan scaffolds for bone tissue engineering. *Biomaterials* **2010**, *31*, 461–466. [CrossRef] [PubMed]
63. Simske, S.J.; Ayers, R.A.; Bateman, T.A. Porous materials for bone engineering. *Mater. Sci. Forum* **1997**, *250*, 151–182. [CrossRef]
64. Dumas, M.; Terriault, P.; Brailovski, V. Modelling and characterization of a porosity graded lattice structure for additively manufactured biomaterials. *Mater. Des.* **2017**, *121*, 383–392. [CrossRef]
65. Sudarmadji, N.; Tan, J.Y.; Leong, K.F.; Chua, C.K.; Loh, Y.T. Investigation of the mechanical properties and porosity relationships in selective laser-sintered polyhedral for functionally graded scaffolds. *Acta Biomater.* **2011**, *7*, 530–537. [CrossRef] [PubMed]

66. De Wild, M.; Zimmermann, S.; Rüegg, J.; Schumacher, R.; Fleischmann, T.; Ghayor, C.; Weber, F.E. Influence of microarchitecture on osteoconduction and mechanics of porous titanium scaffolds generated by selective laser melting. *3D Print. Addit. Manuf.* **2016**, *3*, 142–151. [CrossRef]
67. Taniguchi, N.; Fujibayashi, S.; Takemoto, M.; Sasaki, K.; Otsuki, B.; Nakamura, T.; Matsushita, T.; Kokubo, T.; Matsuda, S. Effect of pore size on bone ingrowth into porous titanium implants fabricated by additive manufacturing: An in vivo experiment. *Mater. Sci. Eng. C* **2016**, *59*, 690–701. [CrossRef] [PubMed]
68. Fukuda, A.; Takemoto, M.; Saito, T.; Fujibayashi, S.; Neo, M.; Pattanayak, D.K.; Matsushita, T.; Sasaki, K.; Nishida, N.; Kokubo, T.; et al. Osteoinduction of porous Ti implants with a channel structure fabricated by selective laser melting. *Acta Biomater.* **2011**, *7*, 2327–2336. [CrossRef] [PubMed]
69. Khan, W.S.; Rayan, F.; Dhinsa, B.S.; Marsh, D. An osteoconductive, osteoinductive, and osteogenic tissue-engineered product for trauma and orthopedic surgery: How far are we? *Stem Cells Int.* **2012**, *2012*, 236231. [CrossRef] [PubMed]
70. Van der Stok, J.; Van der Jagt, O.P.; Amin Yavari, S.; De Haas, M.F.P.; Waarsing, J.H.; Jahr, H.; Van Lieshout, E.M.M.; Patka, P.; Verhaar, J.A.N.; Zadpoor, A.A.; et al. Selective laser melting-produced porous titanium scaffolds regenerate bone in critical size cortical bone defects. *J. Orthop. Res.* **2013**, *31*, 792–799. [CrossRef] [PubMed]
71. Hollister, S.J. Porous scaffold design for tissue engineering. *Nat. Mater.* **2006**, *5*, 590. [CrossRef]

17

Fretting Behavior of SPR Joining Dissimilar Sheets of Titanium and Copper Alloys

Xiaocong He *, Cong Deng and Xianlian Zhang

Innovative Manufacturing Research Centre, Kunming University of Science and Technology, Kunming 650500, China; d_c8745@163.com (C.D.); zxlian4gd@163.com (X.Z.)

* Correspondence: x_he@kmust.edu.cn

Academic Editor: Mark T. Whittaker

Abstract: The fretting performance of self-piercing riveting joining dissimilar sheets in TA1 titanium alloy and H62 copper alloy was studied in this paper. Load-controlled cyclic fatigue tests were carried out using a sine waveform and in tension-tension mode. Scanning electron microscopy and energy-dispersive X-ray techniques were employed to analyze the fretting failure mechanisms of the joints. The experimental results showed that there was extremely severe fretting at the contact interfaces of rivet and sheet materials for the joints at relatively high loads levels. Moreover, the severe fretting in the region on the locked sheet in contact with the rivet was the major cause of the broken locked sheet for the joints at low load level.

Keywords: titanium alloy; copper alloy; fretting; self-piercing riveting; SEM; EDX

---

1. Introduction

Facing a decreasing amount of resources, lightweight design principles continue to prosper rapidly in different engineering fields. As a result, lightweight materials for different applications have been developed. Fusion welding and friction stir welding techniques are normally used to join lightweight metal sheets [1,2]. However, it is not easy to weld dissimilar metal sheets due to the formation of intermetallic phases. Self-piercing riveting (SPR) is a new high-speed mechanical fastening technique suitable for point-joining advanced lightweight sheet materials that are dissimilar, coated, and hard to weld [3].

In the past few years, the mechanical properties of SPR joints and the SPR technique itself have been studied by many scholars. Kang et al. [4] aimed to evaluate the static and fatigue strengths of the joints using coach-peel, cross-tension and tensile-shear specimens with experimental tests and numerical analysis. Su et al. [5] investigated the fracture and fatigue behaviors of SPR and clinch joints in lap-shear specimens of 6111-T4 aluminum sheets based on experimental observations, and examined the optical micrographs of both types of joints before and after failure under quasi-static and cyclic loading conditions. Calabrese et al. [6] conducted a long-time salt spray corrosion test for steel/aluminum hybrid joints obtained by SPR technique to evaluate the mechanical degradation in these critical environmental conditions. The influence of resistance heating on dissimilar SPR joints with unequal thickness was studied systematically by Lou et al. [7]. They reported that SPR joints using rivet-welding could obtain higher tensile-shear strength than with conventional SPR joints. Haque et al. [8] developed a simple geometrical method to calculate rivet flaring without having to cross-section a joint. It is a nondestructive testing method to determine rivet flaring based on the characteristic force-displacement curve, and could be a very useful tool in joint product development and process optimization.

A newly developed solid state joining technique—friction self-piercing riveting (F-SPR)—has been applied for joining high strength aluminum alloy to magnesium alloy [9,10]. The process was performed on a specially designed machine where the spindle can achieve a sudden stop. The effects of rivet rotating rate and punch speed on axial plunge force, torque, joint microstructure, and quality were analyzed systematically. A 3-D thermo-mechanical finite-element (FE) model of F-SPR process was developed using an LS-DYNA code [11]. Temperature-dependent material parameters were utilized to calculate the material yield and flow in the joint formation. A preset crack failure method was used to model the material failure of the top sheet. The calculated joint geometry exhibited a good agreement with the experimental measurement.

In a recent study, Haque and Durandet [12] described a parametric study of the mechanical behavior of SPR joints of steel sheets in two loading conditions (lap-shear and cross-tension). An empirical model was developed to predict the joint strength in cross-tension loading using characteristic joint data determined directly from the SPR process (force-displacement) curve. The tensile and fatigue behavior of SPR in carbon fiber reinforced plastic (CFRP) to aluminum 6111 T82 alloys were evaluated by Kang et al. [13]. The SPR lap-shear joints under fatigue loads failed predominantly due to kinked crack growth along the width of the bottom aluminum sheet. The fatigue cracks initiated in the plastically deformed region of the aluminum sheet close to the rivet shank in the rivet–sheet interlock region. Mucha and Witkowski [14] discussed the strength of riveted joints of various sheet materials: DC01 steel, AW-5754 aluminum alloy, and their hybrid arrangements. The fracture mechanism of riveted joints in unilateral tensile tests of T-shaped specimens made of various sheet materials was also analyzed. Chung and Kim [15] investigated the fatigue strength of SPR joints in tensile-shear specimens with dissimilar Al-5052 and steel sheets. A structural analysis of the specimen was conducted. For this specimen, the upper steel sheet with stood applied load in a monotonic test and played a major role in the low-cycle region. In the high-cycle region, however, the harder surface of the upper steel sheet reduced the fatigue strength by enhancing fretting crack initiation on the opposite softer aluminum surface.

Titanium alloy sheets and copper alloy sheets have been widely used in different engineering fields due to excellent strength, ductility, and corrosion resistance. However, it is not easy to weld titanium sheets with copper alloys, as their melting points and thermal conductivity are very different. It is widely accepted that dissimilar metal sheets of titanium with copper alloys can be jointed well by SPR. In a previous paper by He et al. [16], the static strength of SPR joints in dissimilar sheets of titanium and copper alloys was studied. However, no investigation on fretting behavior of such SPR joints has been reported so far.

The present paper deals with the fretting performance of SPR joints in dissimilar sheets of titanium (TA1) and copper (H62) alloys (defined as STH joints). Fatigue load–fatigue life curves were obtained via tension-tension fatigue tests to characterize the fatigue properties of the joints. The typical fracture interfaces were analyzed by scanning electron microscopy (SEM) and energy-dispersive X-ray spectroscopy (EDX) techniques. The results showed that there was extremely severe fretting at the contact interfaces of rivet and sheet materials for the joints at relatively high load levels. In addition, the severe fretting in the region of the locked sheet in contact with the rivet was the major cause of broken locked sheets for the joints at low load level.

2. Experimental Procedure

2.1. Materials

The materials used in this study were TA1 titanium alloy (TA1) sheets (Baoji Chuangxin Metal Materials Co. Ltd., Baoji, China) and H62 copper alloy (H62) sheets (Baoji Chuangxin Metal Materials Co. Ltd., Baoji, China) with the dimensions 110 mm length × 20 mm width × 1.5 mm thickness. To obtain the sheet mechanical properties, material tests were conducted using an MTS 634.31F-24 extensometer (MTS System Corporation, Eden Prairie, MN, USA) with a 20 mm gauge length on an MTS servo hydraulic test machine (MTS System Corporation, Eden Prairie, MN, USA). The chemical compositions of TA1 and H62 sheets are shown in Table 1, and the stress–strain curves at a constant crosshead speed of 5 mm/min are exhibited in Figure 1.

Table 1. Chemical compositions of TA1 and H62 sheet materials.

Material	Fe	C	N	H	O	Ti
TA1	0.02	0.01	0.01	0.001	0.08	Rest
Material	Zn	Fe	Pb	P	Cu	-
H62	36.8	0.15	0.08	0.01	Rest	-

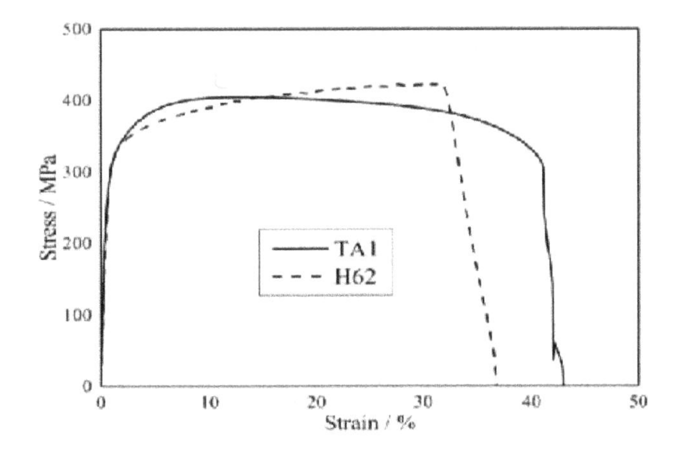

Figure 1. Stress–strain curves for TA1 and H62 sheet materials.

2.2. Specimens Preparation

As mentioned previously, SPR is a new high-speed mechanical fastening technique for point joining of sheet materials. To date, no agreed standard for testing SPR joints exists. The most common test configuration performed in previous studies was the single overlap joint, since it is simple and quick to fabricate. All specimen designs and joint tests referred to China welding standard (GB-2649). As shown in Figure 2, the specimens used for experimental investigation were of the lap-shear type. All specimens were produced with a rivet and a flat bottom die on a RIVSET VARIO-FC servo-driven riveting machine by Böllhoff (Böllhoff Produktion GmbH & Co. KG, Bielefeld and Sonnewalde, Germany). Both the rivet and the die were fabricated from high-strength steel, and were supplied by Böllhoff Produktion GmbH & Co. The mechanical properties of the rivets used in present paper were as follows: Young's modulus 206 GPa, Poisson's ratio 0.3, yield strength 1720 MPa, compressive strength 1595.6 MPa, and Rockwell hardness 44 HRC. The dimensions of the rivet and the die are shown in Figure 3.

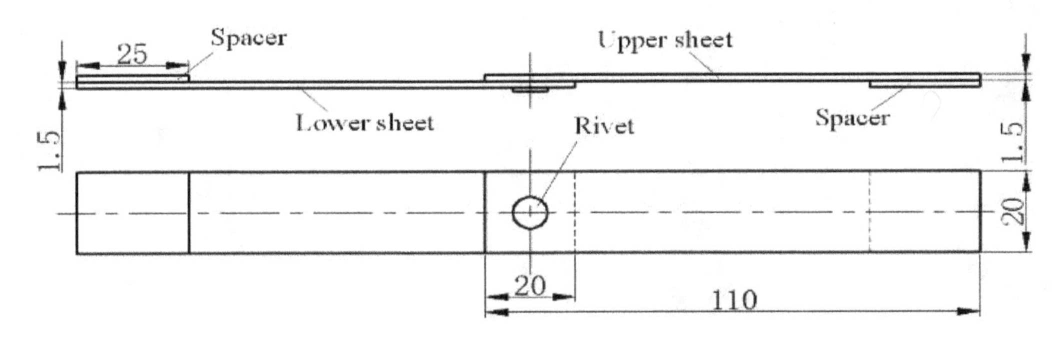

Figure 2. Specimen geometry (dimensions in mm).

To increase the material formability, TA1 sheets in STH joints were heated to 700 °C by an oxyacetylene flame gun. The temperature was controlled by an infrared thermometer (Shenzhen

Jumaoyuan Science And Technology Co., Ltd., Shenzhen, China). The SPR processes were produced immediately after heat treatment. The quality of specimens was monitored by an online window monitoring system (Böllhoff Produktion GmbH & Co. KG, Bielefeld and Sonnewalde, Germany) in the riveting equipment during the SPR process. The monitoring was carried out by measuring the actual SPR setting force through a force sensor and the punch travel through a position sensor, and generating a force-travel curve for one SPR joint [17]. The force-travel curves should be almost identical under the same working conditions for different joints. This indicates that the quality of corresponding joints is good. Through the online window monitoring system, all specimens used were judged as qualified. The satisfied joining parameters of STH joints in terms of punch travel, pre-clam pressure, riveting pressure, and compressing pressure were 131.20 mm, 50 bar, 195 bar, and 110 bar, respectively. Twenty-five STH joints were made, ten of which were selected randomly for the static tests. The rest were prepared for the fatigue tests.

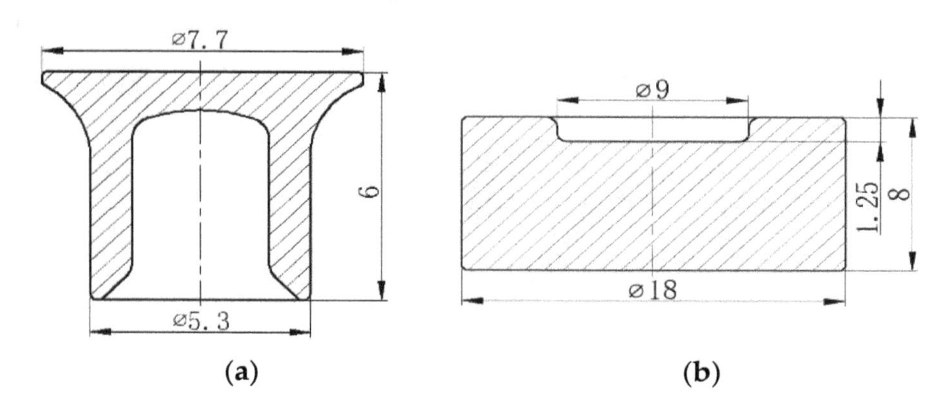

Figure 3. Dimensions of the rivet and the die (dimensions in mm): (**a**) Rivet; (**b**) Die.

3. Static Tests and Results Analysis

The static tests were conducted on an MTS servo hydraulic testing machine using tensile-shear mode. Spacers with dimensions of 25 mm length × 20 mm width × 1.5 mm thickness were glued on both ends of the specimens to reduce the influence of additional bending and to centralize the load. Ten tests for the STH joints were conducted in turn. The tests were performed with a constant displacement rate of 5 mm/min and terminated when the sheets were separated or the force dropped to 5% of the peak force value. Continuous records of the applied force-displacement curves were obtained during each test. The tensile process of the STH joints is exhibited in Figure 4.

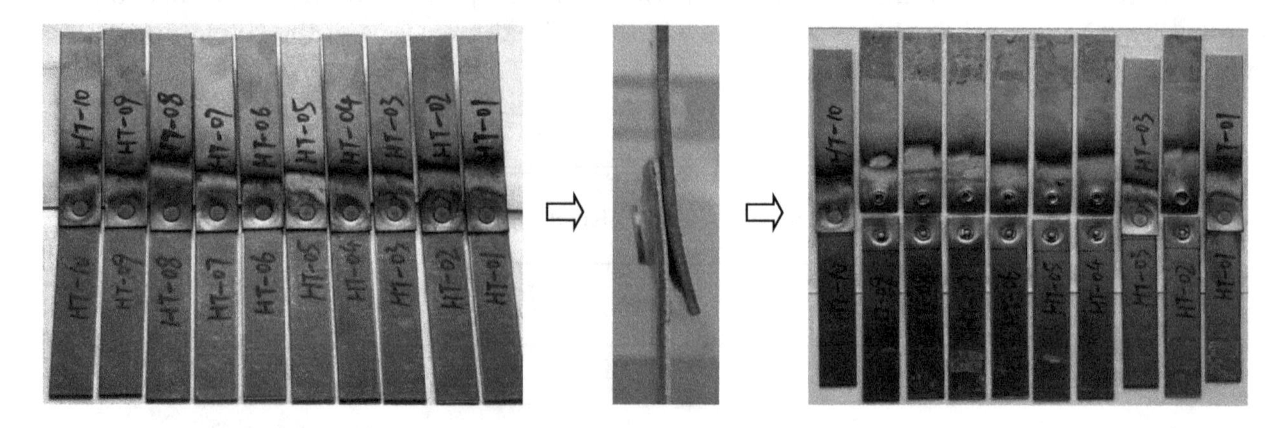

Figure 4. Tensile process of the STH joints.

Figure 5 shows the force-displacement curves and histogram of maximum shear strength of the STH joints. It can be seen from Figure 4 that in static tests of the STH joints, rivets were broken for seven samples, corresponding to the seven relatively low maximum load force-displacement curves in Figure 5. Three joints were not separated completely, which correspond to the force-displacement curve characterized by both high maximum load and high failure deflection in Figure 5.

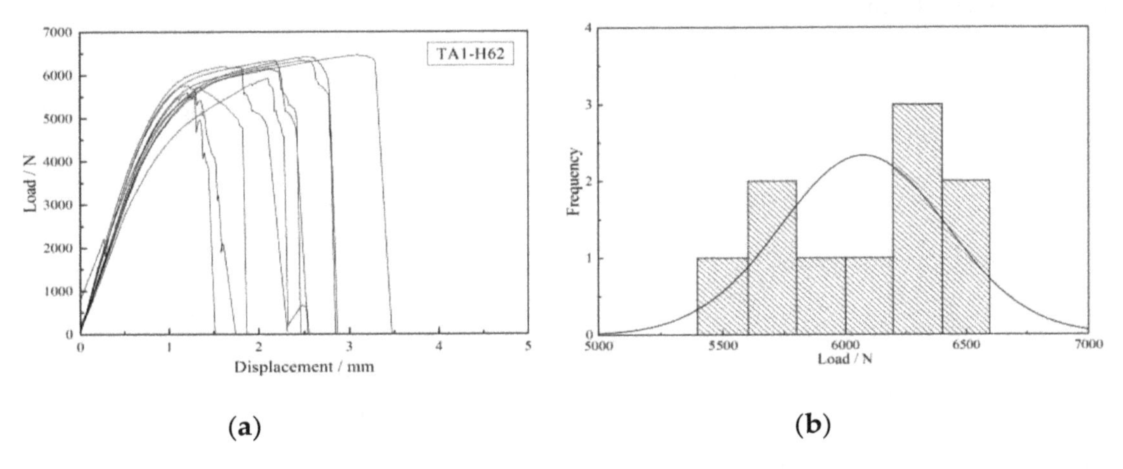

Figure 5. Force-displacement curves and histogram of maximum shear strength of the STH joints. (**a**) Force-displacement curves; (**b**) Histogram of maximum shear strength.

From Figure 5, it can be seen that the data for the failure load and failure displacement of the STH joints were relatively stable, showing that the results were reliable and repeatable. To examine the rationality of the test data, the normal hypothesis tests were performed using MATLAB software (MathWorks Inc., Natick, MA, USA). As shown in Figure 5, the results show that the tensile-shear strength of the STH joints follow normal distributions. All test data fitting the region bounded by the 95% confidence limits.

4. Fatigue Tests and Results Analysis

It can be obtained from Figure 5a that the average peak load of the STH joints was 6.076 kN. The fatigue test parameters were determined based on the average peak load. The load-controlled cyclic fatigue tests were carried out on the MTS servo hydraulic testing machine using a sine waveform and in tension-tension mode. The load ratio $R = 0.1$ and the frequency $f = 10$ Hz were employed for all specimens. Each test was run until 2 million cycles were attained or visible failure occurred. Three load levels of 35% (around 2.1 kN), 30% (around 1.8 kN), and 25% (around 1.5 kN) were performed for the STH joints. Three specimens tested at each load level were randomly selected from the fifteen joints prepared well. To reduce the bending of the sheets and ensure straight-line load paths, spacers were glued to both ends of all specimens in the clamping area. Figure 6 shows the fatigue process of the STH joints.

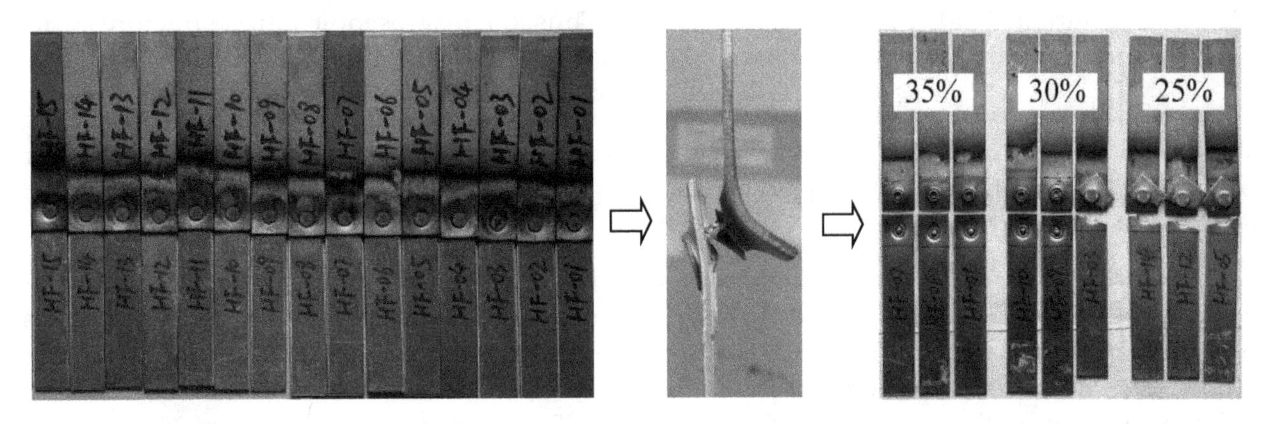

Figure 6. Fatigue process of the STH joints.

The fatigue data and the fatigue load-fatigue life (*F-N*) curve fitted by the least square method are presented in Figure 7. The calculated linear equation of the STH joints is $F = 4.739 - 0.549\ \lg N$. It can be seen from Figure 7 that the STH joints had an average cycle number of 72,102 ($\lg N = 4.9$) at the fatigue load of approximately 2.1 kN.

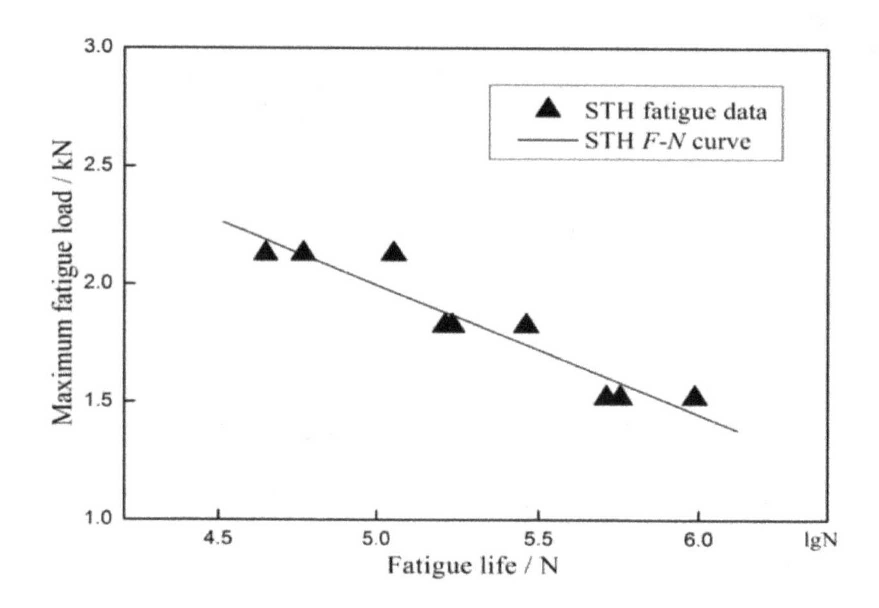

Figure 7. *F-N* curve of the STH joints.

During the static tensile-shear tests, the most STH joints failed with the rivets broken as shown in Figure 4. However, as presented in Figure 6, the failures during the fatigue tests were quite different. At the load level of 35%, all three specimens fractured in the rivet. At the load level of 25%, the locked sheets were broken for all specimens. At the load level of 30%, however, specimens fractured in both failure models.

5. Fretting Failure Mechanism

To analyze the fretting failure mechanisms of the STH joints, SEM (A.S. Tescan Inc., Brno, Czech Republic) and EDX (EDAX Inc., San Diego, CA, USA) techniques were employed to examine the typical fatigue fracture surfaces. As stated previously, the STH joints had an average cycle number of 72,102 at the fatigue load of approximately 2.1 kN. Thus, the macroscopic fatigue fracture surfaces of the STH specimens that failed in the rivet at 2.1 kN were chosen to discuss in detail.

The SEM image of the rivet fracture surface is exhibited in Figure 8, from which the fatigue striations can be obviously observed. This characteristic belongs to the ductile fatigue fracture. It can be seen from the direction marked by an arrow in Figure 8 that fatigue cracks propagated from the outside of the rivet to the inside.

As shown in Figure 9a, the fretting debris could be found in the region near the broken rivet. The corresponding local images of fretting debris are shown in Figure 9b. Based on these features of wear debris, it is inferred that there was extremely severe fretting in the STH joints at load levels of 35% and 30%. Moreover, the severe fretting in the region on the locked sheet in contact with the rivet was the major cause of the broken locked sheet for the STH joints at low load level of 25%.

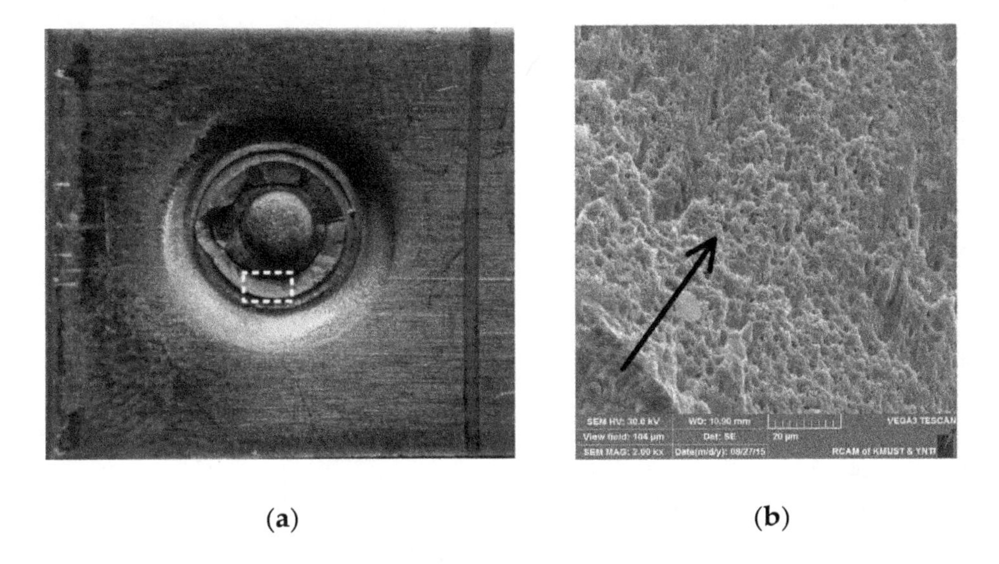

Figure 8. SEM (scanning electron microscopy) images of fracture surface for the STH specimens (load level of 35%). (**a**) Fracture surface of the rivet; (**b**) Enlarged fracture area.

Figure 9. An example of SEM analysis of oxide debris in the STH specimens (load level of 35%). (**a**) Fretting debris; (**b**) Enlarged fretting debris.

Figure 10 shows the spectrum of fretting debris near the broken rivet on the locked sheet for STH joints by EDX (EDAX Inc., San Diego, CA, USA) tests. The main ingredient of the fretting debris was deemed to be CuO, and some zinc, stannum, ferrum, and titanium and other elements in smaller quantities were tested. It could be deduced that the metallic oxide at the contact interface was caused by the continuous cyclic load and micro-movement between the rivet and the sheet. In short, the corresponding elements were fallen off from the sheet or the rivet by the fretting wear processes and oxidized to the relevant oxide: TiO_2 and CuO.

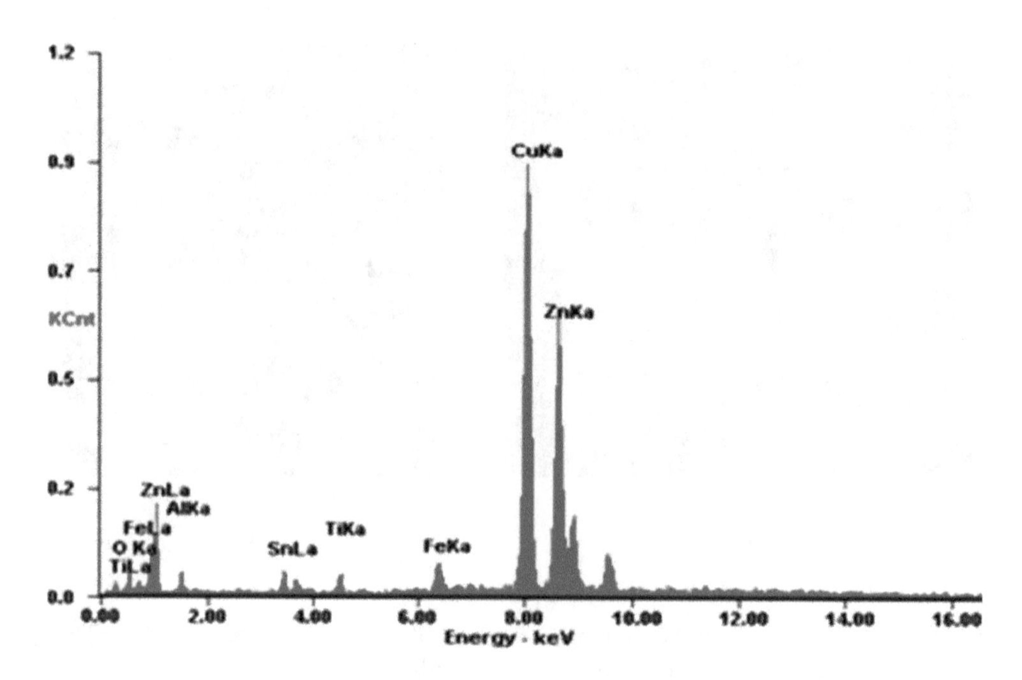

Figure 10. An example of energy-dispersive X-ray spectroscopy (EDX) analysis of oxide debris in the STH specimens.

6. Summary

In this paper, the fretting performance of SPR joining dissimilar sheets in TA1 titanium alloy and H62 copper alloy was studied. The static tests were conducted using tensile-shear mode, and the histograms of maximum shear strength of the STH joints were obtained. Fatigue load–fatigue life curves were obtained via tension-tension fatigue tests to characterize the fatigue properties of the joints. To study the fretting behavior of the STH joints, the typical fracture interfaces were analyzed by SEM and EDX techniques. The results showed that there was extremely severe fretting at the contact interfaces of rivet and sheet materials for the joints at relatively high load levels. Moreover, the severe fretting in the region of the locked sheet in contact with the rivet was the major cause of the broken locked sheets for the joints at low load level. It could be deduced that the metallic oxide at the contact interface was caused by the continuous cyclic load and micro-movement between the rivet and the sheet.

Acknowledgments: This study is supported by the National Natural Science Foundation of China (Grant No. 51565023) and Major Program Foundation of the Education Department of Yunnan Province, China (Grant No. ZD201504).

Author Contributions: Xiaocong He conceived and designed the experiments; Cong Deng and Xianlian Zhang performed the experiments and analyzed the data; Xianlian Zhang contributed reagents/materials/analysis tools; Xiaocong He wrote the paper.

References

1. He, X.; Gu, F.; Ball, A. A review of numerical analysis of friction stir welding. *Prog. Mater. Sci.* **2014**, *65*, 1–66. [CrossRef]
2. He, X. Finite Element Analysis of Laser Welding: A State of Art Review. *Mater. Manuf. Process.* **2012**, *27*, 1354–1365. [CrossRef]
3. Mucha, J. The failure mechanics analysis of the solid self-piercing riveting joints. *Eng. Fail. Anal.* **2015**, *47*, 77–88. [CrossRef]

4. Kang, S.; Kim, H. Fatigue strength evaluation of self-piercing riveted Al-5052 joints under different specimen configurations. *Int. J. Fatigue* **2015**, *80*, 58–68. [CrossRef]
5. Su, Z.; Lin, P.; Lai, W.; Pan, J. Fatigue analyses of self-piercing rivets and clinch joints in lap-shear specimens of aluminum sheets. *Int. J. Fatigue* **2015**, *72*, 53–65. [CrossRef]
6. Calabrese, L.; Proverbio, E.; Pollicino, E.; Galtieri, G.; Borsellino, C. Effect of galvanic corrosion on durability of aluminium/steel self-piercing rivet joints. *Corros. Eng. Sci. Technol.* **2015**, *50*, 10–17. [CrossRef]
7. Lou, M.; Li, Y.; Wang, Y.; Wang, B.; Lai, X. Influence of resistance heating on self-piercing riveted dissimilar joints of AA6061-T6 and galvanized DP590. *J. Mater. Process. Technol.* **2014**, *214*, 2119–2126. [CrossRef]
8. Haque, R.; Williams, N.; Blacket, S.; Durandet, Y. A simple but effective model for characterizing SPR joints in steel sheet. *J. Mater. Process. Technol.* **2015**, *223*, 225–231. [CrossRef]
9. Li, Y.B.; Wei, Z.Y.; Wang, Z.Z.; Li, Y.T. Friction self-piercing riveting of aluminum alloy AA6061-T6 to magnesium alloy AZ31B. *J. Manuf. Sci. Eng. Trans. ASME* **2013**, *135*, 6. [CrossRef]
10. Liu, X.; Lim, Y.C.; Li, Y.B.; Tang, W.; Ma, Y.W.; Feng, Z.L.; Ni, J. Effects of process parameters on friction self-piercing riveting of dissimilar materials. *J. Mater. Process. Technol.* **2016**, *237*, 19–30. [CrossRef]
11. Ma, Y.W.; Li, Y.B.; Hu, W.; Lou, M.; Lin, Z.Q. Modeling of Friction Self-Piercing Riveting of Aluminum to Magnesium. *J. Manuf. Sci. Eng. Trans. ASME* **2016**, *138*, 6. [CrossRef]
12. Haque, R.; Durandet, Y. Strength prediction of self-pierce riveted joint in cross-tension and lap-shear. *Mater. Des.* **2016**, *108*, 666–678. [CrossRef]
13. Kang, J.; Rao, H.; Zhang, R.; Avery, K.; Su, X. Tensile and fatigue behaviour of self-piercing rivets of CFRP to aluminium for automotive application. *IOP Conf. Ser. Mater. Sci. Eng.* **2016**, *137*, 1. [CrossRef]
14. Mucha, J.; Witkowski, W. Mechanical Behavior and Failure of Riveting Joints in Tensile and Shear Tests. *Strength Mater.* **2015**, *47*, 755–769. [CrossRef]
15. Chung, C.-S.; Kim, H.-K. Fatigue strength of self-piercing riveted joints in lap-shear specimens of aluminium and steel sheets. *Fatigue Fract. Eng. Mater. Struct.* **2016**, *39*, 1105–1114. [CrossRef]
16. He, X.; Wang, Y.; Lu, Y.; Zeng, K.; Gu, F.; Ball, A. Self-piercing riveting of similar and dissimilar titanium sheet materials. *Int. J. Adv. Manuf. Technol.* **2015**, *80*, 2105–2115. [CrossRef]
17. Zhao, L.; He, X.; Xing, B.; Lu, Y.; Gu, F.; Ball, A. Influence of sheet thickness on fatigue behavior and fretting of self-piercing riveted joints in aluminum alloy 5052. *Mater. Des.* **2015**, *87*, 1010–1017. [CrossRef]

Permissions

All chapters in this book were first published by MDPI; hereby published with permission under the Creative Commons Attribution License or equivalent. Every chapter published in this book has been scrutinized by our experts. Their significance has been extensively debated. The topics covered herein carry significant findings which will fuel the growth of the discipline. They may even be implemented as practical applications or may be referred to as a beginning point for another development.

The contributors of this book come from diverse backgrounds, making this book a truly international effort. This book will bring forth new frontiers with its revolutionizing research information and detailed analysis of the nascent developments around the world.

We would like to thank all the contributing authors for lending their expertise to make the book truly unique. They have played a crucial role in the development of this book. Without their invaluable contributions this book wouldn't have been possible. They have made vital efforts to compile up to date information on the varied aspects of this subject to make this book a valuable addition to the collection of many professionals and students.

This book was conceptualized with the vision of imparting up-to-date information and advanced data in this field. To ensure the same, a matchless editorial board was set up. Every individual on the board went through rigorous rounds of assessment to prove their worth. After which they invested a large part of their time researching and compiling the most relevant data for our readers.

The editorial board has been involved in producing this book since its inception. They have spent rigorous hours researching and exploring the diverse topics which have resulted in the successful publishing of this book. They have passed on their knowledge of decades through this book. To expedite this challenging task, the publisher supported the team at every step. A small team of assistant editors was also appointed to further simplify the editing procedure and attain best results for the readers.

Apart from the editorial board, the designing team has also invested a significant amount of their time in understanding the subject and creating the most relevant covers. They scrutinized every image to scout for the most suitable representation of the subject and create an appropriate cover for the book.

The publishing team has been an ardent support to the editorial, designing and production team. Their endless efforts to recruit the best for this project, has resulted in the accomplishment of this book. They are a veteran in the field of academics and their pool of knowledge is as vast as their experience in printing. Their expertise and guidance has proved useful at every step. Their uncompromising quality standards have made this book an exceptional effort. Their encouragement from time to time has been an inspiration for everyone.

The publisher and the editorial board hope that this book will prove to be a valuable piece of knowledge for researchers, students, practitioners and scholars across the globe.

List of Contributors

Lin Gao, Guorong Cui, Wenzhen Chen and Wencong Zhang
School of Materials Science and Engineering, Harbin Institute of Technology, Weihai 264209, China

Jiabin Hou
School of Materials Science and Engineering, Harbin Institute of Technology, Weihai 264209, China
Naval Architecture and Marine Engineering College, Shandong Jiaotong University, Weihai 264209, China

Wenguang Tian
Oriental Bluesky Titanium Technology Co., LTD, Yantai 264003, China

Sergey Zherebtsov, Maxim Ozerov, Margarita Klimova, Nikita Stepanov and Gennady Salishchev
Laboratory of Bulk Nanostructured Materials, Belgorod State University, Belgorod 308015, Russia

Dmitry Moskovskikh
Centre of Functional Nanoceramics, National University of Science and Technology, Moscow 119049, Russia

Sever Gabriel Racz, Radu Eugen Breaz, Melania Tera, Claudia Gîrjob, Cristina Biris, Anca Lucia Chicea and Octavian Bologa
Department of Industrial Machines and Equipment, Engineering Faculty, "Lucian Blaga" University of Sibiu, Victoriei 10, 550024 Sibiu, Romania

Geonhyeong Kim, Jae Nam Kim and Chong Soo Lee
Graduate Institute of Ferrous Technology (GIFT), Pohang University of Science and Technology (POSTECH), Pohang 37673, Korea

Taekyung Lee
School of Mechanical Engineering, Pusan National University, Busan 46241, Korea

Yongmoon Lee
Center for Advanced Aerospace Materials, Pohang University of Science and Technology (POSTECH), Pohang 37673, Korea

Seong Woo Choi and Jae Keun Hong
Advanced Metals Division, Korea Institute of Materials Science, Changwon 51508, Korea

André Reck and André Till Zeuner
Institute of Materials Science, Technical University of Dresden, 01062 Dresden, Germany

Martina Zimmermann
Institute of Materials Science, Technical University of Dresden, 01062 Dresden, Germany
Department of Materials Characterization and Testing, Fraunhofer-Institute for Material and Beam Technology IWS, 01277 Dresden, Germany

Pavlo E. Markovsky, Vadim I. Bondarchuk, Oleksandr O. Stasyuk, Dmytro G. Savvakin and Mykola A. Skoryk
G.V. Kurdyumov Institute for Metal Physics of N.A.S. of Ukraine, 36 Academician Vernadsky Boulevard, UA-03142 Kyiv, Ukraine

Jacek Janiszewski, Kamil Cieplak and Piotr Dziewit
Jarosław Dąbrowski, Military University of Technology, 2 gen. Sylwester Kaliski str, 00-908 Warsaw, Poland

Sergey V. Prikhodko
Department of Materials Science and Engineering, University of California Los Angeles, Los Angeles, CA 90095, USA

Shuyu Sun
School of Mechanical Engineering, Taizhou University, Taizhou 318000, China

Weijie Lu
State Key Laboratory of Metal Matrix Composites, Shanghai Jiao Tong University, Shanghai 200240, China

Changzhou Yu, Peng Cao and Mark Ian Jones
Department of Chemical and Materials Engineering, The University of Auckland, Auckland 1142, New Zealand

Dongjun Wang
National Key Laboratory for Precision Hot Processing of Metals, Harbin Institute of Technology, Harbin 150001, China
Key Laboratory of Micro-Systems and Micro-Structures Manufacturing, Ministry of Education, Harbin 150001, China

Hao Yuan and Jianming Qiang
School of Materials Science and Engineering, Harbin Institute of Technology, Harbin 150001, China

Hongbo Ba, Limin Dong, Zhiqiang Zhang and Xiaofei Lei
Institute of Metal Research, Chinese Academy of Sciences, 72 Wenhua Road, Shenyang 110016, China

Elena N. Korosteleva and Maksim G. Krinitcyn
Department of High Technology Physics in Mechanical Engineering, Tomsk Polytechnic University, 30 Lenin av., 634050 Tomsk, Russia
Institute of Strength Physics and Materials Science of Siberian Branch of the Russian Academy of Sciences, 2/4, pr. Akademicheskii, 634055 Tomsk, Russia

Victoria V. Korzhova
Institute of Strength Physics and Materials Science of Siberian Branch of the Russian Academy of Sciences, 2/4, pr. Akademicheskii, 634055 Tomsk, Russia

María Prados-Privado
Department Continuum Mechanics and Structural Analysis, Higher Polytechnic School, Carlos III University, Avenida de la Universidad, 30, 28911 Leganés, Madrid, Spain
Research Department, ASISA Dental, Calle José Abascal, 32, 28003 Madrid, Spain

Henri Diederich
Private Practice, 51 Avenue Pasteur, L2311 Luxembourg, Luxembourg

Juan Carlos Prados-Frutos
Department of Medicine and Surgery, Faculty of Health Sciences, Rey Juan Carlos University, Avenida de Atenas s/n, 28922 Alcorcón, Madrid, Spain

Sergey Mironov
Laboratory of Mechanical Properties of Nanoscale Materials and Superalloys, Belgorod National Research University, Pobeda 85, 308015 Belgorod, Russia

Yutaka S. Sato
Department of Materials Processing, Tohoku University, 6-6-02 Aramaki-aza-Aoba, Sendai 980-8579, Japan

Hiroyuki Kokawa
Department of Materials Processing, Tohoku University, 6-6-02 Aramaki-aza-Aoba, Sendai 980-8579, Japan
Shanghai Key Laboratory of Materials Laser Processing and Modification, School of Materials Science and Engineering, Shanghai Jiao Tong University, 800 Dongchuan Road, Minhang District, Shanghai 200240, China

Satoshi Hirano
Hitachi Research Laboratory, Hitachi Ltd., 7-1-1 Omika-cho, Hitachi 319-1291, Japan

Adam L. Pilchak and Sheldon Lee Semiatin
Air Force Research Laboratory, Materials and Manufacturing Directorate, Wright-Patterson AFB, OH 45433-7817, USA

Ling Ding, Yulei Gu, Danying Zhou, Fuwen Chen, Lian Zhou and Hui Chang
Tech Institute for Advanced Materials & College of Materials Science and Engineering, Nanjing Tech University, Nanjing 210009, China

Rui Hu
State Key Laboratory of Solidification Processing, Northwestern Polytechnical University, Xi'an 710072, China

Khaja Moiduddin, Syed Hammad Mian, Hisham Alkhalefah and Usama Umer
Advanced Manufacturing Institute, King Saud University, Riyadh 11421, Saudi Arabia

Ezgi Onal, Jessica E. Frit, Marten Jurg and Andrey Molotnikov
Department of Materials Science and Engineering, Monash University, Clayton, VIC 3800, Australia

Xinhua Wu
Department of Materials Science and Engineering, Monash University, Clayton, VIC 3800, Australia
Monash Centre for Additive Manufacturing, Monash University, Clayton, VIC 3800, Australia

Xiaocong He, Cong Deng and Xianlian Zhang
Innovative Manufacturing Research Centre, Kunming University of Science and Technology, Kunming 650500, China

Index

O

P

R

S

T

V

Printed in the USA
CPSIA information can be obtained
at www.ICGtesting.com
JSHW050845251023
50683JS00018B/71

9 781647 2665